Biometeorology

Volume 3

More information about this series at http://www.springer.com/series/8101

Heikki Hänninen

Boreal and Temperate Trees in a Changing Climate

Modelling the Ecophysiology of Seasonality

Springer

Heikki Hänninen
Department of Biosciences
University of Helsinki
Helsinki, Finland

Biometeorology
ISBN 978-94-017-7547-2 ISBN 978-94-017-7549-6 (eBook)
DOI 10.1007/978-94-017-7549-6

Library of Congress Control Number: 2016938116

Cover illustration: The cover photo was taken by Juho Aalto.

Printed on acid-free paper

This Springer imprint is published by Springer Nature
The registered company is Springer Science+Business Media B.V. Dordrecht

To
the memory of my mother (1920–2008)
and
the memory of Lisa Lena (1957–2013)

Preface

Boreal and temperate trees grow under climatic conditions in which the ambient air temperature displays pronounced seasonal variation. Unlike herbs and grasses, trees overwinter without a sheltering snow cover, so that they are exposed to all the harsh climatic conditions. That is why their climatic adaptation is based on their annual cycle of development, whereby the frost-hardy dormant phase and the susceptible growth phase are synchronised with the seasonality of the climate. There is a long and rich research tradition focusing on this key adaptive strategy of boreal and temperate trees, not only because trees are major elements of forest ecosystems but also because many of these tree species are economically important for forestry or horticulture. More recently, the projected climate change has dramatically increased scholarly interest in the annual cycle of trees.

The last few decades have seen increasing use of dynamic simulation models in studies of the annual cycle of boreal and temperate trees. In this book I review these studies and introduce some new ideas. Having worked on the subject for more than 30 years, I have no doubt developed a personal approach to the work, and I do not try to hide that in this volume, either. I mainly concentrate on my own research and other work closely related to it. This should not be taken as any disparagement of the studies not reviewed in the same detail or not reviewed at all. Rather, this choice was made in order to keep the presentation concise and uniform throughout the book, as a detailed discussion of all relevant literature would have required a much larger and much more diverse treatise. This is true especially because during the last 10 years or so, I have been very glad to see a dramatic increase in the number of studies published in the field. In some cases I make critical notes on approaches where I see problems in the methods applied. It is my sincere wish that my presentation should facilitate scientific discussion, also in cases where my choices differ from those of some colleagues.

In line with my long-term research strategy, I put the main emphasis on the combining of modelling and empirical studies. Thus, the book is directed not only to modellers but also to experimental tree ecophysiologists, and the bridging of the gap between these two research traditions is a major theme. Consequently, the number of formulae presented is relatively low, and besides modelling studies, I also discuss related empirical research. Because of the vast number of empirical publications about the annual cycle of boreal and temperate trees, a comprehensive review of them is inevitably outside the scope of the book. When discussing the effects of climate change, I mainly concentrate on the overwintering of trees. This choice is a deliberate one, caused not only by the orientation of my own expertise but also by the need to emphasise the ecological importance of the often neglected overwintering period of the trees.

Depending on the background of the reader, the chapters of the book may be readable independently from each other. Even so, the book is planned to form an integrated unity, so that optimally the chapters should be read in sequence. The principles of dynamic modelling are presented in Chap. 2 for those readers who may not be previously familiar with this research tool. Throughout the text, special emphasis is devoted to thorough explanations of the concepts and modelling principles presented. Specialist readers familiar with the topic may find the text even too elementary and lengthy at times, but I have made that choice deliberately, for I wish to serve a wide audience, including advanced graduate and postgraduate students.

Basically, this treatise belongs to the discipline of whole-tree ecophysiology. It seems to me that time is becoming ripe for a synthesis of the whole-tree modelling approach and the physiological and molecular approach, the two of which have largely been pursued in isolation from each other so far. However, rather than trying to make such a synthesis, my aim is to explicate the whole-tree modelling approach for the benefit of forthcoming efforts towards a synthesis. I also briefly discuss the upscaling of ecophysiological information from the whole-tree level to the stand and the ecosystem level. I hope this facilitates the introduction of biologically realistic ecophysiological models of the annual cycle of boreal and temperate trees into the larger-scale ecological models currently applied in climate change research.

I take a historical approach, so that rather than just describing the state of the art at the time of writing, I also discuss the evolution of the art over time into its present form of different concepts, theories, and models. I hope this approach helps the reader to get a deep understanding of the subject, including the perception that in several cases the present theories and models are hardly the final ones. Furthermore, the conceptualisation of many of the models discussed has changed over time; by explaining the changes I hope to help the reader to compare old publications with recent ones. Last but not least, by taking the historical approach, I want to acknowledge the work of many early researchers who are often forgotten. The foundation for projecting the effects of climate change on boreal and temperate trees was laid by these early researchers long before climate change became an issue. Today this heritage is needed more than ever before.

It goes without saying that the 30 years of work summarised here would not have been possible without collaboration with supervisors, colleagues, students, and technicians. Tapani Repo was my nearest colleague in the 1980s and 1990s, when the foundation for this book was laid in our research group at the University of

Joensuu (currently the University of Eastern Finland). My doctoral supervisors Paavo Pelkonen and Veikko Koski conveyed the Finnish research tradition pertaining to the annual cycle of boreal trees to me. Ilkka Leinonen completed his PhD in our group and had a major impact on our research. At the very beginning of my research career, Pertti 'Pepe' Hari familiarised me with the principles of dynamic modelling in tree ecophysiology and with the relevance of the philosophy of science for the practical researcher. I have been glad to continue cooperation with these friends over the years.

The review paper I published in 2007 with Koen Kramer is the backbone of this book in several ways. Two other, more recent joint papers, published with Karen Tanino and Olavi Junttila, were a great help in finalising the manuscript. It is not possible here to mention all my other collaborators over the years, and even the most important ones make quite a long list: Risto Häkkinen, Seppo Kellomäki, Pasi Kolari, Kari Laine, Tarja Lehto, Sune Linder, Tapio Linkosalo, Jaana Luoranen, Annikki Mäkelä, Eero Nikinmaa, Jouni Partanen, Risto Rikala, Outi Savolainen, Heikki Smolander, Marja-Liisa Sutinen, Kari Taulavuori, and Gang Zhang.

After moving to Helsinki in 1997, I have broadened my scope to cover other plant life forms in addition to forest trees. This would not have been possible without my colleagues and students in Helsinki, who have expertise in dwarf shrubs, herbs, and grasses: Timo Saarinen, Robin Lundell, Helena Åström, Sirpa Rasmus, and Friederike Gehrmann. It has been a great pleasure for me to see that the approach developed for trees works in our PECC (Plant Ecophysiology and Climate Change) group for other plant life forms as well. Some of our studies addressing the other plant life forms are mentioned in passing.

Margaret Deignan of Springer invited me to write this book, and she has given me indispensable help throughout the preparation of the manuscript. I am especially grateful for her patience over the long years it took me to complete the book. Valuable technical help was also provided by Timo Bazuin, Takeesha Moerland-Torpey, and Chitra Sundarajan.

The entire manuscript was thoroughly commented on by Olavi Junttila. His comments were especially useful because in addition to providing constructive criticism, he also encouraged me to focus on my cup of tea – the modelling approach – rather than trying to broaden the scope to cover related topics, about which my colleagues have much more to say. I also received comments on the entire manuscript from Tapani Repo, and large parts of it were commented on by Risto Häkkinen, Pepe Hari, Øystein Johnsen, Pasi Kolari, Veikko Koski, Tapio Linkosalo, Robin Lundell, Marja-Liisa Sutinen, Karen Tanino, and Kari Taulavuori. More restricted but nevertheless useful comments were provided by Helena Åström, Koen Kramer, Xavier Morin, and Timo Saarinen. In addition, I received useful hints for literature references from Heikki Henttonen, Øystein Johnsen, Pasi Kolari, Robin Lundell, Jaana Luoranen, Timo Saarinen, Sirpa Rasmus, Karen Tanino, and Jan Weckström.

Ilkka Juuso, Robin Lundell, and Timo Saarinen provided technical help in drawing the figures. The following persons provided me with photographs: Juho

Aalto, Eeva Pudas, Risto Rikala, Timo Saarinen, Annika Saarto, and Sirkka Sutinen. Tapani Repo and Jaana Luoranen provided me with data for redrawing published figures.

My access to the long-term set of data collected by the Finnish Meteorological Institute and digitalised by the Finnish Forest Research Institute (currently Natural Resources Institute Finland) was made possible through cooperation with Pepe Hari, Risto Häkkinen, Heikki Tuomenvirta, and Tapio Linkosalo. Kimmo Ruosteenoja provided me unpublished climatic scenarios for Finland.

The language of the manuscript was carefully revised by Pekka Hirvonen, professor (emer.) of English. He spent a countless number of hours editing the text. During the multi-year process, he also taught me a lot about the nuances of English.

The writing of this book was economically supported by the Kone Foundation, the Academy of Finland (project 122194), and the Finnish Cultural Foundation.

I am grateful to all, including many not mentioned above, who have helped me in various ways over the years. To many of the collaborators named, I am also grateful for their friendship. 'I have been privileged to carry out my studies in a research community where only the joy of life and laughter have overridden scientific enthusiasm'. These words, quoted from the preface to my doctoral thesis, are as true today as they were 25 years ago.

Helsinki, Finland
November 2015

Heikki Hänninen

Contents

Chapter 1
Climatic Adaptation of Boreal and Temperate Tree Species

Contents

Abstract Boreal and temperate trees grow under climatic conditions in which the ambient air temperature displays pronounced seasonal variation. Unlike herbs and grasses, trees overwinter without a sheltering snow cover, so that they are exposed to all the harsh climatic conditions. That is why their climatic adaptation is based on their annual cycle of development, whereby the frost-hardy dormant phase and the susceptible growth phase are synchronised with the seasonality of the climate. The main aspects of this adaptive strategy of trees are briefly discussed, emphasising both the geographical and the year-to-year variation of the seasonal air temperature conditions. Many boreal and temperate tree species have large ranges of geographical distribution, so that their different provenances have adapted to the particular local climate prevailing at their native growing site. The extent of the geographical variation in air temperature crucial for this adaptation is highlighted by examining the climatic records of four locations within the European distribution range of *Pinus sylvestris*. The extent of the year-to-year variation is similarly highlighted by examining a 92-year climatic record from Jyväskylä, central Finland. In the coolest summer, the temperature sum in Jyväskylä was similar to the average temperature sum 600 km north of Jyväskylä; and in the warmest summer it was similar to the average temperature sum 600 km south of Jyväskylä. This limited analysis suffices to reveal the extent of the climatic year-to-year variation that trees need to acclimate to at their native growing site.

Keywords Annual cycle • Boreal trees • Climatic adaptation • Climatic variation • Ecophysiological modelling • Seasonality • Temperate trees

H. Hänninen, *Boreal and Temperate Trees in a Changing Climate*, Biometeorology 3,
DOI 10.1007/978-94-017-7549-6_1

1.1 Thermal Conditions in the Boreal and the Temperate Zone

1.1.1 Seasonal and Geographical Variation

At the highest latitudes of the globe, the climate is too cold to support tree growth, so that a biome of treeless tundra is formed there (Breckle 2002). Although the climate of the forested regions at somewhat lower latitudes is warmer, these regions are also characterized by winter seasons with air temperatures dropping below zero for prolonged periods. The seasonal occurrence of frost sets special requirements on the climatic adaptation of tree species growing in these conditions (Sarvas 1972, 1974; Larcher 2003; Crawford 2008).[1]

The forested regions from the high to the intermediate latitudes of the northern hemisphere are broadly classified into two biomes. South of the arctic tundra, there is a large circumpolar biome of boreal coniferous forest (Breckle 2002). The climate in the boreal zone is typically classified as cool, though temperate and cold climates are also represented (Larcher 2003). Besides the dominating coniferous tree species, most of which are evergreen, broadleaved species are also common in the boreal zone. In the forested regions located further south, broadleaved deciduous tree species become dominant. These regions belong to the biome of temperate broadleaved forest (Breckle 2002).

There is a distinct seasonal pattern of incoming solar radiation in the boreal and the temperate zone, and this seasonality of radiation produces a similar seasonal pattern of air temperature. The summer season, with plenty of incoming solar radiation and relatively high air temperatures, alternates with the winter season, with both the amount of incoming radiation and the air temperatures relatively low (Bonan 2008). Though the seasonal patterns of air temperature in the boreal and the temperate zone are thus generally similar, there are also great geographical differences between the thermal conditions of these zones.

A large part of this geographical variation can be attributed to two main gradients. Firstly, the climate generally gets cooler when one moves from south to north. Thus the mean annual temperature in temperate forests is usually in the range of +5 to +15 °C, depending on the location, whereas in the boreal forests further north the mean annual temperature is typically only a few degrees above zero and, in the coldest part of the zone, even below zero (Whittaker 1975). Secondly, maritime regions are characterized by mild winters and cool summers, whereas continental regions further away from the sea are characterized by cold winters and warm summers (Bonan 2008). Thus the annual range of air

[1] Throughout the present volume, the concept of adaptation is used for the genetic adjustment of the tree populations to their native environments, the adjustment being caused by natural selection working over evolutionary time scales. Correspondingly, the concept of acclimation is used for the physiological adjustment of individual trees to the environmental conditions prevailing at their growing site during their life cycle.

temperatures is wider in continental than in maritime regions. In Yakutsk, Siberia, for instance, the annual range of mean monthly temperatures is about 70 °C, with the wintertime minimum and the summertime maximum of approximately −50 °C and +20 °C, respectively (Archibold 1995).

Many boreal and temperate tree species have large ranges of geographical distribution, so that their different provenances have adapted to the particular local climate prevailing at their native growing site (Langlet 1971; Rehfeldt et al. 1999; Aitken and Hannerz 2001; Savolainen et al. 2007). This important aspect of the climatic adaptation of boreal and temperate tree species is discussed in detail in Chap. 6. An idea of the extent of the geographical climate variation can be formed from an analysis of short-term temperature records gathered at four locations in Europe, representing a restricted part of the geographical distribution range of *Pinus sylvestris* (Fig. 1.1). Among these four locations, the mean annual temperature is the highest in Eskdalemuir, Scotland (+7.1 °C), and the lowest in Murmansk, northwestern Russia (−0.2 °C). Out of the remaining two locations, Jyväskylä, central Finland, has a mean annual temperature of +2.7 °C and Voronez, southern Russia, the second highest at +6.0 °C.

In studies addressing the seasonality and climatic adaptation of boreal and temperate tree species, it is necessary to also examine the seasonal pattern of air temperatures, not just the average level indicated by the mean annual temperature. Among the four locations examined here, for instance, there is only a relatively small difference in the mean annual temperature between Eskdalemuir and Voronez (Fig. 1.1a), but the seasonal pattern of mean monthly temperatures is drastically different at these two locations (Fig. 1.1b). In the maritime climate of Eskdalemuir, the mean monthly temperature remains slightly above zero even in winter, to rise to approximately +13 °C in summer. Thus the annual range of the mean monthly temperature in Eskdalemuir is only slightly over 10 °C (Fig. 1.1b). At the other end of the spectrum is Voronez. Due to its relatively southern location, the mean monthly temperature does not drop much below −10 °C in winter, whereas in summer it approaches +20 °C. Thus the annual range of the mean monthly temperature is about 30 °C (Fig. 1.1b). Accordingly, the air temperature's rate of rising in spring and dropping in autumn is higher in Voronez than in Eskdalemuir.

The other two locations, Jyväskylä and Murmansk, represent intermediate positions in this seasonal comparison. It is noteworthy that despite the northern location and generally cold climate of Murmansk (Fig. 1.1a), its mean temperature in July is almost as high as that of the otherwise much warmer and more southern Eskdalemuir (Fig. 1.1b). Furthermore, with the exception of Eskdalemuir, the mid-winter temperatures are almost equal in the four locations. This would not be the case if a location from the most continental part of the geographical range of *Pinus sylvestris* in Siberia had been included in the comparison (Archibold 1995).

There are also drastic differences among the four locations in the annual accumulation of temperature sum (Fig. 1.1c).[2] In the warm and continental climate

[2] This essential concept is defined in Sect. 2.3, and its ecophysiological interpretation is discussed in Sect. 3.2.1.2.

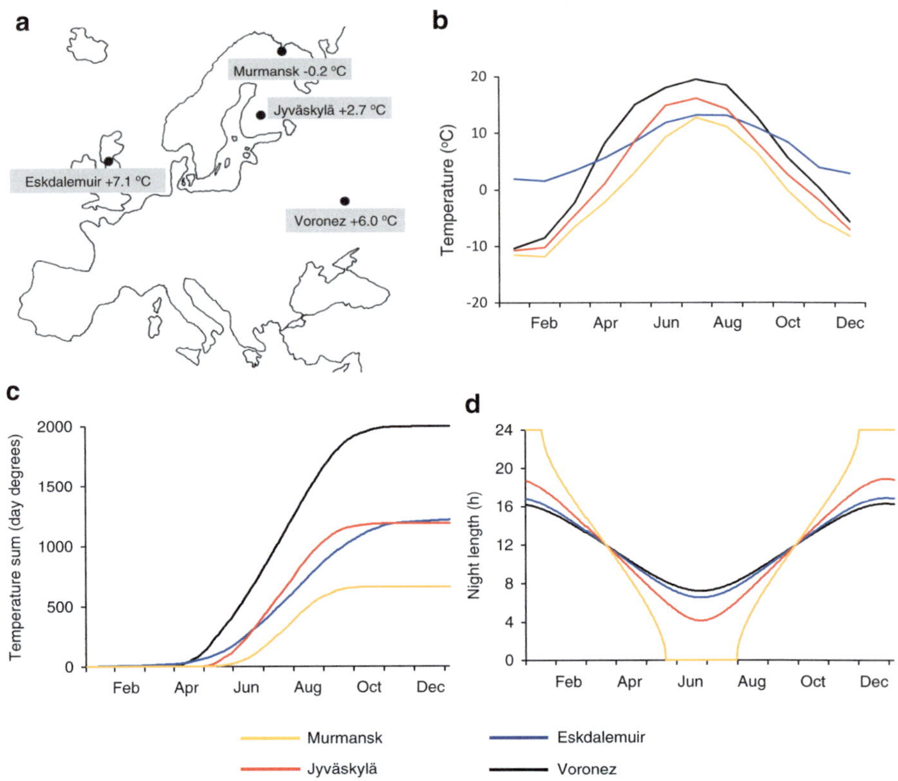

Fig. 1.1 Seasonal and geographical variation of the thermal and night length conditions in the European distribution range of *Pinus sylvestris*. (**a**) The locations of the four meteorological stations, with their respective mean annual air temperatures (Modified from Hänninen and Hari 1996, and published with the permission of the Finnish Forest Research Institute, currently the Natural Resources Institute Finland, and the Finnish Society of Forest Science). (**b**) The seasonal patterns of the mean monthly air temperatures at the four locations. (**c**) Average seasonal patterns of temperature sum accumulation at the four locations. The temperature sum is expressed in day degree units calculated with a threshold temperature of +5 °C (Sect. 2.3). (**d**) The seasonal patterns of night length, i.e., the length of the period during the diurnal cycle when the sun is below the horizon, at the four locations. The temperature indices have been calculated on the basis of standard air temperature records collected by the national meteorological institutes at the four locations during the 15 years of 1966–1980

of Voronez, the temperature sum starts to accumulate in early April, and by the end of the growing season, an average of almost 2000 day degree units has accumulated. In the colder climate of Murmansk, the accumulation starts about 1 month later, and the average total accumulation at the end of the growing season is about one third of the total accumulation in Voronez (Fig. 1.1c).

The remaining two locations, Jyväskylä and Eskdalemuir, display an almost identical average total accumulation of temperature sum, about 1200 day degree units in both. Otherwise, too, the seasonal pattern is quite similar in these two locations, except that in the maritime climate of Eskdalemuir, the accumulation of temperature sum starts earlier and stops later than in the more continental climate of Jyväskylä. In fact, there are many years in Eskdalemuir with the temperature sum accumulation going on even in mid-winter. The difference between Eskdalemuir and Jyväskylä in the wintertime accumulation of temperature sum, though relatively minor in the overall climatological comparison (Fig. 1.1c), has a drastic effect on the climatic adaptation of the tree species. The ecophysiological importance of these climatic geographical differences is discussed in Sect. 6.3.4.

1.1.2 Year-to-Year Variation

The climate of the boreal and temperate zones is characterized by large year-to-year variation (Bonan 2008). In Jyväskylä, central Finland (Fig. 1.1a), the year-to-year range of the daily mean temperature over 92 years was about 35 °C in winter and about 20 °C in summer (Fig. 1.2a). Similarly, there is also large year-to-year variation in the temperature sum accumulated during the growing season (Koski and Sievänen 1985). In Jyväskylä, the average total accumulation of temperature sum over 92 years was about 1150 day degree units, with the minimum and maximum accumulations at about 750 and 1500 units, respectively (Fig. 1.2b). Because of the large year-to-year variation in the accumulation of temperature sum, there is also great year-to-year variation in the dates when a given value of temperature sum is reached. For instance, the temperature sum of 600 day degrees was attained on July 6 at the earliest and on August 17 at the latest, so that the range of the date of attaining this temperature sum in Jyväskylä was 42 days (Fig. 1.2b).

The magnitude of the year-to-year variation in the accumulation of temperature sum is further emphasized by a comparison of the minimum and maximum accumulation of temperature sum in Jyväskylä with the geographical variation in the average accumulation of temperature sum. The minimum Jyväskylä accumulation of 750 day degree units corresponds to the average accumulation in Kittilä, located in Finnish Lapland about 600 km north of Jyväskylä. Similarly, the maximum accumulation of 1500 day degree units corresponds to the average accumulation in southern Estonia, about 600 km south of Jyväskylä (Viherä-Aarnio et al. 2005). In all, then, considering the most extreme years, the year-to-year variation of air temperature sum accumulation in Jyväskylä corresponds to a geographical distance of about 1200 km in the north-south direction as regards the average accumulation of temperature sum.

The dates of both the last frosts in the spring and the first frosts in the autumn are essential for the climatic adaptation of boreal and temperate tree species. In Jyväskylä, the date of the last frost of −5 °C or below varied from March 27 to May 13, i.e., by 47 days, over 92 years, and the date of the first autumn frost from September 25 to December 1, i.e., by 67 days (Fig. 1.2c). Furthermore, there was practically no correlation between the date of the last frost and the first frost in the

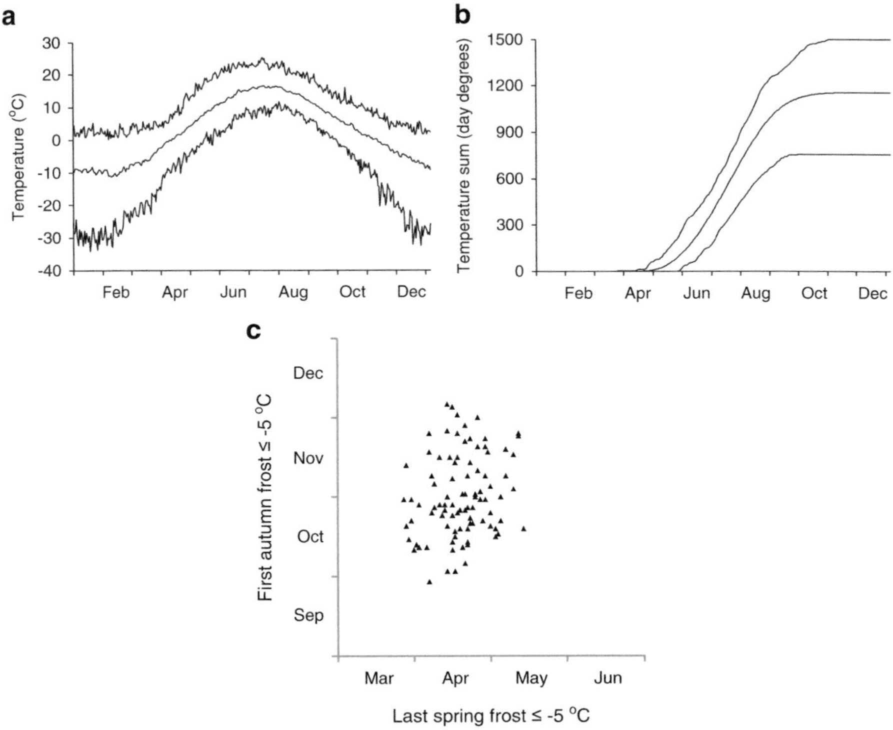

Fig. 1.2 Year-to-year variation of air temperature in Jyväskylä, central Finland (62°14′N, 25°44′E, 115 m asl; Fig. 1.1a). (**a**) The daily mean air temperature, with its year-to-year minimum (*lower curve*), mean (*middle curve*), and maximum (*upper curve*). (**b**) The corresponding three statistics for the accumulation of temperature sum during the growing season. A threshold temperature of +5 °C was used in the calculations (Sect. 2.3). (**c**) The timing of the occurrence of the last spring frost with the daily minimum air temperature T_{min} $-\leq 5$ °C (*horizontal axis*) and the corresponding first autumn frost (*vertical axis*); each symbol corresponds to one year. The results are based on standard air temperature records collected by the Finnish Meteorological Institute in Jyväskylä over the period of 1883–1980. On account of missing observations, the years 1911–1916 were excluded, so that the coverage of the data is 92 years

same year, so that a spring with a late termination of the frost season, for instance, was equally likely to be followed by an early or late onset of the autumn frost season (Fig. 1.2c).

Though only a few air temperature indices were reported here, and for one location only (Fig. 1.2), they may be sufficient to give the reader an idea of the magnitude of the year-to-year variation in the thermal conditions of the boreal and the temperate zone. The large year-to-year variation and the low predictability of the air temperature constitute an additional requirement for the climatic adaptation of the boreal and temperate tree species. The trees are exceptionally long-lived perennials, so that it is not sufficient for them to acclimate to the average seasonal climatic pattern, though it involves a large range of variation in itself (Fig. 1.1b).

Rather, the trees need to acclimate to the climatic conditions of all years, including the most exceptional ones (Koski and Sievänen 1985; Fig. 1.2).

1.2 Night Length in the Boreal and the Temperate Zone

Solar radiation provides the energy source for all green plants, including the trees in the boreal and temperate regions (Larcher 2003). It goes without saying, then, that the seasonality of incoming solar radiation is of major importance for the photosynthetic production of boreal and temperate trees. However, as has been known for long, the seasonality of the light-climate also has other important effects on boreal and temperate trees and many other plants as well (Garner and Allard 1923). These effects are mediated by the relative lengths of the dark part ('night') and the light part ('day') in the diurnal cycle, and they are generally referred to as photoperiodic effects (Thomas and Vince-Prue 1977).

The causal factor in the induction of height growth cessation in boreal and temperate trees is often (Nitsch 1957; Howe et al. 1996), but not always (Håbjørg 1972; Junttila and Kaurin 1985, 1990), night length rather than day length. Regarding several other photoperiodic effects in the annual cycle of the trees, we lack experimental studies that do away with the relationship between day length and night length, so that we do not know which one of them is the causal factor. The modelling studies discussed in the present volume usually refer to night length, so that the concept of night length, NL, is constantly used in the present volume as well.

Night length varies seasonally according to a fixed pattern determined by the latitude. The pattern varies geographically so that the difference between maximum night length during the cold season and minimum night length during the warm season is small at low latitudes and increases as one goes north or south towards the poles. The extreme difference occurs at the highest latitudes, with their periods of polar night (NL = 24 h) and of midnight sun (NL = 0 h) (Bonan 2008). The four locations of the geographical distribution range of *Pinus sylvestris* examined here also evince considerable variation in the seasonal pattern of night length (Fig. 1.1d) though the southernmost regions of the distribution range in Turkey and Spain (Sarvas 1964) are not included in this comparison.

1.3 The Annual Cycle as an Adaptive Strategy

Many plant species are able to tolerate short-term episodic frosts even in the active growth phase. That is especially the case with alpine plant species growing above the tree line in tropical regions (Larcher 2003; Crawford 2008). In these conditions, a heavy night frost may occur at any time of the year, so that the frost survival of the plants cannot be based on seasonal synchronisation of plant development with the climate. The acclimation of these plants involves several biophysical and physiological mechanisms. A crucial one is supercooling, i.e., the lowering of the tissue temperature below the freezing point of the cell sap without forming ice crystals (Sakai and Larcher 1987; Körner 2003). This strategy of frost survival is traditionally referred to as freezing avoidance (Levitt 1980; Larcher 2003).

In boreal and temperate zones, the freezing temperatures occur seasonally rather than episodically (Sakai and Larcher 1987). This implies that overwintering is essential for the frost survival of plants in these zones. Many perennial herbs and grasses overwinter below an insulating and protecting snow cover, and some even in the soil, where not only snow but also the soil itself protects the overwintering plant organs. In Raunkiaer's (1934) classification, the former plants belong to the life form *hemicryptophytes* (Gr. hemi = half; crypto = hidden) and the latter to the life form *cryptophytes*. In contrast, trees belong to the class of *phanerophytes* (Raunkiaer 1934), i.e., except for the roots, they overwinter above the snow cover. During long-lasting seasonal exposure to severe frosts, ice crystals are formed in extracellular spaces in the tree tissues, so that the frost survival of trees is largely based on freezing tolerance (Levitt 1980; Larcher 2003). Due to this extracellular freezing, part of the cellular water is lost to the extracellular ice, which leads to dehydration and lowering of the freezing point of the cell sap (Sutinen et al. 2001). The development of freezing tolerance requires dormancy, i.e., cessation of growth (Weiser 1970; Fuchigami et al. 1982; Junttila 2007). Therefore, as discussed in more detail in Sect. 5.3, studies of the development of freezing tolerance in trees often address growth cessation also.[3]

The seasonally occurring frosts of the boreal and temperate regions, despite their large year-to-year variation (Fig. 1.2c), are still relatively predictable in comparison with the episodic occurrence of frosts in the tropics, and that makes the frost survival strategy based on dormancy possible. Thus, with proper synchronization of the frost hardy dormant phase and the susceptible active growth phase with the seasonality of the air temperature, the trees can grow even at sites where the temperature may drop dozens of degrees below zero in winter (Weiser 1970; Sutinen et al. 1992). In the present volume, the concept of the *annual cycle of development* is used in a broad sense, so that in addition to morphological changes (Sarvas 1972, 1974), it is also used for all biophysical and physiological changes that relate to the seasonal alternation of the dormant and the active phase of the trees (Weiser 1970; Perry 1971; Fuchigami et al. 1982).

The regulation of the annual cycle of development takes place as a result of interaction between genetic and environmental factors, so that each genotype reacts to the environmental cues in its specific way (Sarvas 1972, 1974; Ekberg et al. 1979; Tanino et al. 2010). Understanding this interaction is the key to

[3] In evolutionary terms, there is a trade-off involved in the life form of phanerophytes, i.e., trees have traded off the benefits of a sheltering snow cover in winter for superiority in the competition for light during the growing season. As freezing stress has a lot in common with drought stress at the cellular level, Larcher (1995) put forward the interesting hypothesis that the frost hardiness of plants has its evolutionary origin in drought hardiness.

understanding the climatic adaptation of boreal and temperate tree species and is therefore also crucial for any attempt to project the effects of climate change on these trees. Accordingly, the interaction between genetic and environmental factors in the regulation of the annual cycle of boreal and temperate trees is the leading theme throughout the present volume.

1.4 Organisation of the Book

Today, mathematical dynamic models are an essential part of studies of the annual cycle of boreal and temperate trees, especially where the effects of the projected climate change on trees are addressed (Cannell 1985; Murray et al. 1989; Hänninen and Kramer 2007; Kramer and Hänninen 2009; Morin et al. 2009; Rammig et al. 2010; Caffarra et al. 2011a, b; Chuine et al. 2013). The models have also been used in studies addressing other woody plants (Pop et al. 2000; Van Wijk et al. 2003; Lundell et al. 2008; Jones et al. 2015) and in studies using remote sensing to investigate the phenology of the entire plant community (Picard et al. 2005; Delbart and Picard 2007).

In this book, modelling studies of the annual cycle are discussed with the emphasis on integrating modelling with empirical research. To grasp the message of the book, the reader needs to understand some underlying principles of the dynamic modelling of the annual cycle. To this end, a tutorial on the principles of dynamic modelling, starting from some general principles of the philosophy of science, is provided in Chap. 2.

In the next three chapters, various aspects of the annual cycle of boreal and temperate trees are discussed. Chap. 3 addresses phenological phenomena, such as the timing of vegetative bud burst and height growth cessation. Due to the multitude of phenomena addressed and the exceptionally emphasized need for conceptual clarification, Chap. 3 is longer than the next two chapters. Chapter 4 is devoted to the seasonality of photosynthesis and Chap. 5 to that of frost hardiness. The leading theme of the whole volume, i.e., the interaction of environmental and genetic factors in the regulation of the annual cycle, is re-examined in Chap. 6, where evolutionary aspects of trees are specifically discussed. Next, the scaling up of the implications from the seasonal phenomena of trees to higher organisational levels is briefly discussed in Chap. 7 on the basis of the whole-tree considerations presented in Chaps. 3, 4, 5, and 6. In Chap. 7, the considerations range all the way from the level of the individual tree to the stand level and even to the global level.

The ultimate aim of this volume is to examine the effects of the projected anthropogenic climate change on boreal and temperate trees. That is the theme of Chap. 8, so that the first seven chapters serve as an *ouverture* of sorts to Chap. 8. Much simpler and more straightforward approaches, such as the 'niche-based' or 'climatic envelope' modelling, are available for evaluating the effects of climate change on various biota, including boreal and temperate trees (Bakkenes et al. 2002; Thomas et al. 2004; Hijmans and Graham 2006). However, despite the useful first approximations of the effects of climate change obtained with these approaches, it is the main postulate of the present volume that the projections obtained with these methods need to be augmented with more detailed approaches, where the pertinent biophysical and ecophysiological phenomena are explicitly addressed. It is for this reason that the lengthy *ouverture* is needed. But facilitating the applied research related to climate change is not the only motivation for this chapter and Chaps. 2, 3, 4, 5, 6, and 7. Rather, understanding the climatic adaptation of boreal and temperate tree species is a scientifically fascinating theme for basic research as such. Furthermore, such an understanding already has several important practical applications in the present climate, especially in horticulture and forestry (Campbell 1974; Cannell et al. 1985; Saure 1985; Hänninen et al. 2009; Halaly et al. 2011). Finally, the conclusions for the entire volume are presented in Chap. 9.

References

Aitken, S. N., & Hannerz, M. (2001). Genecology and gene resource management strategies for conifer cold hardiness. In F. J. Bigras & S. J. Colombo (Eds.), *Conifer cold hardiness* (pp. 23–53). Dordrecht: Kluwer Academic Publishers.

Archibold, O. W. (1995). *Ecology of world vegetation*. London: Chapman & Hall. 510 p.

Bakkenes, M., Alkemade, J. R. M., Ihle, F., Leemans, R., & Latour, J. B. (2002). Assessing effects of forecasted climate change on the diversity and distribution of European higher plants for 2050. *Global Change Biology, 8*, 390–407.

Bonan, G. (2008). *Ecological climatology. Concepts and applications*. Cambridge: Cambridge University Press. 550 p.

Breckle, S.-W. (2002). *Walter's vegetation of the Earth. The ecological systems of the geo-biosphere* (4th ed.). Berlin: Springer-Verlag. 527 p.

Caffarra, A., Donnelly, A., Chuine, I., & Jones, M. B. (2011a). Modelling the timing of *Betula pubescens* budburst. I. Temperature and photoperiod: a conceptual model. *Climate Research, 46*, 147–157.

Caffarra, A., Donnelly, A., & Chuine, I. (2011b). Modelling the timing of *Betula pubescens* budburst. II. Integrating complex effects of photoperiod into process-based models. *Climate Research, 46*, 159–170.

Campbell, R. K. (1974). Use of phenology for examining provenance transfers in reforestation of Douglas-fir. *Journal of Applied Ecology, 11*, 1069–1080.

Cannell, M. G. R. (1985). Analysis of risks of frost damage to forest trees in Britain. In P. M. A. Tigerstedt, P. Puttonen, & V. Koski (Eds.), *Crop physiology of forest trees* (pp. 153–166). Helsinki: Helsinki University Press.

Cannell, M. G. R., Murray, M. B., & Sheppard, L. J. (1985). Frost avoidance by selection for late budburst in *Picea sitchensis*. *Journal of Applied Ecology, 22*, 931–941.

Chuine, I., de Cortazar-Atauri, I. G., Kramer, K., & Hänninen, H. (2013). Plant development models. In: M. D. Schwartz (Ed.), *Phenology: An integrative environmental science, Second Edition* (pp. 275–293). Dordrecht: Springer.

Crawford, R. M. M. (2008). *Plants at the margin. Ecological limits and climate change*. Cambridge: Cambridge University Press. 478 p.

Delbart, N., & Picard, G. (2007). Modeling the date of leaf appearance in low-arctic tundra. *Global Change Biology, 13*, 2551–2562.

Ekberg, I., Eriksson, G., & Dormling, I. (1979). Photoperiodic reactions in conifer species. *Holarctic Ecology, 2*, 255–263.

Fuchigami, L. H., Weiser, C. J., Kobayashi, K., Timmis, R., & Gusta, L. V. (1982). A degree growth stage (°GS) model and cold acclimation in temperate woody plants. In P. H. Li & A. Sakai (Eds.), *Plant cold hardiness and freezing stress. Mechanisms and crop implications* (Vol. 2, pp. 93–116). New York: Academic Press.

Garner, W. W., & Allard, H. A. (1923). Further studies in photoperiodism, the response of the plant to relative length of day and night. *Journal of Agricultural Research, 23*, 871–920.

Håbjørg, A. (1972). Effects of light quality, light intensity and night temperature on growth and development of three latitudinal populations of *Betula pubescens* Ehrh. *Scientific Reports of the Agricultural University of Norway, 51*(26), 1–17.

Halaly, T., Zion, B., Arbel, A., Regev, R., Barak, M., & Or, E. (2011). Short exposure to sublethal heat shock facilitates dormancy release in grapevines. *American Journal of Enology and Viticulture, 62*, 106–112.

Hänninen, H., & Hari, P. (1996). The implications of geographical variation in climate for differentiation of bud dormancy ecotypes in Scots pine. In P. Hari, J. Ross, M. Mecke (Eds.), Production process of Scots pine; Geographical variation and models. *Acta Forestalia Fennica, 254*, 11–21.

Hänninen, H., & Kramer, K. (2007). A framework for modelling the annual cycle of trees in boreal and temperate regions. *Silva Fennica, 41*, 167–205.

Hänninen, H., Luoranen, J., Rikala, R., & Smolander, H. (2009). Late termination of freezer storage increases the risk of autumn frost damage to Norway spruce seedlings. *Silva Fennica, 43*, 817–830.

Hijmans, R. J., & Graham, C. H. (2006). The ability of climate envelope models to predict the effect of climate change on species distributions. *Global Change Biology, 12*, 2272–2281.

Howe, G. T., Gardner, G., Hackett, W. P., & Furnier, G. R. (1996). Phytochrome control of short-day-induced bud set in black cottonwood. *Physiologia Plantarum, 97*, 95–103.

Jones, H. G., Gordon, S. L., & Brennan, R. M. (2015). Chilling requirement of Ribes cultivars. *Frontiers in Plant Science, 5, Article 767*. doi:10.3389/fpls.2014.00767.

Junttila, O. (2007). Regulation of annual shoot growth cycle in northern tree species. In E. Taulavuori & K. Taulavuori (Eds.), *Physiology of northern plants under changing environment* (pp. 177–210). Kerala: Research Signpost.

Junttila, O., & Kaurin, Å. (1985). Climatic control of apical growth cessation in latitudinal ecotypes of *Salix pentandra* L. In Å. Kaurin, O. Junttila, & J. Nilsen (Eds.), *Plant production in the north* (pp. 83–91). Tromsø: Norwegian University Press.

Junttila, O., & Kaurin, Å. (1990). Environmental control of cold acclimation in *Salix pentandra. Scandinavian Journal of Forest Research, 5*, 195–204.

Körner, C. (2003). *Alpine plant life. Functional plant ecology of high mountain ecosystems* (2nd ed.). Berlin: Springer-Verlag. 344 p.

Koski, V., & Sievänen, R. (1985). Timing of growth cessation in relation to the variations in the growing season. In P. M. A. Tigerstedt, P. Puttonen, & V. Koski (Eds.), *Crop physiology of forest trees* (pp. 167–193). Helsinki: Helsinki University Press.

Kramer, K., & Hänninen, H. (2009). The annual cycle of development of trees and process-based modelling of growth to scale up from the tree to the stand. In: A. Noormets (Ed.), *Phenology of ecosystem processes* (pp 201–227). Dordrecht: Springer.

Langlet, O. (1971). Two hundred years genecology. *Taxon, 20*, 653–721.

Larcher, W. (1995). *Physiological plant ecology. Ecophysiology and stress physiology of functional groups* (3rd ed.). Berlin: Springer. 506 p.

Larcher, W. (2003). *Physiological plant ecology. Ecophysiology and stress physiology of functional groups* (4th ed.). Berlin: Springer-Verlag. 513 p.

Levitt, J. (1980). *Responses of plants to environmental stresses* (Chilling, freezing, and high temperature stresses 2nd ed., Vol. I). New York: Academic Press. 497 p.

Lundell, R., Saarinen, T., Åström, H., & Hänninen, H. (2008). The boreal dwarf shrub *Vaccinium vitis-idaea* retains its capacity for photosynthesis through the winter. *Botany, 86*, 491–500.

Morin, X., Lechowicz, M. J., Augspurger, C., O'Keefe, J., Viner, D., & Chuine, I. (2009). Leaf phenology in 22 North American tree species during the 21st century. *Global Change Biology, 15*, 961–975.

Murray, M. B., Cannell, M. G. R., & Smith, R. I. (1989). Date of budburst of fifteen tree species in Britain following climatic warming. *Journal of Applied Ecology, 26*, 693–700.

Nitsch, J. P. (1957). Photoperiodism in woody plants. *Proceedings of the American Society for Horticultural Science, 70*, 526–544.

Perry, T. O. (1971). Dormancy of trees in winter. *Science, 171*, 29–36.

Picard, G., Quegan, S., Delbart, N., Lomas, M. R., Le Toan, T., & Woodward, F. I. (2005). Bud-burst modelling in Siberia and its impact on quantifying the carbon budget. *Global Change Biology, 11*, 2164–2176.

Pop, E. W., Oberbauer, S. F., & Starr, G. (2000). Predicting vegetative bud break in two arctic deciduous shrub species, *Salix pulchra* and *Betula nana*. *Oecologia, 124*, 176–184.

Rammig, A., Jönsson, A. M., Hickler, T., Smith, B., Bärring, L., & Sykes, M. T. (2010). Impacts of changing frost regimes on Swedish forests: Incorporating cold hardiness in a regional ecosystem model. *Ecological Modelling, 221*, 303–313.

Raunkiaer, C. (1934). *The life forms of plants and statistical plant geography*. Oxford: Clarendon Press. 632 p.

Rehfeldt, G. E., Ying, C. C., Spittlehouse, D. L., & Hamilton, D. A., Jr. (1999). Genetic responses to climate in *Pinus contorta*: Niche breadth, climate change, and reforestation. *Ecological Monographs, 69*, 375–407.

Sakai, A., & Larcher, W. (1987). *Frost survival of plants. Responses and adaptation to freezing stress*. Berlin: Springer-Verlag. 321 pp.

Sarvas, R. (1964). *Havupuut*. Porvoo: Werner Söderström Osakeyhtiö. 531 p.

Sarvas, R. (1972). Investigations on the annual cycle of development of forest trees. Active period. *Communicationes Instituti Forestalis Fenniae, 76*(3), 1–110.

Sarvas, R. (1974). Investigations on the annual cycle of development of forest trees. II. Autumn dormancy and winter dormancy. *Communicationes Instituti Forestalis Fenniae, 84*(1), 1–101.

Saure, M. C. (1985). Dormancy release in deciduous fruit trees. *Horticultural Reviews, 7*, 239–300.

Savolainen, O., Pyhäjärvi, T., & Knürr, T. (2007). Gene flow and local adaptation in trees. *Annual Review of Ecology, Evolution, and Systematics, 38*, 595–619.

Sutinen, M.-L., Palta, J. P., & Reich, P. B. (1992). Seasonal differences in freezing stress resistance of needles of *Pinus nigra* and *Pinus resinosa*: Evaluation of the electrolyte leakage method. *Tree Physiology, 11*, 241–254.

Sutinen, M.-L., Arora, R., Wisniewski, M., Ashworth, E., Strimbeck, R., & Palta, J. (2001). Mechanisms of frost survival and freeze-damage in nature. In F. J. Bigras & S. J. Colombo (Eds.), *Conifer cold hardiness* (pp. 89–120). Dordrecht: Kluwer Academic Publishers.

Tanino, K. K., Kalcsits, L., Silim, S., Kendall, E., & Gray, G. R. (2010). Temperature-driven plasticity in growth cessation and dormancy development in deciduous woody plants: A working hypothesis suggesting how molecular and cellular function is affected by temperature during dormancy induction. *Plant Molecular Biology, 73*, 49–65.

Thomas, B., & Vince-Prue, D. (1977). *Photoperiodism in plants* (2nd ed.). San Diego: Academic Press. 428 p.

Thomas, C. D., Cameron, A., Green, R. E., Bakkenes, M., Beaumont, L. J., Collingham, Y. C., Erasmus, B. F. N., de Siqueira, M. F., Grainger, A., Hannah, L., Hughes, L., Huntley, B., van Jaarsveld, A. S., Midgley, G. F., Miles, L., Ortega-Huerta, M. A., Peterson, A. T., Phillips, O. L., & Williams, S. E. (2004). Extinction risk from climate change. *Nature, 427*, 145–148.

Van Wijk, M. T., Williams, M., Laundre, J. A., & Shaver, G. R. (2003). Interannual variability of plant phenology in tussock tundra: Modelling interactions of plant productivity, plant phenology, snowmelt and soil thaw. *Global Change Biology, 9*, 743–758.

Viherä-Aarnio, A., Häkkinen, R., Partanen, J., Luomajoki, A., & Koski, V. (2005). Effects of seed origin and sowing time on timing of height growth cessation of *Betula pendula* seedlings. *Tree Physiology, 25*, 101–108.

Weiser, C. J. (1970). Cold resistance and injury in woody plants. *Science, 169*, 1269–1278.

Whittaker, R. H. (1975). *Communities and ecosystems* (2nd ed.). New York: Macmillan Publishing Co. 385 p.

Chapter 2
Dynamic Modelling of the Annual Cycle

Contents

Abstract A hypothetico-deductive framework for dynamic ecophysiological modelling of the annual cycle of boreal and temperate trees is described and discussed. In the framework used, the modelling is closely associated with experimental and observational empirical studies, so that inductive ('empirical') and deductive ('theoretical') phases alternate in the research. Computer simulations are deductive, and their results therefore contain nothing but implications of the assumptions of the model used in the simulations. Empirical data are used either for inductive formulation of the models or for testing the models after deducing their predictions by means of simulations. In dynamic modelling, the time courses of seasonal ecophysiological processes are simulated by first calculating the momentary rate of development on the basis of the input data of environmental factors, such as air temperature and night length. After that, the time course of the state of development is obtained by mathematical integration of the rate of development with respect to time. A unifying notation is described for the rate and state variables of different aspects of the annual cycle. Diverse model categories, based on major differences in the ecophysiological phenomena addressed, are described and compared. Finally, the realism, accuracy, and generality of the models are discussed. A novel concept of coverage is introduced for use in the assessment of the realism of the models. Biological levels of organisation, vertical reduction, and emergent properties are briefly discussed in relation to the realism of the models. The framework described and discussed in this chapter forms the basis for the subsequent chapters, in which different aspects of the annual cycle in boreal and temperate trees are addressed.

H. Hänninen, *Boreal and Temperate Trees in a Changing Climate*, Biometeorology 3,
DOI 10.1007/978-94-017-7549-6_2

Keywords Annual cycle • Boreal trees • Computer simulations • Day degrees • Dynamic models • Ecophysiological modelling • Modelling framework • Model realism • Seasonality • Temperate trees

2.1 A Hypothetico-Deductive Modelling Framework

In the philosophy of science, research is generally divided into two main categories, which are referred to as inductive and deductive research. In broad terms, induction refers to empirical research: whenever consistent results are obtained in empirical data, they are generalised into a rule, model, or theory (Niiniluoto 1983). Historically, the development of any scientific discipline is typically inductive at the early stages, when the research has not yet produced many generalised concepts and theories. In tree ecophysiology, inductive research is exemplified by determining the environmental responses of ecophysiological phenomena, such as photosynthesis or frost hardening/dehardening, on the basis of experiments carried out in controlled conditions (Hänninen and Lundell 2007).

At a later stage in the history of the scientific discipline, when generalised concepts and theories have accumulated, the research often adopts a deductive (or 'theoretical') approach and starts applying the hypothetico-deductive method (HDM; Haila 1982; Niiniluoto 1983). Now the starting point is a generalised theory, which is to be tested against data from a particular study specifically designed for the purpose. First, predictions to be tested by means of various experimental designs are derived (deduced) from the theory. The research takes an "if x then y" approach: if the theory or model x is true, then a given experiment will yield the results y. With such logical inference (or calculations), maximally efficient experimental designs are planned (Hänninen 1990a, b). Second, the experiment is carried out and its results are compared with the predictions of the theory to be tested. In this way, the theory is either falsified or corroborated, i.e., supported, by the experimental data (Niiniluoto 1983). Deductive inference is also applied whenever a simulation model is run on a computer (Hänninen and Lundell 2007). In this case, the computer calculates the implications that the generalised model used in the simulations produces for the particular case specified by the input data (Fig. 2.1).

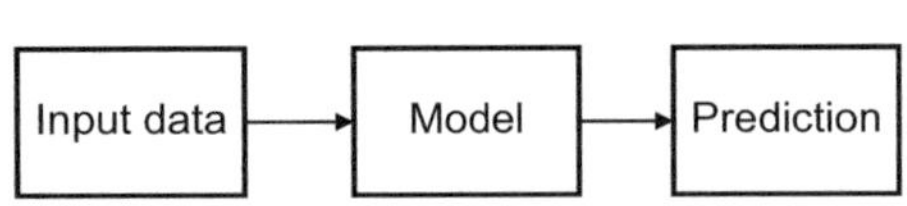

Fig. 2.1 The deductive principle of using mathematical models for research calculations. The formulae of the model, together with the input data, determine the prediction of the model. In this way the prediction of the model is a logical necessity implied by the model for the particular set of input data used, so that the prediction contains nothing new as such. Even so, the calculations are needed in practical research, since it is generally not possible to see the implications directly from the structure of the model and the input data

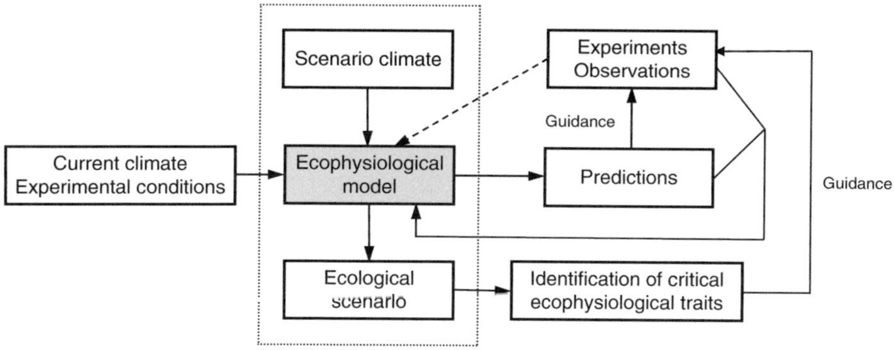

Fig. 2.2 The modelling framework for the study of the annual cycle of boreal and temperate trees adopted in the present volume, based on the hypothetico-deductive method (HDM). Simulations with various types of input data are carried out in order to reveal the implications of ecophysiological models of the annual cycle and thus to guide the planning of informative experimental designs for further testing and development of the models. Besides the HDM approach, models are also formulated directly on the basis of previous empirical results (*dashed arrow*). The *box* drawn with the *dotted line* represents the studies where scenarios for the ecological effects of the projected climate change are obtained by means of computer simulations using scenario climates as input for the ecophysiological models. Like the other implications of the models, the ecological scenarios thus obtained are not taken as the final truth but are used to guide experimental studies aiming at improving the models (Modified from Hänninen and Lundell (2007), and published with the permission of Research Signpost)

This volume makes use of a hypothetico-deductive modelling framework designed for plant ecophysiological studies (Fig. 2.2; Hänninen 1990a, b; Hänninen and Lundell 2007; Hänninen et al. 2013). Following an approach that is commonly used in ecological research on climatic change nowadays, the implications of the projected climatic change are deduced by means of computer simulations with ecophysiological models. In this case, the deduction carried out in the simulation reveals the implications of the ecophysiological generalisations written into the model for the particular case of assumed climatic scenario (Fig. 2.2, box drawn with a dotted line). It should be noted that the ecological scenario thus obtained contains nothing new as such because it is totally dictated by the assumptions inherent in the model and by the input climatic scenario (Fig. 2.1). However, deduction with a computer simulation is needed because it is not possible to see the results directly from the structure of the model applied. Even though the results are altogether logical implications of the model used, it is not rare that even counterintuitive results, i.e., results that seem contradictory to the assumptions of the model and/or input data used, are obtained as a result of the simulations. This is also the case in the paradoxical hypothesis that climatic warming will increase frost damage to boreal and temperate trees in the future (Cannell 1985; Hänninen 1991). The hypothesis is discussed in detail in Chap. 8.

The box drawn with a dotted line in Fig. 2.2 provides ecological projections for the future, which is the kind of information requested by decision-makers and other end-users of scientific information. However, it is of utmost importance that the ecological scenarios obtained are interpreted with care. They will be realized only (i) if the climatic scenario is realized and (ii) if the ecophysiological model applied provides a sufficient description of the ecophysiological phenomena involved (Hänninen 1995a, b). It goes without saying that we do not have the option of waiting for the formulation of perfect models for the scenario simulations, so that we have to carry out the simulations with the best possible knowledge, i.e., with the best models available. At the same time, however, it also necessary to continue the critical testing and improving of the models. The simulations mercilessly reveal the implications of the models, so that they are also useful for the further testing and improving of the models (Fig. 2.2). In this way, inductive ('empirical') and deductive ('theoretical') phases alternate in the research, with the ultimate aim of understanding the climatic adaptation of the boreal and temperate tree species and the effects of climatic change on them.

2.2 Basic Concepts[1]

Dynamic models are commonly used in numerous fields of research nowadays. When they are used, regardless of the discipline, the processes are studied over time (Mesarovic and Takahara 1975; Landsberg 1986; Thornley and Johnson 1990). The variables of the models are therefore presented as a function of time, and in figures representing the predictions of the models, the time dimension is usually seen on the horizontal axis. The variables of dynamic models fall into two categories. The value of a *state variable* indicates the status of the system studied at any given moment, and the value of the corresponding *rate variable* indicates the rate of change of the state. Mathematically, the rate variable is the first derivative of the state variable with respect to time, and the state variable is the integral of the rate variable with respect to time (Jeffers 1978; Landsberg 1986; Thornley and Johnson 1990; Hänninen and Lundell 2007).

Dynamic models have been used in studies related to plant ecophysiology since the time of Réaumur (1735), so that the history of the approach is much longer than the history of plant ecophysiology as a formal scientific discipline, let alone the history of using computers in plant ecophysiological studies. The explicit introduction of dynamic modelling into studies of the annual cycle of boreal and temperate trees took place in 1970s (Hari et al. 1970; Hari 1972; Sarvas 1972, 1974; Winter 1973; Landsberg 1974; Richardson et al. 1974). Different researchers have used different terminologies and symbols in their publications. Throughout this volume, the terminology and notation introduced in Hänninen and Kramer's (2007) review is used.

[1] This section is based on Hänninen and Kramer's (2007) review.

In dynamic models of the annual cycle of boreal and temperate trees, the state and rate variables are specified as the *state of development*, S(t), and the *rate of development*, R(t), both having their specific value at each time instant t (Hari 1968, 1972; Hari et al. 1970; Landsberg 1977).[2]

In each modelling study, this pair of variables is further defined according to the ecophysiological phenomenon addressed in that particular study. In studies addressing frost hardiness, for instance, the state variable is the state of frost hardiness, $S_h(t)$, indicating the frost hardiness of the tree (or a given tissue of the tree) at each time instant t. The value of the variable $S_h(t)$ is expressed in a unit of temperature (°C), and by definition it indicates, for each time instant t, the threshold temperature under which frost damage will occur. Correspondingly, the rate of development in this case is specified as the rate of change of frost hardiness, $R_h(t)$, i.e., the rate of hardening or dehardening, and it is expressed in units of air temperature per unit of time, in most cases as °C day^{-1} (Kellomäki et al. 1992, 1995).

Most models of the annual cycle are limited to one aspect of the cycle, but there are also more comprehensive models, where several attributes of the annual cycle are simulated at the same time. In these models, both the rate of development and the state of development involve several attributes, so that $S(t) = S_i(t)$, $i = 1,.., n$ and $R(t) = R_i(t)$, $i = 1,..., n$; where i is the index of the attribute and n is the number of attributes addressed. Both types of models are exemplified in the present volume.

The concept of the *environmental response of the rate of development* links the annual development of the tree to its environment and is thus a crucial concept in the ecophysiological modelling approach (Hänninen and Kramer 2007). Mathematically, the environmental response of the rate of development is a function f that indicates the dependence of the value of R(t) on the values of the environmental factors E(t) (Sarvas 1972, 1974; Hari 1972; Landsberg 1974):

$$R(t) = f(E(t)) \tag{2.1}$$

The environmental factors E(t) may involve one (e.g., air temperature) or several factors (e.g., air temperature and night length). In the latter case $E(t) = E_i(t)$, $i = 1,..., n$; where i is the index of the environmental factor and n is the number of factors considered.

Equation 2.1 describes the simplest case, where the response of the rate of development to one or several environmental factors does not change over time. Such models are frequently used when a limited part of the annual cycle is considered. In trees, however, the responses change during the annual cycle (Pelkonen and Hari 1980; Kobayashi and Fuchigami 1983a, b; Leinonen

[2] The concept "development" is used here in a broad sense, including all biophysical, physiological, and morphological changes involved in the annual cycle of the boreal and temperate trees (Hänninen and Kramer 2007). In textbooks of plant physiology and plant developmental biology, the concept "development" is classically used in a more restricted sense, referring only to the irreversible anatomical and organological changes (ontogenesis) taking place in plants.

et al. 1997), so that in more comprehensive models the environmental response also changes during the development. In some models, feedback is assumed, i.e., the environmental response of the rate of development changes along the development of the simulated attribute itself:

$$R_i(t) = f_i(E(t), S_i(t)) \tag{2.2}$$

where f_i is the environmental response of the rate of development of attribute i. In this case the prevailing state of development of the attribute i, $S_i(t)$, affects its own rate of development, $R_i(t)$. This approach was introduced by Pelkonen and Hari (1980) in their modelling of the springtime recovery of photosynthetic capacity in *Pinus sylvestris*. In their model, high air temperatures (T(t)) cause the recovery of the photosynthetic capacity (E(t) = T(t)), but the effect of any given air temperature changes as a function of the prevailing photosynthetic capacity. This phenomenon, crucial for the photosynthetic production of boreal coniferous trees, is dealt with in detail in Chap. 4.

In some models the rate of development of a given attribute, $R_i(t)$, is affected by the state of development of another attribute, $S_j(t)$. In this case the rate of development is calculated as follows:

$$R_i(t) = f_i\left(E(t), S_j(t)\right) \tag{2.3}$$

This approach has been taken especially in the modelling of the annual phenological cycle and the annual cycle of frost hardiness. In this approach, the environmental responses of the rate of ontogenetic development towards bud burst (Chap. 3) and the environmental responses of the rate of change in frost hardiness (Chap. 5) are treated as dependent on other attributes of the annual cycle. Accordingly, in the present volume the environmental regulation of a given attribute i of the annual cycle is divided into direct and indirect parts. The environmental regulation is regarded as direct whenever no other attribute of the annual cycle is referred to, whereas indirect environmental regulation of the attribute i takes place via the effects of the environmental factors on another attribute j, whose state of development affects the rate of development of the examined attribute i, as formulated in Eq. 2.3.

Regardless of how the rate of development is modelled as a function of the environmental factors (Eqs. 2.1, 2.2, and 2.3), the state of development, S(t), is by definition obtained by integrating the rate of development with respect to the time from the starting day of the simulation, t_0, till the time instant t in question (Hari et al. 1970; Hari 1972; Landsberg 1977):

$$S(t) = \int_{t_0}^{t} R(t)dt \tag{2.4}$$

Mathematically, both the rate of development, R(t), and the state of development, S (t), are continuous variables, which means that theoretically their values are given for any time instance, and the value of S(t) is obtained as an analytical integral of R (t) (Eq. 2.4). In ecophysiological applications, however, the continuous analytical formulae are replaced by numerical approximations, which means that the mean rate of development is calculated for a short time step, typically 1 day (or 1 h) in models of the annual cycle, using the mean value of the environmental variable (s) for the time step. Subsequently, the analytical integration presented in Eq. 2.4 is replaced by summing up the mean values obtained for the short time steps (Landsberg 1977, 1986; see the example in Sect. 2.3).

In order to further facilitate the explication and comparison of the models of the annual cycle of boreal and temperate trees, Hänninen and Kramer (2007) divided the models and the corresponding phenomena into two categories. The phenomena belonging to the category of *fixed sequence development* are irreversible. Accordingly, in models addressing developmental phenomena belonging to this category, the state of development, S(t), can only increase or stay constant but not decrease (resetting the value at the beginning of a new annual cycle is, of course, an exception to this rule). This implies that the rate of development, R(t), is always either positive (increasing S(t)) or zero (constant S(t)). The irreversible ontogenetic development of trees belongs to the category of fixed sequence development (Sarvas 1972, 1974; Fuchigami et al. 1982). The best known examples of fixed sequence modelling are the various models predicting the timing of the bud burst or the flowering of trees in spring (Wang 1960; Boyer 1973; Kramer 1994a, b; Hänninen 1995a; Häkkinen et al. 1998; Chuine 2000; Linkosalo et al. 2008). These are discussed in Sects. 3.2, 3.3, and 3.4.

Some physiological phenomena of the annual cycle are reversible. That is the case, at least partially, in the annual cycle of frost hardiness (Repo et al. 1990; Repo 1991) and that of the photosynthetic capacity of conifers (Pisek and Winkler 1958; Pelkonen 1980, 1981a, b; Bergh et al. 1998). In Hänninen and Kramer's (2007) classification, these phenomena and the corresponding models are classified into the category of *fluctuating development*. Contrary to the fixed sequence models, the rate of development R(t) can be negative and the value of S(t) can correspondingly decrease in models of fluctuating development. Furthermore, the driving force in fluctuating development is a change in the environmental factors. In any constant environmental conditions, fluctuating development will attain a stationary state, i.e., the state of development will stay constant (Pelkonen and Hari 1980; Repo et al. 1990; Repo 1993). This approach is discussed in detail in Chap. 4, in connection with the annual cycle of photosynthesis. Integrated models, i.e., models in which both fixed sequence and fluctuating phenomena are included, are discussed in Chap. 5, in connection with the annual cycle of frost hardiness.

2.3 The Modelling Principle

The formulation of the environmental responses is a crucial phase in ecophysiological modelling, not only because the responses link the tree to its environment but also because all primary biological information is introduced into the study in the formulation of the responses (Hänninen and Lundell 2007). For this reason, the empirical researchers who determine the responses make an important contribution to the modelling, too, even though they do not necessarily present their results with mathematical formulae or carry out computer simulations themselves. The computer simulations carried out after the responses have been formulated are pure mathematical deductions, so that all results of the simulation studies are actually determined when the responses are formulated. Any new hidden information revealed by the simulations is a logical and necessary implication of the formulated responses (Fig. 2.1).

The day degree – or temperature sum – model is the best known example of models used in studies of the annual cycle of boreal and temperate trees. The temperature sum model is often used for predicting the timing of phenological events such as vegetative bud burst or flowering in spring, not only with trees but also with other plants (Katz 1952; Arnold 1959; Wang 1960; Boyer 1973; Diekman 1996). The model is introduced here in order to illustrate the modelling approach in general; the ecophysiological interpretation of the approach is discussed in more detail in Sect. 3.2.1.2.

In the temperature sum model, too, the crucial point is the environmental response of the rate of development, or, more precisely, the air temperature response of the accumulation rate of arbitrarily chosen developmental units (day degrees, dd), $R_{dd}(t)$. When the daily mean air temperature $T_{mean}(t)$ is below a threshold T_{thr}, the rate is assumed to be zero, and when above the threshold, the rate is assumed to increase linearly with rising air temperature (Fig. 2.3a):

$$R_{dd}(t) = \begin{cases} 0 & T_{mean}(t) < T_{thr} \\ T_{mean}(t) - T_{thr} & T_{mean}(t) \geq T_{thr} \end{cases} \tag{2.5}$$

It is a common practice, especially in Finland, to use the threshold value of $T_{thr} = +5$ °C in the temperature sum model (Sarvas 1967; Hari and Häkkinen 1991; Fig. 2.3a).

When the response is fixed in the above form, then the result of the simulation is determined for any given input temperature data (Fig. 2.1). Assuming the hypothetical data presented in Fig. 2.3b for the daily mean air temperature, T_{mean}, the daily accumulation rate of dd units, $R_{dd}(t)$, is determined as follows: during the first 5 days, T_{mean} fluctuates below the threshold of +5 °C (Fig. 2.3b), so that the daily accumulation rate of dd units is zero during that time (Fig. 2.3c), despite the fluctuation of the air temperature. On day 6, T_{mean} rises to 8 °C (Fig. 2.3b), so that according to the air temperature response, the accumulation rate of dd units is 3 dd day^{-1} (Eq. 2.5; Fig. 2.3a); so this value is plotted for day 6 in the graph presenting the daily accumulation rate of dd units (Fig. 2.3c). Correspondingly, on day 7, T_{mean} is 10 °C (Fig. 2.3b), so that the accumulation rate of dd units is 5 dd day^{-1} (Eq. 2.5; Fig. 2.3a, c).

The state variable, i.e., the accumulated temperature sum, $S_{dd}(t)$, is obtained by integrating the rate of development (Fig. 2.3c) with respect to the time from the beginning of the simulation till the time instant t in question. In the numerical

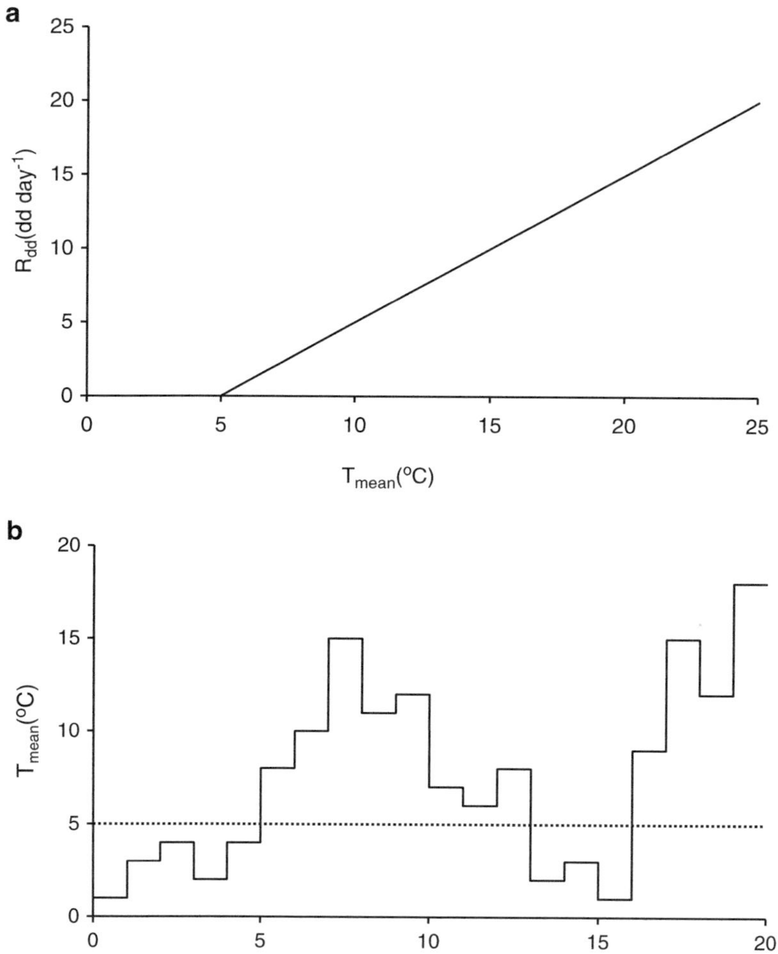

Fig. 2.3 The principle of computer simulations with dynamic models in studies of the annual cycle of boreal and temperate trees. The classical temperature sum model is used as an example. (**a**) Air temperature response of the accumulation rate of arbitrary developmental units, (day degrees, dd), R_{dd}. (**b**) A hypothetical pattern of daily mean air temperatures, T_{mean}. The horizontal dotted line indicates the threshold temperature used in the temperature sum model applied (Eq. 2.5). (**c**) The pattern of the accumulation rate of day degrees, R_{dd}, implied by the model presented in (**a**) for the particular air temperature pattern presented in (**b**). (**d**) The corresponding pattern implied for the temperature sum, S_{dd}, i.e., the accumulation of the dd units. According to the prediction of the model, any developmental event with a high temperature requirement, H_{crit}, of 70 dd occurs on day 20 (*dashed lines*). The *dotted bar* and the *bar with oblique lines* in (**c**) illustrate the numerical integration carried out (see text for explanation)

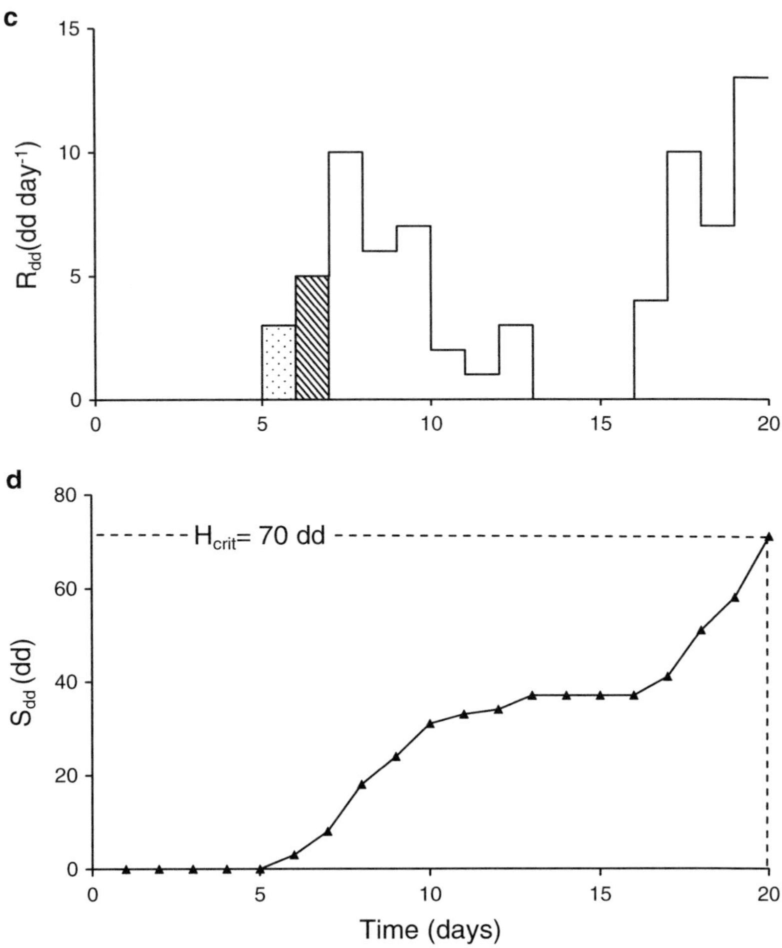

Fig. 2.3 (continued)

approximation, however, the analytical integral of Eq. 2.4 is replaced by a summing up of the average rates calculated for each day. Geometrically, the summation is identical to the calculation of the area delineated by the x and y axes, the curve of the accumulation rate of dd units, R_{dd}, and a vertical line indicating the time instant t till which the numerical integration is carried out. The area of a rectangle representing 1 day is obtained by multiplying the width of the rectangle (= 1 day in each case) by its height. Thus, the calculation produces the accumulation of 1 day $\times$ 3 dd day^{-1} = 3 dd for day 6. This is the area corresponding to day 6 (Fig. 2.3c, the dotted bar); so this value is plotted as the value of the temperature sum, $S_{dd}(t)$, at the end of day 6 (Fig. 2.3d). For the area corresponding to day 7, the value of 1 day $\times$ 5 dd day^{-1} = 5 dd is obtained (Fig. 2.3c, the bar with oblique lines). Thus the numerical integration (=summation) produces the value of 3 dd + 5 dd = 8 dd for the value of the temperature sum, $S_{dd}(t)$, at the end of day 7 (Fig. 2.3d).

This procedure is repeated for all days, and the curve for the accumulated temperature sum is obtained as a result (Fig. 2.3d). For each day, the value of the rate variable from Fig. 2.3c is identical to the slope of the line representing the increase in S_{dd} during that day in Fig. 2.3d. Thus the state variable S_{dd} stays constant during days 1–5 and again during days 14–16 (Fig. 2.3d), when the rate variable R_{dd} has the value of zero (Fig. 2.3c).

In the example illustrated in Fig. 2.3, an arbitrary meteorological scale (day degrees, dd) is used for the state of development and the rate of development (dd day^{-1}). Thus the result obtained in the simulation (Fig. 2.3d) carries no ecophysiological information as such. If ecophysiological predictions, e.g., for the timing of growth onset, are to be deduced from the model, then the information represented by the response curve (Fig. 2.3a) needs to be augmented with the value of one ecophysiological parameter, i.e., the *high temperature requirement of growth onset*, H_{crit} (Hänninen and Kramer 2007), often also referred to as the day degree requirement (Olsen et al. 2003) or critical temperature sum threshold (Linkosalo et al. 2008; Lappalainen 2010). When the value of this parameter is known, then the timing of bud burst can be predicted by using the accumulated day degrees as the dimension selected for the state variable (Fig. 2.3d). This is a standard procedure in using temperature sum models, and that is why the procedure is described here, too. The transformation of the meteorological approach into an ecophysiologically explicit one is presented in Sect. 3.2.1.2.

The simulation example in Fig. 2.3 demonstrates an essential point that was made above already: Once the environmental response of the rate of development is determined, then the result of any simulation with a given input is actually determined also; even so, the simulation is needed to reveal hidden implications (Fig. 2.1). Thus, even though the calculations produce no new information as such, they are needed to reveal the predictions they imply. Without making the calculations it is not possible, for instance, to see directly from Fig. 2.3a, b that according to the prediction of the model, any developmental event requiring 70 dd units will occur on day 20 (Fig. 2.3d) in all cases where the daily mean air temperature fluctuates as portrayed in Fig. 2.3b.

The classical temperature sum model is an example of fixed sequence modelling, i.e., the rate of development is either zero or positive (Fig. 2.3c), so that the state of development either stays constant or increases (Fig. 2.3d). In fluctuating development, the rate of development can also be negative (Hänninen and Kramer 2007). When it is negative, the area corresponding to the rate of development is formed below the x axis (cf. Fig. 2.3c) and is accordingly equipped with a minus sign in the summation taking place in the numerical integration. The value equipped with the minus sign indicates the negative slope of the decreasing curve for the state variable during the respective days. The fluctuating development approach is illustrated in Chaps. 4 and 5, where the annual cycle of photosynthesis and frost hardiness, respectively, is discussed.

2.4 Model Realism and the Discipline of Whole-Tree Ecophysiology

Levins (1966, 1968) introduced the concepts of realism, accuracy, and generality to the evaluation of biological models.[3] This 'Modeller's Triangle' has become an important tool for examining and comparing the properties of different biological models (Sharpe 1990; Hänninen 1995b). In brief, the concept of *model realism* refers to the degree to which the model describes and explains real-life phenomena. However, as no model provides a full description of real-life phenomena, the degree of model realism is evaluated on two main criteria in the present volume. The first one is causality. The relationships described by a realistic model have to be causal relationships, not just occasional correlations between variables (Hänninen and Lundell 2007). The second criterion, introduced in the present volume, is the coverage of the model. A realistic model has to cover most, if not all, of the relevant aspects of the object of study.

However, the relevance of the different aspects, and thus also the coverage of the model, is context-dependent. For instance, in the context of using ecophysiological models of plant photosynthesis in short-term simulations, one can neglect the effects of atmospheric concentration of carbon dioxide because that aspect can be regarded as a constant. It goes without saying, however, that this cannot be done in long-term simulations projecting the effects of climate change over decades.

The concept of *model accuracy* is about how well the output of the model corresponds to empirical observations. It can be evaluated by fitting the model to empirical data, i.e., by estimating the values of the model parameters (Kramer 1994a, b; Chuine et al. 1998; Mäkelä et al. 2004) or, better still, by testing the already parameterized model with independently gathered data (Hari et al. 2008, 2009; Linkosalo et al. 2008). In either case, the model accuracy is examined by means of statistical indices such as the root mean squared error (RMSE, Linkosalo et al. 2008). In the evaluation of the model by interpreting the obtained value of the RMSE, the precision of measurement should be taken into account. When the precision of measurement is low, any difference between the observation and the model prediction may be partly caused by shortcomings inherent in the data. In such cases the model may be corroborated even by relatively high values of the RMSE.

This notion is particularly crucial for the modelling of the annual cycle of boreal and temperate trees, because the measurements of the different aspects of the annual cycle have different degrees of precision. Phenological data, such as data concerning the timing of bud burst in spring (Chap. 3), are usually based on visual inspection and therefore have relatively low precision of measurement, so that a

[3] Levins originally used the concept of "precision" instead of "accuracy", and that nomenclature has been used later by several other authors (e.g., Sharpe 1990; Hänninen 1995b). However, in the present volume the concept of "precision" is used only when referring to the exactness of the data with the concept of "precision of measurement", so that in the comparisons of model output with empirical data, the concept of "accuracy" is used.

relatively high RMSE can be considered sufficient for corroborating the model under inspection. In contrast, several physiological attributes of the annual cycle, such as photosynthetic capacity (Chap. 4), are determined by means of exact physiological measurements, so that data concerning these aspects of the annual cycle usually have relatively high precision of measurement, which implies that a relatively low RMSE is required for corroborating the model.

The concept of *model generality* refers to the width of the application area of the model (Sharpe 1990). The use of a model with low generality is restricted to the particular conditions in which the model was developed, whereas a model of high generality can be applied to several other conditions as well (Hänninen and Lundell 2007). The generality of ecophysiological models can be evaluated in several respects. We can determine, for instance, whether a model of frost hardiness predicts the annual cycle of frost hardiness for years with relatively average climatic conditions only or whether it is able to predict the frost hardiness for years with exceptional climatic conditions also.

Though the concepts of realism, accuracy, and generality refer to different aspects of the model, they are interlinked. Accuracy is a technical concept, so that it can readily be given an exact mathematical definition, such as the RMSE. Realism is a philosophical concept referring to the ultimate aim of science to understand natural phenomena and the causal factors associated with them. Accuracy is necessarily used as one indicator of realism, though high accuracy does not automatically imply high realism (see next paragraph). Generality, when measured as the consistency of the model's accuracy in different tests of the model, provides a useful measure of the realism of the model. As discussed in Sect. 8.3.2, this methodological principle is crucial for studies addressing the ecological effects of climate change (Hänninen 1995a, b; Hänninen et al. 1996).

Basically, all the three model properties, i.e., realism, accuracy, and generality, are measures of the quality of the model. Theoretically, then, it would be desirable to maximize all the three in the development of a biological model. However, experience tells us that this is not possible. Rather, one has to sacrifice one of the desired properties in order to improve the other two. Which one is sacrificed depends on the general aim of the research project (Levins 1966, 1968). When using a model for applied purposes, such as predicting the frost hardiness of nursery seedlings (Timmis et al. 1994; Hänninen et al. 2009) or the atmospheric concentrations of allergenic pollen (Sofiev et al. 2013), one is often well advised to sacrifice model realism to accuracy and generality. In basic research, on the other hand, realism is by far the most important property of the model, so that model accuracy in particular can be sacrificed as long as the model describes and explains phenomena existing in reality. For instance, when the population dynamics of animals is modelled, it is often of interest whether the model predicts a fluctuating or a relatively constant population size (Hanski and Henttonen 2002). In such cases, quantitative models of population dynamics are actually used for producing qualitative predictions, so that the accuracy of the model, i.e., the actual fit between the predicted and the observed population size at any moment, is of secondary interest.

Sacrificing model realism to model accuracy in applied cases, though often advisable, is not without its risks. In a new application situation, an occasional correlation of a completely non-realistic but accurate model may disappear, and if it does, so will the accuracy of the model. A realistic model, in contrast, always addresses the key phenomena involved in the real object of study, so that even though the model is often not especially accurate, it retains its moderate accuracy in new situations, at least by predicting the direction of the change in new situations correctly. In the long run, then, model realism is important for applications also. As discussed in detail in Chap. 8, this notion is especially important when models, either ecophysiological or other ones, are applied to the assessing of the ecological implications of the projected climate change. Therefore, the importance of model realism and of continued critical testing of the models (Fig. 2.2) cannot be overemphasized (Hänninen 1995b; Hänninen and Lundell 2007).

As mentioned above, one of the two aspects of model realism, i.e., the coverage of the model, is context-dependent. Thus, even the most realistic models do not address all issues known to be involved in the real-life phenomena studied by means of the model (Levins 1966). This notion has important implications, especially at the biological levels of organisation. Biological research is carried out at several organisation levels, from the molecular to the ecosystem one (Thornley and Johnson 1990). Thus, it goes without saying that a model constructed with concepts belonging to a given level, e.g., the population level, will fail to address lower-level entities and phenomena, such as stomatal conductance or the activities of various enzymes in individual plants, which nevertheless affect the population-level phenomena addressed by the model.

According to the reductionist philosophy of science, the real explanations are always found at the lower levels of organisation, so that in order to be realistic, the model needs to address the lower-level phenomena explicitly (Ayala and Dobzhansky 1974; Thornley and Johnson 1990). The present volume is based on the contrary notion that each level of organisation, besides reflecting properties of the lower levels, also evinces properties characteristic of itself, so-called emergent properties, i.e., properties specific to the organisational level examined, which cannot be reduced to the lower-level properties only (Ayala and Dobzhansky 1974). In other words, the phenomena discussed in the present volume at the whole tree level are emergent manifestations of the phenomena occurring at the lower levels of organisation. According to this non-reductionist philosophy of science, phenomena such as the inter-specific competition of plants or the frost hardening of the tissues of a given tree individual are as real as the cell divisions or the activation of genes in plants or phenomena related to the elementary particles studied in physics. Thus, in order to obtain a comprehensive understanding of natural phenomena, nature should be studied at all organisational levels, not just at the levels of elementary particles and molecules.

Emergent properties are sometimes considered to be generated by complicated interactions of the lower-level phenomena (Thornley and Johnson 1990; Hari 2013). According to this notion, no strictly emergent properties actually exist, because the seemingly emergent properties can be reduced, at least in principle, to lower-level phenomena. It is not possible to resolve this classical controversy in this volume. Rather, the adoption of the non-reductionist philosophy of science is motivated here on the basis of common-sense reasoning. For instance, despite the recent advances of molecular biology it would not be fruitful to try and adopt it – any more than quantum physics – to replace all the biological research carried out at the higher levels of biological organisation.

Even though physiological phenomena are discussed throughout the present volume, the *Fragestellung* is ultimately ecological: How do boreal and temperate tree species adapt to the seasonal climate prevailing at their growing sites and how will the climate change affect these trees? Accordingly, in the discussion of various aspects of tree seasonality, the most emphasis will be at the whole-tree level, i.e., the treatment belongs essentially to the discipline of whole-tree ecophysiology. Therefore the coverage and realism of the models addressed are also discussed at the whole-tree level, in most cases without explicit consideration of phenomena at the anatomical, cellular, or molecular levels.[4]

However, this notion and the adoption of the ecological approach and the non-reductionist philosophy of science should not obscure the importance of the interrelationships among the phenomena appearing at the different levels of organisation. Despite the emergent properties involved, the ecophysiological quantities measured at the whole-tree level reflect in several ways the physiological properties existing and the phenomena occurring at the anatomical, cellular, and molecular levels. Because of these interrelationships, phenomena belonging to different levels of organisation should not be studied in isolation from each other. This is especially important in the discipline of tree ecophysiology, which by definition synthesizes the ecology and the physiology of trees. Indeed, it appears that it is beginning to be time for a synthesis of the whole-tree modelling approach, with the physiological and molecular approaches, to the annual cycle of boreal and temperate trees; however, rather than trying to achieve such a synthesis, the present volume aims at explicating the whole-tree modelling approach for the purposes of forthcoming efforts towards a synthesis.

2.5 Summary

In this volume, a hypothetico-deductive (HDM) modelling approach is adopted for a study of the annual cycle in boreal and temperate trees, with the projection of the effects of the projected climate change to these trees as the ultimate aim. The approach is based on a close combination of modelling with empirical studies, both experimental and observational. Accordingly, computer simulations are carried out not only for projecting the ecological effects of climate change, but also for

---

[4] Following its standard usage, the concept of whole-tree ecophysiology is used in the present volume when different tree organs, such as buds or leaves, are discussed.

developing effective and informative experimental designs for testing and further development of the models.

When using the models in computer simulations deductive research is carried out. Accordingly, the results of the computer simulations are logical and necessary implications of the models used, so that in this sense these results contain nothing new as such. The computer simulations, however, are necessary because without them the implications of the models cannot be revealed.

The trustworthiness of the ecological scenarios obtained with computer simulations using ecological models stands or falls according to the degree of realism of the model applied. All biological models are simplifications of Nature, but an ecological model with high realism fulfils two criteria. First the model represents causal relationships, not occasional correlations between variables. Second, the model covers most, if not all, relevant aspects of the phenomenon described by the model. As will be demonstrated in Chap. 8, the importance of model realism calls for further critical empirical tests of the model even in cases where the model has already passed several tests.

In the dynamic modelling of the annual cycle, the rate of development at each moment is calculated on the basis of the environmental factors prevailing at that moment. In this way the *environmental response of rate of development* determines the development of the tree predicted by the model for any given environmental conditions, including those represented by climatic scenarios for climate change. Once the environmental responses of the rate of development included in the model have been formulated, the rest of the modelling is pure mathematical deduction. For this very reason the close combination of theoretical modelling with experimental studies is crucial for the advancement of this field of research.

The present volume belongs to the discipline of whole tree ecophysiology. This means that the various state variables of the models applied describe such physiological attributes which are in most cases measured at relatively high organisational levels, e.g. at the levels of leaves or buds, without addressing the underlying physiological mechanisms at the cellular and molecular levels. However, as the understanding of the physiological mechanisms of various whole tree attributes of the annual cycle has been improved considerably during last decades, a synthesis of the whole-tree modelling approach with the mechanistic approaches is called for in the future.

The developmental phenomena and the corresponding models are classified into two categories. First, phenomena belonging to the category of *fixed-sequence development* are irreversible. The annual phenological cycle discussed in Chap. 3 belongs to that category. Second, phenomena belonging to the category of *fluctuating development* are reversible. The annual cycle of photosynthetic capacity discussed in Chap. 4 represents that category. In *integrated models,* both fixed-sequence and fluctuating aspects of the annual cycle are included. As discussed in Chap. 5, integrated models are frequently used for the annual cycle of tree frost hardiness.

List of Symbols

t time

Time-Dependent Environmental Variables

$E(t)$ environmental factor(s)
$T(t)$ air temperature (°C)
$T_{mean}(t)$ daily mean air temperature (°C)

Time-Dependent Rate and State Variables

$R(t)$ rate of development
$S(t)$ state of development
$R_i(t)$ rate of development of ith attribute of the annual cycle
$S_i(t)$ state of development of ith attribute of the annual cycle
$R_{dd}(t)$ accumulation rate of day degree units (dd day^{-1})
$S_{dd}(t)$ accumulated temperature sum (dd)

Model Parameters

T_{thr} air temperature threshold for accumulation of day degree units
H_{crit} the high temperature requirement of growth onset (dd)

Other Symbols

dd day degree unit
f environmental response of the rate of development (function)

References

Arnold, C. Y. (1959). The determination and significance of the base temperature in a linear heat unit system. *Proceedings of the American Society of Horticultural Science, 74*, 430–445.
Ayala, F. J., & Dobzhansky, T. (1974). *Studies in the philosophy of biology. Reduction and related problems*. London: The Macmillan Press Limited. 390 p.

Bergh, J., McMurtrie, R. E., & Linder, S. (1998). Climatic factors controlling the productivity of Norway spruce: A model-based analysis. *Forest Ecology and Management, 110*, 127–139.

Boyer, W. D. (1973). Air temperature, heat sums, and pollen shedding phenology of longleaf pine. *Ecology, 54*, 420–426.

Cannell, M. G. R. (1985). Analysis of risks of frost damage to forest trees in Britain. In P. M. A. Tigerstedt, P. Puttonen, & V. Koski (Eds.), *Crop physiology of forest trees* (pp. 153–166). Helsinki: Helsinki University Press.

Chuine, I. (2000). A unified model for budburst of trees. *Journal of Theoretical Biology, 207*, 337–347.

Chuine, I., Cour, P., & Rousseau, D. D. (1998). Fitting models predicting dates of flowering of temperate-zone trees using simulated annealing. *Plant, Cell and Environment, 21*, 455–466.

de Réaumur, M. (1735). *Observations du thermomètre faites à Paris pendant l'année 1735 comparées avec celles qui ont été faites sous la ligne à l' Ile de France, à Alger et en quelques-unes de nos îles de l'Amérique*. Académie royale des sciences – Année 1735.

Diekman, M. (1996). Relationship between flowering phenology of perennial herbs and meteorological data in deciduous forests of Sweden. *Canadian Journal of Botany, 74*, 528–537.

Fuchigami, L. H., Weiser, C. J., Kobayashi, K., Timmis, R., & Gusta, L. V. (1982). A degree growth stage (°GS) model and cold acclimation in temperate woody plants. In P. H. Li & A. Sakai (Eds.), *Plant cold hardiness and freezing stress* (Mechanisms and crop implications, Vol. 2, pp. 93–116). New York: Academic Press.

Haila, Y. (1982). Hypothetico-deductivism and the competition controversy in ecology. *Annales Zoologici Fennici, 19*, 255–263.

Häkkinen, R., Linkosalo, T., & Hari, P. (1998). Effects of dormancy and environmental factors on timing of bud burst in *Betula pendula. Tree Physiology, 18*, 707–712.

Hänninen, H. (1990a). Modelling the annual growth rhythm of trees: Conceptual, experimental, and applied aspects. In H. Jozefek (Ed.), Modelling to understand forest functions. *Silva Carelica, 15*, 35–45.

Hänninen, H. (1990b). Modelling bud dormancy release in trees from cool and temperate regions. *Acta Forestalia Fennica, 213*, 1–47.

Hänninen, H. (1991). Does climatic warming increase the risk of frost damage in northern trees? *Plant, Cell and Environment, 14*, 449–454.

Hänninen, H. (1995a). Effects of climatic change on trees from cool and temperate regions: An ecophysiological approach to modelling of bud burst phenology. *Canadian Journal of Botany, 73*, 183–199.

Hänninen, H. (1995b). Assessing ecological implications of climatic change: Can we rely on our simulation models? *Climatic Change, 31*, 1–4.

Hänninen, H., & Kramer, K. (2007). A framework for modelling the annual cycle of trees in boreal and temperate regions. *Silva Fennica, 41*, 167–205.

Hänninen, H., & Lundell, R. (2007). Dynamic models in plant ecophysiology. In E. Taulavuori & K. Taulavuori (Eds.), *Physiology of northern plants under changing environment* (pp. 157–175). Kerala: Research Signpost.

Hänninen, H., Leinonen, I., Repo, T., & Kellomäki, S. (1996). Overwintering and productivity of Scots pine in a changing climate. *Silva Fennica, 30*, 229–237.

Hänninen, H., Luoranen, J., Rikala, R., & Smolander, H. (2009). Late termination of freezer storage increases the risk of autumn frost damage to Norway spruce seedlings. *Silva Fennica, 43*, 817–830.

Hänninen, H., Zhang, G., Rikala, R., Luoranen, J., Konttinen, K., & Repo, T. (2013). Frost hardening of Scots pine seedlings in relation to the climatic year-to-year variation in air temperature. *Agricultural and Forest Meteorology, 177*, 1–9.

Hanski, I., & Henttonen, H. (2002). Population cycles of small rodents in Fennoscandia. In A. Berryman (Ed.), *Populations cycles: The case for trophic interactions* (pp. 44–68). New York: Oxford University Press.

Hari, P. (1968). A growth model for a biological population, applied to a stand of pine. *Communicationes Instituti Forestalis Fenniae, 66*(7), 1–16.
Hari, P. (1972). Physiological stage of development in biological models of growth and maturation. *Annales Botanici Fennici, 9*, 107–115.
Hari, P. (2013). The approach to construct and test the theory of forest ecology. In P. Hari, K. Heliövaara, & L. Kulmala (Eds.), *Physical and physiological forest ecology* (pp. 7–25). Dordrecht: Springer.
Hari, P., & Häkkinen, R. (1991). The utilization of old phenological time series of budburst to compare models describing annual cycles of plants. *Tree Physiology, 8*, 281–287.
Hari, P., Leikola, M., & Räsänen, P. (1970). A dynamic model of the daily height increment of plants. *Annales Botanici Fennici, 7*, 375–378.
Hari, P., Kolari, P., Bäck, J., Mäkelä, A., & Nikinmaa, E. (2008). Photosynthesis. In P. Hari & L. Kulmala (Eds.), *Boreal forest and climate change* (pp. 231–242). Dordrecht: Springer.
Hari, P., Hänninen, H., Berninger, F., Kolari, P., Nikinmaa, E., & Mäkelä, A. (2009). Predicting boreal conifer photosynthesis in field conditions. *Boreal Environment Research, 14*(suppl. A), 19–28.
Jeffers, J. N. R. (1978). *An introduction to systems analysis: With ecological applications*. London: Edward Arnold (Publishers) Limited. 198 p.
Katz, Y. H. (1952). The relationship between heat unit accumulation and the planting and harvesting of canning peas. *Agronomy Journal, 44*, 74–78.
Kellomäki, S., Väisänen, H., Hänninen, H., Kolström, T., Lauhanen, R., Mattila, U., & Pajari, B. (1992). A simulation model for the succession of the boreal forest ecosystem. *Silva Fennica, 26*, 1–18.
Kellomäki, S., Hänninen, H., & Kolström, M. (1995). Computations on frost damage to Scots pine under climatic warming in boreal conditions. *Ecological Applications, 5*, 42–52.
Kobayashi, K. D., & Fuchigami, L. H. (1983a). Modelling temperature effects in breaking rest in red-osier dogwood (*Cornus sericea* L.). *Annals of Botany, 52*, 205–215.
Kobayashi, K. D., & Fuchigami, L. H. (1983b). Modeling bud development during the quiescent phase in red-osier dogwood (*Cornus sericea* L.). *Agricultural Meteorology, 28*, 75–84.
Kramer, K. (1994a). Selecting a model to predict the onset of growth of *Fagus sylvatica*. *Journal of Applied Ecology, 31*, 172–181.
Kramer, K. (1994b). A modelling analysis of the effects of climatic warming on the probability of spring frost damage to tree species in The Netherlands and Germany. *Plant, Cell and Environment, 17*, 367–377.
Landsberg, J. J. (1974). Apple fruit bud development and growth; analysis and an empirical model. *Annals of Botany, 38*, 1013–1023.
Landsberg, J. J. (1977). Effects of weather on plant development. In J. J. Landsberg & C. V. Cutting (Eds.), *Environmental effects on crop physiology* (pp. 289–307). London: Academic Press.
Landsberg, J. J. (1986). *Physiological ecology of forest production*. London: Academic Press. 198 p.
Lappalainen, H. (2010). *Role of temperature in the biological activity of a boreal forest* (Helsinki: Finnish Meteorological Institute Contributions, Vol. 84, 52 p).
Leinonen, I., Repo, T., & Hänninen, H. (1997). Changing environmental effects on frost hardiness of Scots pine during dehardening. *Annals of Botany, 79*, 133–138.
Levins, R. (1966). The strategy of model building in population biology. *American Scientist, 54*, 421–431.
Levins, R. (1968). *Evolution in changing environments. Some theoretical explorations*. Princeton: Princeton University Press. 120 p.
Linkosalo, T., Lappalainen, H. K., & Hari, P. (2008). A comparison of phenological models of leaf bud burst and flowering of boreal trees using independent observations. *Tree Physiology, 28*, 1873–1882.
Mäkelä, A., Hari, P., Berninger, F., Hänninen, H., & Nikinmaa, E. (2004). Acclimation of photosynthetic capacity in Scots pine to the annual cycle of temperature. *Tree Physiology, 24*, 369–376.

Mesarovic, M. D., & Takahara, Y. (1975). *General systems theory: Mathematical foundations*. New York: Academic Press. 268 p.

Niiniluoto, I. (1983). *Tieteellinen päättely ja selittäminen*. Helsinki: Kustannusosakeyhtiö Otava. 416 p.

Olsen, A., Bale, J. S., Leadbeater, B. S. C., Callow, M. E., & Holden, J. B. (2003). Developmental thresholds and day-degree requirements of *Paratanytarsus grimmii* and *Corynoneura scutellata* (Diptera: Chironomidae): Two midges associated with potable water treatment. *Physiological Entomology, 28*, 315–322.

Pelkonen, P. (1980). The uptake of carbon dioxide in Scots pine during spring. *Flora, 169*, 386–397.

Pelkonen, P. (1981a). *Recovery and cessation of CO_2uptake in Scots pine at the beginning and at the end of the annual photosynthetic period* (Research notes, Vol. 30. 95 p). Helsinki: University of Helsinki, Department of Silviculture.

Pelkonen, P. (1981b). Investigations on seasonal CO_2 uptake in Scots pine. I CO_2 exchange in Scots pine after wintering period. II The autumn fall in CO_2 uptake in Scots pine. *Communicationes Instituti Forestalis Fenniae, 99*(5), 1–40.

Pelkonen, P., & Hari, P. (1980). The dependence of the springtime recovery of CO_2 uptake in Scots pine on temperature and internal factors. *Flora, 169*, 398–404.

Pisek, A., & Winkler, E. (1958). Assimilationsvermögen und Respiration der Fichte (Picea excelsa Link) in verschiedener Höhenlage und der Zirbe (Pinus cembra L.) an der alpinen Waldgrenze. *Planta, 51*, 518–543.

Repo, T. (1991). Rehardening potential of scotch pine seedlings during dehardening. *Silva Fennica, 25*, 13–21.

Repo, T. (1993). Impedance spectroscopy and temperature acclimation of forest trees (Research notes, Vol. 9. 53 p). Joensuu: University of Joensuu, Faculty of Forestry.

Repo, T., Mäkelä, A., & Hänninen, H. (1990). Modelling frost resistance of trees. In H. Jozefek (Ed.), Modelling to understand forest functions. *Silva Carelica, 15*, 61–74.

Richardson, E. A., Seeley, S. D., & Walker, D. R. (1974). A model for estimating the completion of rest for 'Redhaven' and 'Elberta' peach trees. *HortScience, 9*, 331–332.

Sarvas, R. (1967). The annual period of development of forest trees. *Proceedings of the Finnish Academy of Science and Letters, 1995*, 211–231.

Sarvas, R. (1972). Investigations on the annual cycle of development of forest trees. Active period. *Communicationes Instituti Forestalis Fenniae, 76*(3), 1–110.

Sarvas, R. (1974). Investigations on the annual cycle of development of forest trees. II. Autumn dormancy and winter dormancy. *Communicationes Instituti Forestalis Fenniae, 84*(1), 1–101.

Sharpe, P. J. H. (1990). Forest modeling approaches: Compromises between generality and precision. In R. K. Dixon, R. S. Meldahl, G. A. Ruark, & W. G. Warren (Eds.), *Process modeling of forest growth responses to environmental stress* (pp. 180–190). Portland: Timber Press.

Sofiev, M., Siljamo, P., Ranta, H., Linkosalo, T., Jaeger, S., Rasmussen, A., Rantio-Lehtimaki, A., Severova, E., & Kukkonen, J. (2013). A numerical model of birch pollen emission and dispersion in the atmosphere. Description of the emission module. *International Journal of Biometeorology, 57*, 45–58.

Thornley, J. H. M., & Johnson, I. R. (1990). *Plant and crop modelling. A mathematical approach to plant and crop physiology*. Oxford: Clarendon Press. 669 p.

Timmis, R., Flewelling, J., & Talbert, C. (1994). Frost injury prediction model for Douglas-fir seedlings in the Pasific Northwest. *Tree Physiology, 14*, 855–869.

Wang, J. Y. (1960). A critique of the heat unit approach to plant response studies. *Ecology, 41*, 785–790.

Winter, F. (1973). Ein Simulationsmodell über die Phänologie und den Verlauf der Frostresistenz von Apfelbäumen. *Ecologia Plantarum, 8*, 141–152.

Chapter 3
The Annual Phenological Cycle

Contents

H. Hänninen, *Boreal and Temperate Trees in a Changing Climate*, Biometeorology 3,
DOI 10.1007/978-94-017-7549-6_3

Abstract The hypothetico-deductive modelling framework introduced in Chap. 2 is applied to the modelling of the annual phenological cycle of boreal and temperate trees. The ecophysiological models of the annual phenological cycle predict the timing of discontinuous developmental events, such as bud burst and height growth cessation. With such events only one, or maximally a few, empirical observations per year are available for testing these models; but as in all other models of the annual cycle, the values of the state variables are nevertheless calculated for each day of the simulation period. The methodological problems caused by this discrepancy are discussed, and an ecophysiological explication of the models is introduced. Most effort is devoted to modelling the springtime developmental events, such as bud burst. The direct environmental regulation by air temperature is first discussed by examining the classical temperature sum (or day degree) models. The ecophysiological interpretation of these models, which use the arbitrary unit of day degree, is explicated, and the experimental research aimed at determining the real ecophysiological air temperature responses of the trees and thus supplementing the day degree approach is discussed. Subsequently, effects of dormancy are introduced into the modelling. Most models for growth cessation are conceptually more straightforward, so that they are not discussed at the same length as the models of springtime development. The chapter is concluded with a discussion of models for the entire annual phenological cycle.

Keywords Annual phenological cycle • Boreal trees • Chilling requirement • Computer simulations • Critical night length • Day degrees • Dormancy • Growth • High temperature requirement • Model realism • Ontogenetic competence • Rest break • Seasonality • Temperate trees

3.1 The Phenomena, the Observations, and the Measurements Involved

3.1.1 Tree Phenology: The Study of the Timing of Seasonal Point Events

3.1.1.1 The Concept of Phenology

The concept of phenology is often defined as the study of the timing of seasonally recurring phenomena in plants and animals and of the relationship of that timing to climatic factors (Schwartz 2013). Other aspects than the timing of the phenomena are outside the scope of phenology. Thus the timing of the bolting of a given crop species, for instance, falls in the scope of phenology, whereas the amount of crop produced does not. In other respects the definition of the concept is not fully unequivocal. Sometimes it is understood in the broad sense of covering the timing of all seasonal phenomena in plants and animals. The seasonal changes in the needle frost hardiness of a coniferous tree, for instance, would then be classified as phenological events (Kramer et al. 2002). More often, however, the concept is understood in a more restricted sense, referring to such seasonal phenomena as are readily visible to the naked eye. Accordingly, the bud burst of deciduous trees in spring and their leaf colouration in autumn are regarded as phenological events, as is the appearance of migratory birds or insect hatching in spring (Schwartz 2013). This restricted definition is also used in the present volume addressing boreal and temperate trees.

Sarvas (1972) introduced the concepts of *point event* and *segment event* for studying the annual cycle of trees. A point event is a developmental event of so short a duration that in each inspection one can classify it into two categories only: (i) the event has not yet occurred or (ii) the event is over. A segment event is of longer duration, allowing three potential classifications in each inspection: the event (i) has not yet occurred, (ii) is going on, or (iii) is over. It goes without saying that the distinction between point events and segment events is not absolute but depends on the methods used in the inspection and the time interval applied between two successive inspections. Nevertheless, the concepts as defined by Sarvas (1972) have turned out to be useful in studies of tree phenology. Using Sarvas's (1972) terminology, tree phenology is defined in the present volume as the study of the timing of point events during the annual cycle of trees.

3.1.1.2 Point Events Frequently Addressed in Modelling Studies

In broad terms, the annual phenological cycle of boreal and temperate trees consists of two major point events, i.e., the onset of growth in spring and its cessation in late summer or autumn (Perry 1971). These two point events mark the borderline between the periods of growth and dormancy, which can both be understood as major segment events of relatively long duration. In research, however, more exact specifications are needed, and these vary according to the tree species studied (Sarvas 1972, 1974; Fuchigami et al. 1982).

One point event frequently addressed in modelling studies is the activation of apical meristems in spring.[1] In deciduous tree species, this is done by visually inspecting the bursting, and often also the subsequent leaf unfolding, of the vegetative buds (Fig. 3.1a; Murray et al. 1989; Kramer 1994a). A similar procedure is often used with several evergreen conifer species, such as *Picea abies* (Fig. 3.1b; Partanen et al. 1998; Slaney et al. 2007). In these cases the visual classification can be carried out quite precisely, provided that standard instructions, often including photographs, are made available to the inspectors (Kubin et al. 2007). In some other coniferous species, such as *Pinus sylvestris*, the new needles become visible from the bud scales later, when the shoot has already started elongation growth. In these cases the elongation growth is often measured simply by means of a ruler, for instance, and the point event of spring activation (i.e., the onset of elongation growth) is subsequently determined by means of the resulting growth curve. The date of the onset of growth can be defined to correspond to the date when the slope of the growth curve exceeds a given threshold value, for instance, or to the date when 10 % of the final growth increment has been attained (Fig. 3.1c; Hänninen and Pelkonen 1988).

Besides the vegetative cycle discussed above, springtime phenological studies often examine the generative cycle, i.e., the development of flower buds. Similarly to the vegetative buds, the phenological observations can be done visually (Chung 1981; Kubin et al. 2007; Tooke and Battey 2010). In the case of wind-pollinated species, male flowering can also be studied indirectly, by measuring pollen concentrations in the air with equipment specially developed for the purpose (Fig. 3.1d). In this case the phenology of male flowering, i.e., the release of pollen grains into the atmosphere, is studied at the population level (Boyer 1973; Sarvas 1972, 1974; Chuine et al. 1998, 1999). At that level, flowering is a segment event, but by means of daily pollen data and statistical methods, various point events, such as the median day of flowering, can be determined (Sarvas 1972, 1974). The long distances travelled by pollen may cause considerable errors in the inferences made about the timing of male flowering in local populations, but there are ways of accounting for this in the processing of the data (Hänninen 1984).

The cessation of growth and related late-summer and early-autumn phenomena have also been addressed frequently in modelling studies, though not as frequently as phenomena related to the onset of growth and flowering in spring. In studies carried out in natural conditions, leaf colouration and abscission (Fig. 3.1e) in deciduous trees have been the main objects (Sarvas 1972; Menzel and Fabian 1999; Pudas et al. 2008; Delpierre et al. 2009). Like the springtime phenological events, these late-season events, too, are often examined visually, using standard instructions to guide the observations. In experimental studies under controlled

[1] Only the activation visible to the naked eye is discussed here. At the anatomical level, the activation starts earlier (Fig. 2.2).

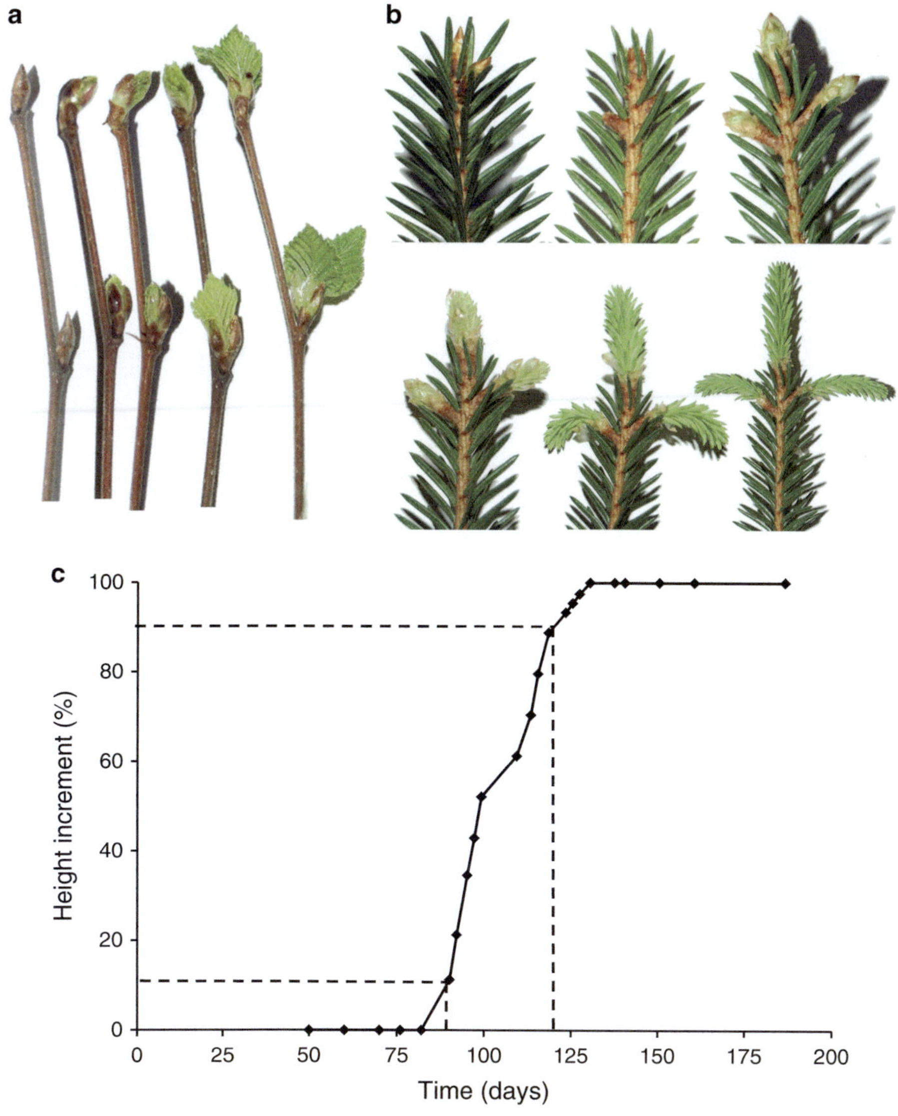

Fig. 3.1 The phenomena addressed and the measurements applied in studies of the annual phenological cycle in boreal and temperate trees. (**a**) Vegetative bud burst and leaf unfolding in *Betula pubescens*. (**b**) Vegetative bud burst and height growth in *Picea abies*. (**c**) Shoot elongation in second-year *Pinus sylvestris* seedlings. The time instants of growth onset (day 89) and cessation (day 120) are defined as the dates corresponding to 10 % and 90 %, respectively, of the final height increment. (**d**) Equipment for measuring the pollen concentration in the air. The data obtained are used for determining the timing of male flowering in the population studied. The unpublished photograph was provided by Annika Saarto. (**e**) Autumn colouration and leaf abscission in *Betula pubescens* (**a**, **b**, and **e** from Kubin et al. 2007, photos by Eeva Pudas, published with the permission of the Natural Resources Institute Finland)

d

e

Fig. 3.1 (continued)

conditions, the cessation of growth and the formation of the overwintering bud are studied frequently. The cessation of height growth is usually determined by means of height measurements, using the resulting growth curve in ways basically similar to those used in studies of the onset of height growth in spring (Fig. 3.1c). Bud set is usually determined by means of visual inspection (Mikola 1982).

The dormant phase of the annual cycle evinces no phenomena visible to the naked eye. However, because of our limited physiological information about the mechanisms of dormancy and the methodological difficulties involved in studying them, the induction and release of bud dormancy are usually studied by means of indirect regrowth tests, using observations basically similar to those used in studies of the onset of growth (Nienstaedt 1966; Worrall and Mergen 1967; Jones et al. 2015). For this reason, the bud dormancy of trees is discussed as part of the annual phenological cycle in Sect. 3.3.

It goes without saying that the seasonal phenomena of tree organs other than the vegetative and generative buds, such as secondary xylem or fine roots, are also crucial for tree growth and the functioning of the ecosystem. Despite their importance, they have only rarely been addressed in modelling studies and are therefore not discussed at any depth in the present volume, either. The reader is referred to a recent review by Delpierre et al. (2015).

3.1.1.3 Large Temporal and Spatial Scales

Phenological observations in field conditions, whether carried out in spring or in autumn, are frequently characterised by large temporal and spatial scales. Large temporal scales are made possible by the long historical time series that are available for several species and phenological events (Chuine et al. 1998, 1999). The time series typically cover several decades (Lappalainen and Heikinheimo 1992; Häkkinen et al. 1995; Linkosalo et al. 1996). Long time series have been used for testing phenological models (Chuine et al. 1998, 1999; Häkkinen et al. 1998; Salminen and Jalkanen 2015) and for examining whether the effects of climatic warming are already observable in nature (Parmesan and Yohe 2003; Menzel et al. 2006; Linkosalo et al. 2009).

Large spatial scales are currently used in two types of research. First, large international programmes have been launched to observe the phenology of predefined groups of species in various geographical locations. These studies have devoted a great deal of effort to standardising the instructions in order to facilitate the collection of comparable phenological data across all locations in all the participating countries (Koch et al. 2009).

Second, phenological studies, both spring and autumn ones, make increasing use of remotely sensed data. In spring, the object of study is the greening of the landscape, i.e., bud burst and subsequent leaf development (Fisher et al. 2007; Delbart and Picard 2007), whereas in autumn it is leaf senescence that has been observed by means of satellites (Keenan and Richardson 2015). In most cases it is impossible to collect species-specific data with this approach, but satellites provide a powerful tool for studies at the ecosystem level (Polgar and Primack 2011). Remotely sensed phenological data are often used in modelling the fluxes of carbon between the atmosphere and vegetation (Picard et al. 2005). Furthermore, remotely sensed data have also been used similarly to long-term land-based observations for detecting potential changes in the phenology of plants and in the length of the growing season that may already have been caused by climatic warming (Myneni et al. 1997).

3.1.2 The Anatomical and Physiological Basis

Most studies of tree phenology are characterised by the discontinuity and scarcity of observations. Bud burst, for instance, is a point event taking place only once during the annual cycle, so that in studies of bud burst phenology, no more than one observation, i.e., the date of bud burst, per annual cycle is often obtained. This is in striking contrast to studies of the height growth or frost hardiness of trees, for instance (Hänninen and Kramer 2007). In such studies the ecophysiological process can be measured continuously, so that the number of successive measurements is determined on the basis of practical issues such as the economic cost of the measurements, not by the ecophysiological phenomenon under scrutiny.

However, the discontinuous phenological point events involved are results of continuous cellular and anatomical development inside the bud. Long before any signs of bud burst are visible to the naked eye, phenomena such as division, growth and differentiation of cells are initiated (Lindfors and Kupila-Ahvenniemi 1988; Sutinen et al. 2009, 2012; Viherä-Aarnio et al. 2014; Clark et al. 2014; Fig. 3.2). These developmental phenomena belong to the general domain of *ontogenetic*

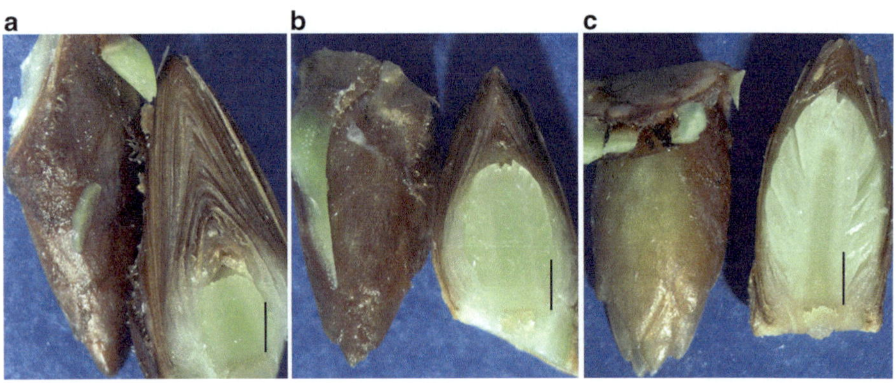

Fig. 3.2 The stages of the ontogenetic development in *Picea abies* buds occurring prior to the bud burst visible to the naked eye, as classified by Sutinen et al. (2009). Each of the photographs (**a–c**) represents one stage in the development, showing a fresh-cut longitudinal section of the bud on the *right-hand side* and a corresponding intact bud on the *left-hand side*, both photographed through a stereo microscope. (**a**) Overwintering bud, no ontogenetic development towards bud burst observable, (**b**) microscopic phase 2 attained, and (**c**) microscopic phase 3 attained. In each photograph, the length of the vertical line is 1 mm. In addition to the growth of the primordial shoot and primordial needles shown in the stereomicroscopic photographs (**b**, **c**), Sutinen et al. (2009) based their classification of the stages on changes within the microscopic phases, such as the turning of the primordial needles from rounded to pointed ones, observed with a light microscope. The first externally visible changes occur upon attainment of microscopic phase 3, as shown in (**c**) (The unpublished photographs were provided by Sirkka Sutinen)

development, or *ontogenesis*, which means the irreversible formation of plant structures, whether at the anatomical or the whole-plant level (Lawrence 2011). In modelling studies of tree phenology, cellular and anatomical phenomena have rarely been addressed, obviously due to the large amount of work and the special skills required. In most modelling studies, only phenological point events visible to the naked eye, such as bud burst in spring, are considered, but in these studies, too, the ontogenetic development is actually modelled implicitly (Hänninen and Kramer 2007). Accordingly, the modelling of the spring phenology of trees would be facilitated if observations on point events occurring in the buds before the visible bud burst were available. To this end, microscopic studies of bud development are especially useful (Fig. 3.2; Sutinen et al. 2012; Viherä-Aarnio et al. 2014). This point is further elaborated in Sect. 3.2.1.3, when discussing Sarvas's (1972) modelling work, which was based on large-scale microscopic observations on the development of flower buds in boreal forest trees.

Over several decades, considerable research effort has been directed to the physiological mechanisms regulating the cellular and anatomical phenomena (Fig. 3.2) behind the annual phenological cycle of boreal and temperate trees. Some aspects of the physiological regulation, such as the role of phytochromes in mediating the effects of day and night length on growth cessation (Sect. 3.5.2), have been known for a long time. More recently, considerable progress has been achieved in research on the regulation of growth and dormancy in boreal and temperate trees by the use of modern molecular approaches. Even so, several aspects of this regulation remain to be revealed in future research. However, as this extensive and rapidly developing field of science is outside the scope of the present volume, the reader is referred to the several reviews available (e.g., Horvath et al. 2003; Arora et al. 2003; Rinne and van der Schoot 2003; Junttila 2007; Rohde and Bhalerao 2007; Cooke et al. 2012; Brunner et al. 2014).

3.1.3 Implications for Modelling

In this Chapter the modelling approach introduced in Chap. 2 is applied to modelling the annual phenological cycle of boreal and temperate trees. Among the seasonal phenomena addressed in the present volume, the annual phenological cycle constitutes a special case for the modelling (Hänninen and Kramer 2007). For instance in the models predicting the timing of springtime point events, such as bud burst, two ecophysiological processes are often simulated by their respective two specifications of the variable state of development, S(t). The pattern of the two S(t) variables is predicted for a prolonged period, lasting typically several months. However, the values of S(t) predicted by the models of the annual phenological cycle do not usually have a continuous counterpart in empirical studies. Thus, when testing the models with field data, the only test is usually provided by comparing the predicted day of bud burst with the observed day. As will be discussed in subsequent sections of this chapter, these peculiarities in modelling the annual phenological cycle have several important implications. Most importantly, these peculiarities emphasize the need for conceptual clarification and critical and diverse testing of the models.

3.2 Direct Environmental Regulation of Spring Phenology[2]

3.2.1 *The High Temperature Requirement*

3.2.1.1 Day Degree Models

In the boreal and temperate zones, the timing of springtime phenological events in trees and other plants is regulated to a large extent by air temperature. Rather than the temperature on any one day, it is the long-term pattern of air temperature that matters. In years when air temperatures rise early in the spring and stay at a relatively high level, plants are relatively early in their phenological events, and conversely, in years with a cool and long spring, plants lag behind in their development (Polgar and Primack 2011). This general pattern is so obvious that it is common knowledge among the public.

In scientific research, the cumulative effects of air temperature on the spring phenology of plants were addressed as early as the eighteenth century by Réaumur (1735), who introduced the concept of temperature sum. According to the temperature sum model, the development of plants is initiated when the air temperature reaches a given threshold, above which the rate of development increases linearly (Fig. 3.3). Any given point event, such as bud burst in trees, occurs when the temperature sum attains the critical value corresponding to the event (Fig. 2.3d in Sect. 2.3). In the present volume this critical value is referred to as the parameter *high temperature requirement of growth onset*, H_{crit} (Hänninen and Kramer 2007). The value of H_{crit} is specific to each species (Kramer 1994b; Chuine et al. 1998) and even to different genotypes within the species (Sarvas 1967, 1972; Rousi and Heinonen 2007).[3]

Because of the linear relationship between above-threshold air temperatures and the accumulation rate of the temperature sum (Fig. 3.3), the unit used in the model was subsequently named *day degree* or *degree day* (dd) (Madariaga and Knott 1951; Kasap and Alten 2005). Correspondingly, when the time step used in the model is an hour, i.e., when the accumulation rate is calculated for each hour on the basis of the hourly mean air temperature, the unit is referred to as *degree hour* (dh) (Lana and Haber 1952; Sarvas 1972). Mathematically, both ways of

[2] As defined in Sect. 2.2, the environmental regulation of any attribute of the annual cycle is regarded as direct whenever no other attribute of the annual cycle is referred to in the modelling of the environmental regulation. Indirect environmental regulation of spring phenology, taking place through regulation of the dormancy status (rest break) in the buds, will be discussed in Sects. 3.3 and 3.4.

[3] In addition to height growth onset, the concept "high temperature requirement" and the corresponding acronym H_{crit} are used in the present volume for a wide range of spring phenological events, such as the bud burst, leaf unfolding, and flowering of the trees. In addition to these phenological events, the concept is also used for microscopic phases of meiosis prior to flowering.

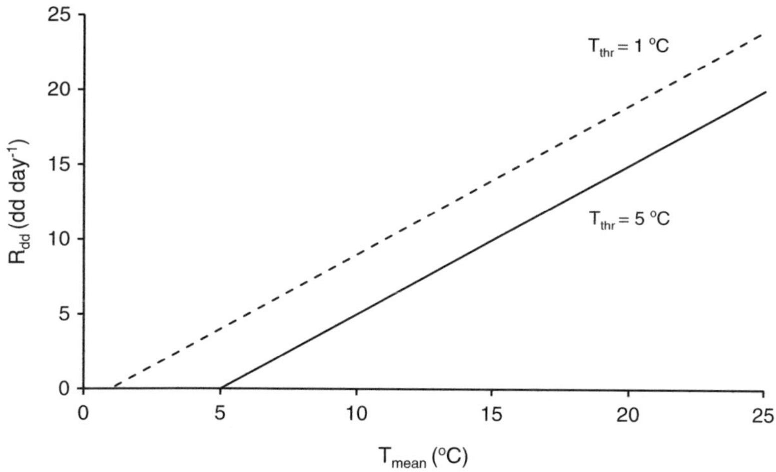

Fig. 3.3 The air temperature response of the accumulation rate of arbitrary developmental units (day degrees, dd), R_{dd}, in two temperature sum models with two different values of the threshold temperature T_{thr}. T_{mean} daily mean air temperature. Note that each of the two temperature sum models has its particular dd unit on the horizontal axis and that these two dd units are incomparable

calculating the temperature sum are numerical approximations of the continuous relationship between instantaneous air temperature and instantaneous rate of development. However, as the environmental responses of plants are characterised by delays, it may well be that a numerical approximation, averaging the air temperature over the time step applied, may even give a more realistic idea of the real developmental phenomenon than the analytical continuous presentation.

After its early introduction by Réaumur (1735), the temperature sum model later became the standard method for describing the spring phenology of plants. The model is perhaps best known under the name "day degree model." Accordingly, it has become a general praxis to speak of the "day degree relations" of a variety or species. The model has had particularly frequent use in agricultural and horticultural sciences, where the suitability of a given cultivar or variety to a given climate is often examined by comparing the number of day degrees required by the variety and the climatological accumulation of day degrees at the growing site. Reviews of the use of the concept of day degree have been provided, among others, by Arnold (1959), Wang, (1960), Sarvas (1967), Tuhkanen (1980), and Chuine et al. (2013).

The concept of day degree has also been used frequently in studies of the spring phenology of boreal and temperate trees. For instance, Hari and Häkkinen (1991) found that the year-to-year variation in air temperature described in terms of the traditional day degree model accounted for a large part of the variation in the timing of the bud burst of *Betula pendula* in central Finland (Fig. 3.4).

The day degree model involves three parameters (Linkosalo et al. 2008): the first is the starting date of the simulations, t_0, i.e., the day when the accumulation

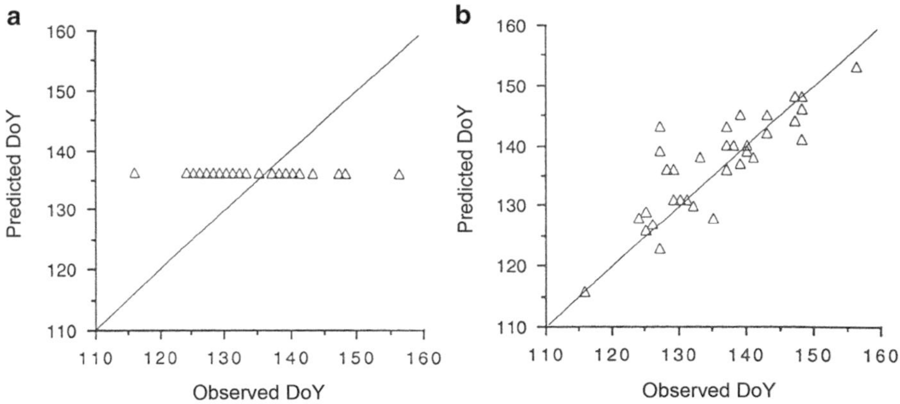

Fig. 3.4 Test of the temperature sum model with 36-year bud burst data of *Betula pendula* growing in central Finland in 1907–1950 (with 8 years excluded on account of missing observations). For comparison's sake, panel (**a**) presents a test of the time model, where bud burst is predicted to occur each year on the mean Day of the Year (DoY) calculated from the observed dates of bud burst over the 36 years. (**b**) Each year, bud burst is predicted to occur on the date when the temperature sum attains the mean of the empirical temperature sums calculated up to the observed day of bud burst. The *straight line* represents a 1:1 relation, indicating a perfect fit between prediction and observation (Modified from Hari and Häkkinen 1991; published with the permission of Oxford University Press)

of temperature sum is initiated. In northern countries such as Finland, 1 January is quite often deliberately set as the value of t_0 (Heikinheimo and Lappalainen 1997; Leinonen and Hänninen 2002). The second is the threshold air temperature, T_{thr} (Fig. 3.3). In Finland, the value of $T_{thr} = +5$ °C has been used as the standard (Sarvas 1967; Leinonen and Hänninen 2002). However, in Heide's (1993a) experimental test the estimated threshold temperature for predicting the timing of bud burst in six deciduous northern tree species varied from −4 °C in *Prunus padus* to +1 °C in *Populus tremula*. On the basis of these experimental results, Heide (1993a) suggested 0 °C for use as the standard threshold with these northern trees. The third parameter is the number of day degrees required for the point event to occur, i.e., the high temperature requirement, H_{crit}, of the event (Hari and Häkkinen 1991; Linkosalo et al. 2008). The value of this parameter has a great effect on the timing of the predicted point event: the higher the value of the H_{crit} of the tree genotype, the later the point event.

Day degree models are fitted to empirical data by estimating the values of the three parameters. Sometimes the value of T_{thr} and/or t_0 is fixed on the basis of a priori information, but the value of H_{crit} is always estimated.[4] In the estimation, iterative calculations with different parameter values are used for minimising the

[4] This procedure is followed when the model is fitted to the data, which is the case discussed here. When the data are used for independent testing of the model, the value of H_{crit} is also fixed on the basis of a priori information.

difference between the observed and the modelled timing of the phenological event, e.g., bud burst (Häkkinen et al. 1998; Linkosalo et al. 2000). Here the procedure of model-fitting is illustrated with the case of a 10-year subset of a longer data set for the bud burst of the *Betula* species in Jyväskylä, central Finland (Häkkinen 1999a). The simplest case, estimating the value of H_{crit} only, is assumed, with the values of the other two parameters set at the standard values of $t_0 = 1$ January and $T_{thr} = +5$ °C (Fig. 3.3). In this case, the procedure of model-fitting consists of estimating the value of H_{crit} as follows (Table 3.1):

1. Using the fixed value of the parameter T_{thr} (here +5 °C), calculate the temperature sum for each year from the fixed starting date, t_0 (here 1 January), until the observed date of bud burst.
2. Calculate the mean of these empirical values of the temperature sums and take this for an estimate of H_{crit}. In this case, the value of $H_{crit} = 47.3$ dd was attained.
3. Starting the accumulation of temperature sum from t_0 (here 1 January) again each year, calculate the date when the accumulated temperature sum attains the estimated value of H_{crit} (here 47.3 dd). For each year, the date of attaining the value of H_{crit} is the predicted date of bud burst.
4. In order to measure the goodness-of-fit of the prediction of the model, compare the predicted and the observed timing of bud burst by calculating the value of the root mean squared error (RMSE; Linkosalo et al. 2000, 2008). In the present case, the value of RMSE = 4.0 days was obtained (Table 3.1). In 3 of the 10 years examined, the predicted time of bud burst was equal to the observed one, but in 1931 the prediction was 7 days early and in 1938, 7 days late (Table 3.1).

In more complicated cases, where the values of one or both of the parameters t_0 and T_{thr} are estimated, the whole sequence of operations (1)–(4) is repeated for each value of the estimated parameter or combinations of the values of the two parameters, depending on whether only t_0 or T_{thr} or both are to be estimated. The parameter value (or combination of values) that minimises the residual error (RMSE) in phase (4) is taken for the estimated value of the parameter.

The fitting of the temperature sum model is sometimes carried out in an erroneous way, so that after step (1), the variance of the obtained empirical temperature sums is calculated and the parameter values that minimise the variance are selected for estimates of the parameters (e.g., Boyer 1973). This procedure is based on the intuitive idea that with the correct model, the accumulated temperature sum at the observed time of bud burst stays constant over different years. Though the intuitive idea is correct, the procedure is not. This is because the scales of the different temperature sum models are not comparable (Sarvas 1972; Fig. 3.3) and therefore, neither are their variances.

It is clear, for instance, that degree hour models generally produce much higher variances than day degree models do because the averages are also much higher in the degree hour models (where the summation is carried out 24 times rather than once a day). Addressing this by calculating the coefficient of variation of the empirical temperature sum (Arnold 1959) does not help, because it artificially favours models that increase the average values without increasing their variance. For instance, an

Table 3.1 The procedure of fitting a temperature-sum model to phenological and air-temperature data

	Observation				Prediction				
Year	Month	Day	DoY	TS (dd)	Month	Day	DoY	TS (dd)	Error (days)
1931	5	13	133	55.0	5	6	126	47.3	−7
1932	5	20	141	51.9	5	20	141	47.3	0
1933	5	27	147	46.2	5	28	148	47.3	1
1934	5	6	126	49.5	5	6	126	47.3	0
1935	5	31	151	38.7	6	6	157	47.3	6
1936	5	11	132	51.8	5	11	132	47.3	0
1937	5	8	128	48.6	5	6	126	47.3	−2
1938	5	19	139	27.3	5	26	146	47.3	7
1939	5	19	139	36.2	5	22	142	47.3	3
1940	5	14	135	68.0	5	10	131	47.3	−4
				Mean $=H_{crit}=47.3$ dd					RMSE = 4.0 days

Bud burst data for *Betula* at Jyväskylä, central Finland, was obtained from Häkkinen (1999a); published with the permission of the Finnish Forest Research Institute, currently the Natural Resources Institute Finland, and the Finnish Society of Forest Science. Daily mean air temperatures based on standard air-temperature records collected by the Finnish Meteorological Institute at Jyväskylä were used in the calculations. Two of the model parameters were fixed a priori, so that the starting day of the simulation was set at $t_0 = 1$ January and the threshold air temperature at $T_{thr} = +5\,°C$. *DoY* Day of Year, *TS* temperature sum, *dd* day degree unit, H_{crit} high-temperature requirement of bud burst in dd units, *RMSE* root mean squared error of the predicted timing of bud burst

imaginary model where t_0 is set at 1 January 10 years before the year for which bud burst is predicted would produce a very low coefficient of variation for the empirical values of the temperature sum accumulated until the observed day of bud burst, but this is no indication that the model is accurate, of course.

This example, though an extreme and artificial one, reveals that examining the models by comparing the values of the coefficients of variation in the empirical temperature sums will not work, either. On account of the incomparability of the scales used by the different temperature sum models, the only way to compare the predictive power of the models is the method including a double calculation of the temperature sums (Table 3.1).[5] In this way, the models are compared in terms of a common scale, i.e., that of time (Häkkinen et al. 1998; Linkosalo et al. 2000).

Another statistical pitfall in the fitting of a temperature sum model to a set of data is the correlation of the parameters t_0 and H_{crit} (Linkosalo et al. 2000, 2008). Due to this correlation, practically identical accuracy is obtained with several parameter combinations, so that in the near-optimal combination, the value of H_{crit} decreases with increasing value of t_0. In other words, if the calculation of the temperature sum is initiated late, then a small accumulation of temperature sum provides the best fit to the data, and vice versa. Thus, when fitting a temperature sum model to a set of 57-year data on the bud burst of *Sorbus aucuparia*, Linkosalo et al. (2008) found that any starting date between the beginning of January and early April attained a similarly optimal fit to the data. As discussed in detail in Sect. 3.4.6.1, these findings call for experimental approaches to the modelling of the springtime phenology of boreal and temperate trees.

As the day degree units obtained with different threshold temperatures, T_{thr}, are not comparable with each other (Sarvas 1972), there is no multiplier to convert a temperature sum calculated with one threshold into one calculated by another threshold. This point is readily substantiated by the different starting times of the accumulation of dd units obtained with different values of the threshold temperature. The accumulation of dd units in spring starts earlier when the threshold has a low rather than a high value (Fig. 3.5).

In plant geography, the concept of day degree has been used as an index of the climatic conditions of a given location by calculating the average total accumulation of temperature sum for the growing season (Kalliola 1973; Tuhkanen 1980). The longer and warmer the growing season, the higher the value of the temperature sum. The obvious reason for using the temperature sum as a rough climatic index lies in the ecophysiological background of the model: the air temperature response of the accumulation rate of dd units approximates that of the rate of plant development. The temperature sum of the growing season also belongs to the essential variables in the niche-based, or climatic envelope, models, which are nowadays frequently used for assessing the effects of climatic change on the geographical range of plant and animal species (Bakkenes et al. 2002; Thomas et al. 2004;

[5] Except in the case where the value of the H_{crit} is known a priori. In that case, the prediction of the model can be calculated by starting directly from step (3).

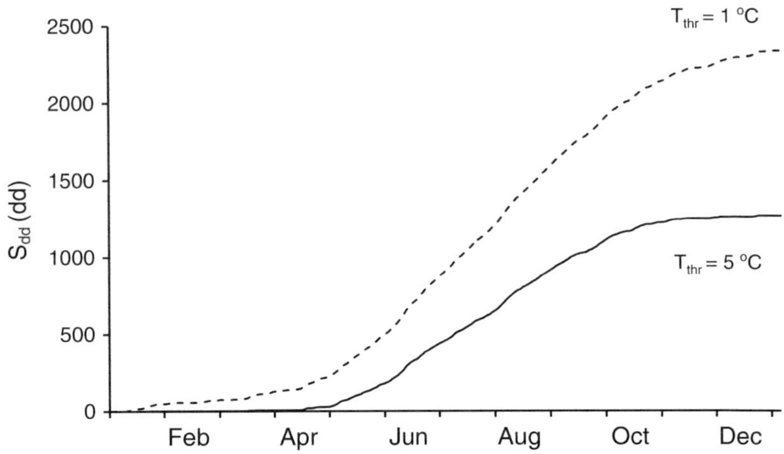

Fig. 3.5 Accumulation of temperature sum, S_{dd}, calculated with two values of the threshold temperature (T_{thr}; Fig. 3.3) for Eskdalemuir in Scotland; dd day degree unit. In each case, the calculations were initiated on 1 January 1970 with the daily mean air temperature as input. Note that each of the two temperature sum models has its particular dd unit on the horizontal axis and that these two dd units are incomparable

Thuiller et al. 2005). This use of the climatic envelope models will be briefly discussed in Sect. 8.2.3.2.

3.2.1.2 Ecophysiological Explication of the Day Degree Models

Most work related to the effects of air temperature on the spring phenology of trees and other plants has been carried out in the framework of the day degree model until recently. Though the concept of day degree has considerably improved our understanding of the spring phenology of trees and other plants as well as some insects and other invertebrate animals, the ecophysiological interpretation of the model requires special attention. This is because rather than the visible point event (e.g., bud burst) as such, it is actually the microscopic ontogenetic development leading to that point event that is modelled with the day degree model (Hänninen and Kramer 2007; Sect. 3.1.2). The accumulation of the day degree units describes the cumulative progress of that development in the same way, basically, as the accumulation of millimetres describes the progress of the ecophysiological process in the modelling of the height growth of plants (Hänninen and Lundell 2007). This comparison highlights two interrelated problems that call for an ecophysiological interpretation of the day degree model:

First, the response variable in day degree models is climatic rather than ecophysiological. This notion is readily substantiated when the units of the vertical axes of various ecophysiological models are compared with the corresponding unit of day degree models. For example, the typical unit in models of height growth is mm day^{-1} (Junttila and Nilsen 1993) and in models of photosynthesis, μmol CO_2 m^{-2} s^{-1} (Larcher 2003), whereas in the day degree models the unit is day degree day^{-1}. In all the three cases, the unit of time is expressed in the denominator, revealing that all of them are units of rate variables. Otherwise the climatological unit of the day degree model, obtained as a multiplication of the unit of time ('day') and an unofficial unit of temperature ('degree'), does not facilitate a direct ecophysiological interpretation, which is straightforward in the other two cases. Second, contrary to the direct experimentation applied to most ecophysiological responses, such as those of height growth or photosynthesis, the response of the temperature sum model is usually examined by means of indirect fitting to empirical data, often gathered in natural conditions, in which the temperature fluctuates (Table 3.1; Wielgolaski 1999).

The ecophysiological interpretation of the day degree model is introduced when the accumulation rate of the day degree units is divided by the value of the parameter H_{crit}, i.e., the high temperature (day degree) requirement of the event modelled (Campbell 1978; Hänninen 1995a; Hänninen et al. 2007):

$$R_o(T(t)) = 100\frac{R_{dd}(T(t))}{H_{crit}} \tag{3.1}$$

where t = time, T(t) = air temperature at time instant t, $R_o(T(t))$ = air temperature response of the rate of ontogenetic development, and $R_{dd}(T(t))$ = air temperature response of the accumulation rate of day degrees. The high temperature requirement, H_{crit}, is expressed in the specific day degree units (i.e., a particular threshold temperature, T_{thr}; Sarvas 1972; Fig. 3.3) used in the particular day degree model for calculating the value of $R_{dd}(T(t))$.

Rather than the accumulation rate of the arbitrary dd-units included in the original day degree model, Eq. 3.1 models the rate of the biological process addressed, i.e., the microscopic ontogenetic development towards the visible point event, such as bud burst or growth onset. This modelling is carried out by means of a relative scale. As an illustrative example, consider the day degree model with T_{thr} = +5 °C. At the temperature of T = 10 °C, for instance, the accumulation rate $R_{dd}(10)$ is 5 dd day^{-1} (Fig. 3.3). Assuming a tree genotype with H_{crit} = 100 dd, the quotient of Eq. 3.1 gets the value of 5 dd day^{-1}/100 dd = 0.05 day^{-1}. Thus the quotient has a dimensionless decimal number in the numerator and a unit of time in the denominator, so that the day degree unit cancels out. This quotient is the ecophysiological variable for the rate of development. It tells us that at the air temperature of T = 10 °C, 5 of the 100 dd units required for the point event to occur are accumulated in 1 day; so, the proportion of 5/100 = 0.05 of the developmental phenomena required for the point event to occur takes place during 1 day. The number of days required for the point event to occur is obtained as the reciprocal value of the quotient of Eq. 3.1, so that the number of days required is 1/0.05 day^{-1} = 20 days. Correspondingly, in the air temperature of T = 15 °C, the accumulation rate of day degrees is $R_{dd}(15)$ = 10 dd day^{-1} (Fig. 3.3), so that for the genotype H_{crit} = 100 dd, the quotient of Eq. 3.1 gets the value of 10 dd day^{-1}/100 dd = 0.1 day^{-1} with the reciprocal of 1/0.1 day^{-1} = 10 days.

In the original day degree model, each genotype has its own [0, H_{crit}] scale for the description of the progress of ontogenetic development from its starting point to the modelled point event, e.g., bud burst. By dividing the accumulation rate of day degrees by the value of H_{crit}, a dimensionless relative [0, 1] scale common for all the genotypes is created. By multiplying this quotient in Eq. 3.1 by 100, the [0, 1] scale is further transferred into a percentage scale [0, 100] (Hänninen 1995a; Hänninen et al. 2007). In this way, the rate of ontogenetic development gets the dimension of % day^{-1}, and any given value of R_o calculated with Eq. 3.1 indicates the percentage of the development taking place during 1 day at a constant temperature T(t), out of the total development required for the modelled point event to occur (Fig. 3.6a; Campbell 1978). Then, by definition, the point event is predicted to occur when the integral of Eq. 3.1 attains the value of $S_o(t) = 100$ % (Fig. 3.6b).

Consider the day degree model for two tree genotypes with the values of $H_{crit} = 50$ dd and $H_{crit} = 150$ dd for bud burst (Fig. 3.6). The shape of the air temperature response of the rate of ontogenetic development is identical for the two genotypes, but the scaling is different, so that at any given temperature above the threshold of +5 °C, the rate of development for the genotype with $H_{crit} = 150$ dd is one third of the corresponding rate for the genotype with $H_{crit} = 50$ dd (Fig. 3.6a). In both cases, the simulated ontogenetic development progresses whenever the daily mean air temperature is above +5 °C, but on each day, the slope of the S_o curve for the genotype with $H_{crit} = 50$ dd is three times as steep as the corresponding slope for the other genotype with $H_{crit} = 150$ dd. In this way, the model predicts that in the spring of 1970 in Jyväskylä, the bud bursts of these two genotypes take place on 19 May and 5 June, respectively (Fig. 3.6b).

Mathematically, Eq. 3.1 only involves the replacement of one scale with another one. Ecophysiologically, however, the equation involves a more profound idea. Contrary to any given value of the temperature sum, S_{dd}, a given value of the variable state of ontogenetic development, S_o, predicts that a given ecophysiological status will occur in real trees. Not only is bud burst predicted to occur, then, when $S_o(t) = 100$ % (Fig. 3.6b), but analogically, any discernible microscopic point event preceding the bud burst is predicted to occur when its specific value of $S_o(t)$ has been attained, regardless of the tree genotype (value of H_{crit}) or the air temperature conditions (input air temperature data). This prediction can be tested by means of microscopic examination of the buds (Fig. 3.2), provided that air temperature data for the period preceding the bud burst is available (Viherä-Aarnio et al. 2014).

With the ecophysiological percentage scale, one is also not committed to any specific form of the air temperature response of the rate of development, so that day degree models with different values of the threshold temperature, T_{thr}, and non-linear models become all comparable when transformed onto a percentage scale (Hänninen and Lundell 2007). A similar idea of an ecophysiologically explicit scale was already used in the Degree Growth State (°GS) model, where the whole annual cycle is described by means of one ecophysiological scale for the state of development (Fuchigami et al. 1982; Sect. 3.6.3).

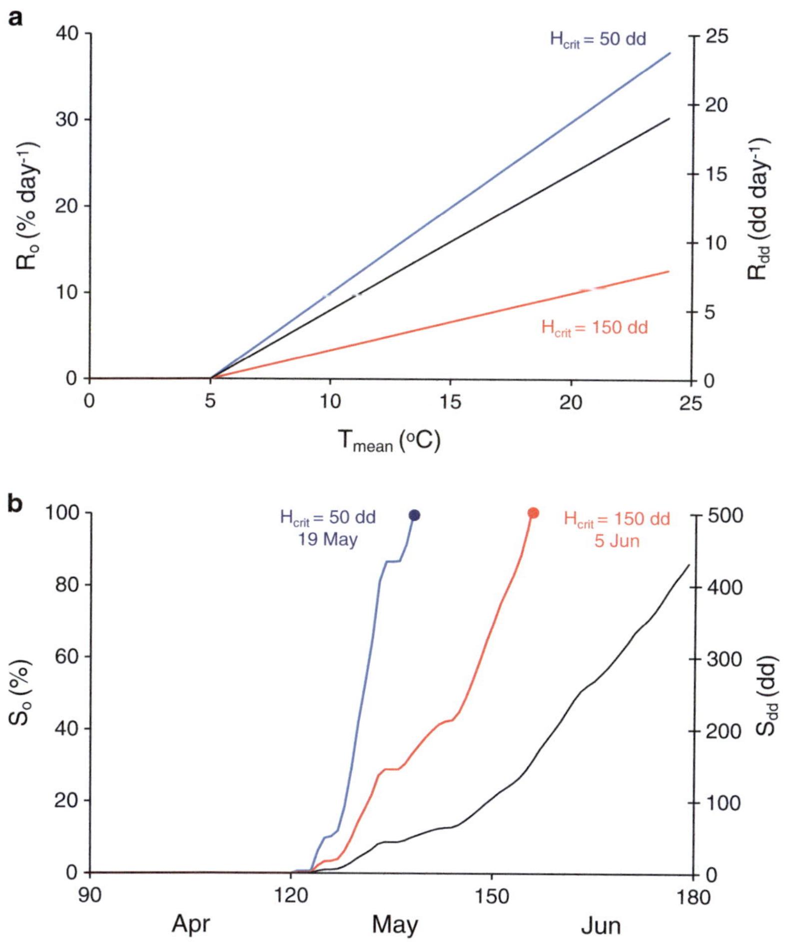

Fig. 3.6 An ecophysiological interpretation of the modelling of the springtime phenological development of trees, e.g., the timing of bud burst, within the traditional day degree approach. In both (**a**) and (**b**), the *black line* and the *vertical axis on the right* denote the meteorological accumulation of arbitrary developmental units (day degrees, dd), and the *coloured lines* and the *vertical axis on the left* denote the simulated ontogenetic development of the buds of two tree genotypes characterised by their respective values for the high temperature requirement of growth onset (here specified as bud burst), H_{crit}. (**a**) The air temperature response of the accumulation rate of dd units, R_{dd}, and the corresponding air temperature responses of the rate of ontogenetic development, R_o, for the two tree genotypes, T_{mean} daily mean temperature. (**b**) Accumulation of temperature sum, S_{dd}, and the predicted state of ontogenetic development, S_o, of the two tree genotypes. Starting from 1 January 1970, the calculations were carried out with daily mean air temperature data from Jyväskylä, central Finland, as input. The predicted date of bud burst for each genotype is indicated at the end of the respective lines of S_o

3.2.1.3 Experimental Studies

The transformation onto the percentage scale not only explicates the ecophysiological interpretation of the day degree models and makes different response functions comparable with each other but also elaborates the experimental design for determining the air temperature response of the rate of ontogenetic development, $R_o(T)$, empirically. For a given constant air temperature, T, $R_o(T)$ is defined as the reciprocal of the developmental time, $\Delta t(T)$, i.e., the time required for the developmental point event to occur in air temperature T, multiplied by 100 (Campbell 1978):

$$R_o(T) = 100\frac{1}{\Delta t(T)} \tag{3.2}$$

Thus, when studying the air temperature response of the rate of development of tree seedlings towards bud burst, for instance, samples of the seedlings are transferred into a set of growth chambers. In each chamber, a constant predetermined air temperature T is maintained, with the temperature varying among the chambers. The timing of the bud burst of the seedlings in each chamber is recorded and the number of days required for bud burst, $\Delta t(T)$, is thus determined for each particular air temperature used in the chambers. The rate of development corresponding to each applied temperature is then determined by multiplying the reciprocal of the days required in that temperature by 100 (Eq. 3.2). The air temperature response of the rate of development is obtained by expressing the rate of development thus determined as a function of the air temperature (Campbell 1978; Caffarra 2007).

Since no observations are available on the development before the occurrence of bud burst, the method described above yields the average rate of development at the particular temperature. It is possible that the rate of development within the period from the start of the experiment until the observed bud burst varies. For this reason Campbell (1974) and Campbell and Sugano (1975) suggested that a concept of DARD (Daily Average Rate of Development) be applied to the percentage scale. Though informative, this concept has not been widely used.

There is a crucial methodological difference between experimental studies addressing the air temperature response of the rate of ontogenetic development and other studies addressing the environmental responses of other ecophysiological phenomena, such as height growth or photosynthesis. When determining the air temperature response of the rate of height growth, for instance, one simply monitors the height at predetermined time intervals, and the rate of the ecophysiological process (height growth) is then obtained as the height increment observed between two successive measurements divided by the time lapsed between the measurements (Junttila and Nilsen 1993). Similarly, one obtains the rate of leaf photosynthesis by measuring the amount of carbon dioxide taken up by the leaf during the measurement and dividing this quantity by the time lapsed in the measurement (Larcher 2003).

Such a straightforward methodology cannot be applied to measuring the rate of ontogenetic development, for the state of development cannot be measured continuously at predetermined time intervals. Rather, there is only one or, at best, a few point events whose occurrence determines the rate of development. Thus, in contrast to the standard ecophysiological method of fixing the time interval and measuring the amount of the ecophysiological process (e.g., height growth or uptake of CO_2) attained, one needs to determine the rate of ontogenetic development the other way round: the amount of development is fixed, e.g., by requiring that the stage of bud burst be attained, and the measured variable is the time it takes for this to occur. Thus, in order to obtain the value of the rate variable, one needs to calculate the reciprocal of the time required.

Furthermore, as there is no scale expressed in SI units available for phenological development, the approach yields a relative value for the rate of development (Eq. 3.2). This reasoning also reveals that the principle of phenological models is based on an analogy with Newtonian physics, where the average velocity (rate variable) of an object is determined by dividing the distance the object has moved by the time lapsed. Since distance cannot be measured with the SI unit of length in phenological models, the concept of distance is replaced with a corresponding relative 'distance' on an imaginary developmental road towards bud burst (Hänninen 2009).

Sarvas (1967, 1972, 1974) carried out large-scale studies on the annual cycle of boreal forest trees, concentrating mainly on the generative meristems of the trees. In his early studies addressing the spring phenology of trees, he applied the traditional day degree model, following the Finnish standard of using the value of $T_{thr} = +5$ °C for the threshold temperature (Sarvas 1967). According to a story often told (Koski 2009; Hänninen 2009) but obviously not documented in any of Sarvas's publications, he was taken by surprise one spring when *Alnus incana* flowered before a single day degree was accumulated at the growing site of the observed trees. According to the story, this observation led Sarvas to start his large-scale experimental studies in connection with his large-scale field studies, which had already been initiated. Some of the results were published, together with a modelling synthesis, in the 1970s (Sarvas 1972, 1974), but many of them were published by a colleague after Sarvas had passed away (Luomajoki 1977, 1982, 1984, 1986a, b, 1993a, b, 1999).

Due to their relatively obscure publishing forum, Sarvas's (1972, 1974) studies remained relatively unknown to the international research community. Even so, his work was discussed early on by Landsberg (1977) and Campbell (1978) and later by Cannell (1989). Fuchigami et al. (1982) acknowledged his studies and his model as an important foundation of their Degree Growth State (°GS) model.[6] As a result of these and some other publications, Sarvas's (1972, 1974) studies gradually became more familiar to the international research community, and during the last 20 years or so, his work has gained so much more attention that it is no exaggeration to speak of a Renaissance of his studies. This was first and foremost because his model of the annual cycle became a key reference in modelling studies for assessing the effects of climate change.

Sarvas (1972, 1974) carried out his experiments mainly with twigs sampled from mature trees growing in natural conditions. A sample of twigs was transferred into a set of growth chambers, each chamber having its predefined constant air temperature. The development of generative meristems was monitored in the chambers. Microscopic investigations of the progress of meiosis in the pollen mother cells of

[6] This model is discussed in Sect. 3.6.3.

several tree species were carried out. Regardless of the phenomenon examined, the principle of the experiment was always the same. Frequently repeated sampling and statistically careful analysis were used to determine the median time instant when a given state of development, say Metaphase I in the pollen mother cells, was attained. The same procedure was repeated in order to determine the median for the time instant when the next point event, in this case Metaphase II, was attained. The developmental time, Δt, was then determined as the difference between these two time instants and the rate of development was calculated as the reciprocal of Δt (Eq. 3.2) (Sarvas 1972, 1974).

With this detailed experimentation, where several sequential microscopic point events were observed by repeating the sampling on a time scale of hours, much more precise information was obtained than is possible in the more common experiment, where only the final point event, which is visible to the naked eye, e.g., bud burst, is observed. However, Sarvas (1972, 1974) also carried out coarser sorts of experiments. In these experiments, the final visible point event was the time instant of the opening of the male catkins, which was determined by observing the release of pollen from the catkins.

Though Sarvas (1972, 1974) determined the air temperature response with exceptionally detailed experimentation, he did not present his results on the ecophysiological percentage scale of Eq. 3.2. Rather, he introduced a new arbitrary scale, referred to as the period unit, PU. This was done by introducing a multiplier into the calculation of the rate of development (Table 3.2):

$$R_{PU}(T) = \frac{5 \cdot \Delta t_{10}}{\Delta t(T)} \tag{3.3}$$

where T = air temperature, $R_{PU}(T)$ = air temperature response of the accumulation rate of period units, Δt_{10} = developmental time required at the reference air temperature T = 10 °C, and $\Delta t(T)$ = developmental time required at air temperature T. Thus, with reference to Eqs. 3.2 and 3.9 (the latter in Sect. 3.2.1.4), the introduced multiplier $5\Delta t_{10}$ is the value of the high temperature requirement, H_{crit}, now expressed in the new period units. Using this scaling, Sarvas (1972) found that the points representing the air temperature response of the rate of development in the generative meristems fell on a common sigmoidal curve (Fig. 3.7). Considering that the experimental data consisted of several tree species, both angiosperms and gymnosperms (results obtained with the *Larix* species are not presented here), and several phases of ontogenetic development, this finding can be regarded as a major scientific achievement.

However, the fact that the accumulation rate of period units is identical at a given temperature with all species and phenomena should not be interpreted to mean that the rate of development is also identical in all examined cases. Rather, the curve indicates that the rate of development responds to the air temperature in a similar way regardless of the species and the phenomenon examined. This means, for instance, that the developmental time required in the air temperature of 10 °C is about twice the developmental time required in 15 °C (Fig. 3.7). To obtain a comparable rate of development, i.e., the reciprocal of the developmental time Δt, one must divide the

Table 3.2 Experimental determination of the air-temperature response of the rate of ontogenetic development in the flower buds of three boreal tree species

Experimental object	T (°C)	Δt_{10} (h)	Δt (T) (h)	$R_{PU}(T)$ (PU h^{-1})
Populus tremula Meiosis in microspore mother cells MI → MII	0.0	6.5	62.0	0.527
	0.3	6.5	55.0	0.590
	1.2	6.5	44.0	0.736
	4.2	6.5	18.5	1.75
	6.1	6.5	12.0	2.77
	8.2	6.5	9.3	3.49
	10.6	6.5	5.0	6.50
Betula pendula Opening of the male catkins	7.9	122.6	172.8	3.54
	10.3	122.6	111.6	5.49
	11.9	122.6	96.0	6.40
	15.7	122.6	54.2	11.40
	18.2	122.6	43.7	14.10
	24.1	122.6	29.3	20.9
	29.1	122.6	25.2	25.5
Betula pubescens Opening of the male catkins Series 1	7.7	73.7	135.1	2.74
	9.2	73.7	89.0	4.15
	11.7	73.7	57.6	6.38
	15.9	73.7	34.1	10.8
	18.2	73.7	26.4	13.8
	23.8	73.7	18.2	20.0
	25.3	73.7	16.3	22.4
	29.3	73.7	15.1	24.3
Betula pubescens Opening of the male catkins Series 2	18.2	50.9	17.0	15.0
	23.8	50.9	12.0	21.2
	35.1	50.9	9.40	27.0

T the experimental temperature applied, Δt_{10} the time required for the developmental event to occur in the reference temperature of 10 °C, and $\Delta t(T)$ the corresponding time required in each temperature T. The rate of development is expressed as the accumulation rate of arbitrary developmental units (period unit *PU*), $R_{PU}(T)$, which is obtained by scaling the reciprocal of Δt (T) as indicated by Eq. 3.3 in the text. *MI* metaphase I, and *MII* metaphase II (Sarvas 1972; published with the permission of the Finnish Forest Research Institute, currently the Natural Resources Institute Finland, and the Finnish Society of Forest Science)

accumulation rate of period units by the multiplier $5\Delta t_{10}$ (Eq. 3.3). Thus, as the value of Δt_{10} varies considerably among the species and phenomena examined (Table 3.2), so does the rate of development. In other words, the PU curve indicates the shape of the air temperature response, but in order to get a comparable rate of development and, for instance, to calculate the predicted timing of a given event, the value of the high temperature requirement $H_{crit} = 5\Delta t_{10}$ must also be known.

In this way the PU model is analogical to the classical degree hour model: there is an arbitrary unit on the vertical axis in both, and the critical number of these units required for the point event to occur must be known before the curve can be used for predicting the occurrence of the event. Despite this similarity, however, one should

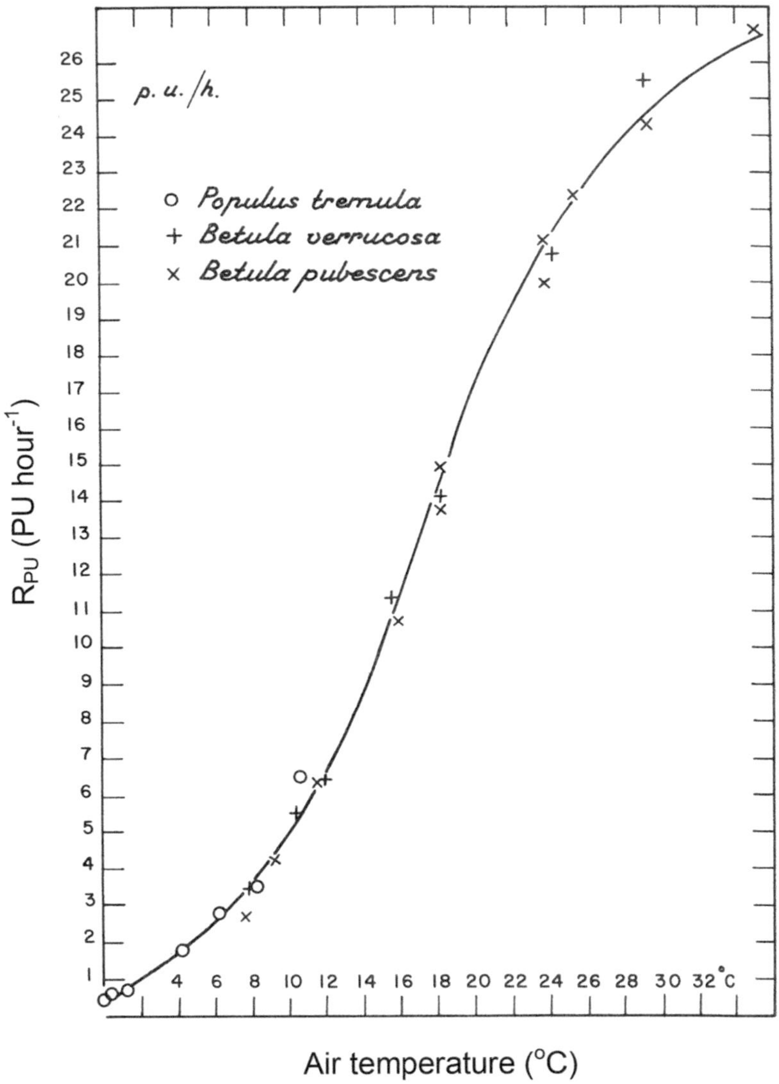

Fig. 3.7 The air temperature response of the rate of ontogenetic development in the generative meristems of three boreal tree species, as expressed with the accumulation rate of arbitrary developmental units (period units, PU). For the experimental design and the calculation of the PU units, see the text and Table 3.2. *Betula verrucosa* is the former scientific name of *Betula pendula* (Sarvas 1972; redrawn with the permission of the Finnish Forest Research Institute, currently the Natural Resources Institute Finland, and the Finnish Society of Forest Science)

not forget the crucial difference between these two models: the degree hour (and the day degree) model is based on a simple rule of thumb developed on the basis of indirect and quite circumstantial information, whereas the PU curve is based on direct and solid experimental evidence (Sarvas 1972; Table 3.2). A comparison of the degree hour and the period unit model reveals that with the exception of the low

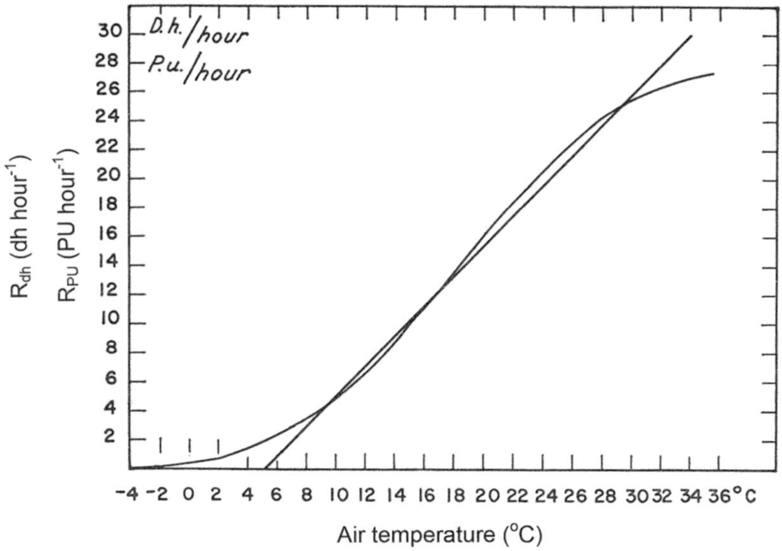

Fig. 3.8 A comparison of the period unit model (*sigmoidal curve*; see also Fig. 3.7) with the classical degree hour model (*straight line*) with the threshold temperature of +5 °C. R_{dh} accumulation rate of dh units (degree hours), R_{PU} accumulation rate of PUs (period units) (Sarvas 1972; redrawn with the permission of the Finnish Forest Research Institute, currently the Natural Resources Institute Finland, and the Finnish Society of Forest Science)

end of the temperature range, i.e., temperatures between zero and 10 °C, the real ecophysiological response can be quite closely approximated by the classical degree hour model (Fig. 3.8).[7] Another exception is seen at the upper end of the temperature range, i.e., above 30 °C, but these temperatures are so rare in natural conditions that this deviation is negligible.

The experimental results synthesized in the PU curve also provide an explanation for the observation that was the impetus, according to the legend (Koski 2009; Hänninen 2009), for Sarvas to start his experimental work. On the basis of the PU curve (Fig. 3.7) it is readily understood that in some years, early trees such as *Alnus incana* can flower before the accumulation of day degrees is initiated, i.e., before the daily mean temperatures rise above +5 °C. It is also a priori obvious that Sarvas's (1972) experimental results concerning the development taking place at temperatures between zero and 10 °C are essential for assessing the effects of climate change: according to the climatic scenarios, climate change will considerably increase the frequency of these temperatures in winter and spring at several northern locations. These issues will be dealt with in detail in Sect. 8.3.

[7] Sarvas (1972) actually indicated that the multiplier 5 Δt_{10} was introduced in order to obtain this similarity.

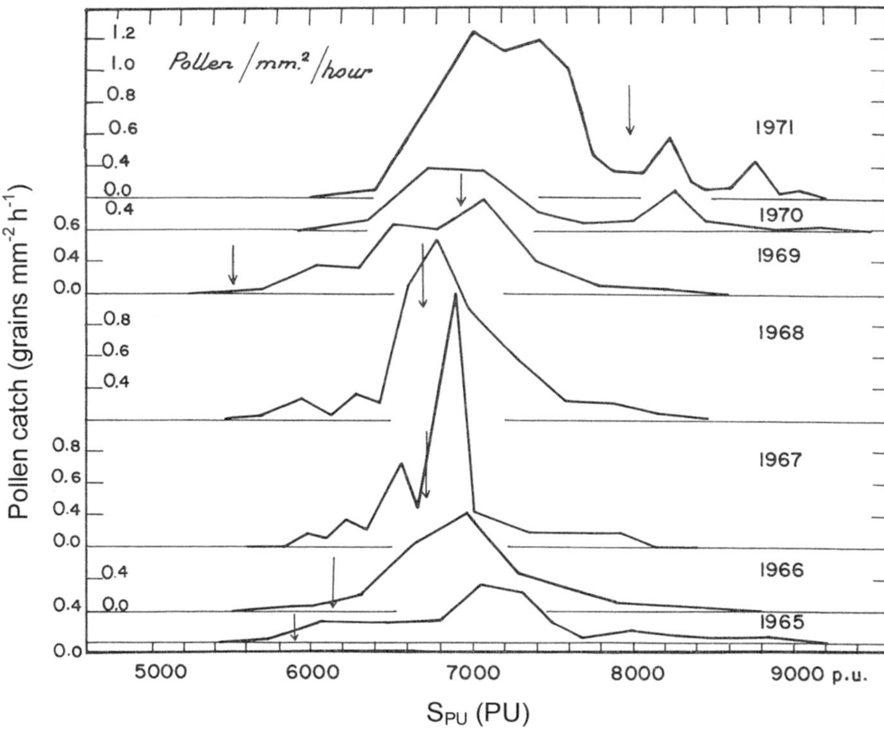

Fig. 3.9 A field test of the period unit model (Fig. 3.7) with observations of the male flowering of *Pinus sylvestris* in Bromarv, southern Finland. The *horizontal axis* indicates the accumulation of period units calculated from the beginning of the year, S_{PU}, on the basis of air temperature records gathered in a screen located above the canopy. The pollen catch on the vertical axis was measured as the number of pollen grains impacted on the surface of a sampler, expressed as grains per mm^{-2} h^{-1}, so that it provides an index of pollen concentration in the air. The *arrows* indicate the accumulation of period units on 10 June each year. There is considerable variation in the timing of peak flowering with respect to calendar time, but as predicted by the period unit model developed on the basis of growth chamber measurements (Table 3.2; Fig. 3.7), the peak flowering occurs at a relatively constant accumulation of period units each year (Sarvas 1972; redrawn with the permission of the Finnish Forest Research Institute, currently the Natural Resources Institute Finland, and the Finnish Society of Forest Science)

Sarvas (1972) tested the period unit model with long-term large-scale studies carried out in natural conditions. The timing of the male flowering of several tree species was determined by measuring the pollen concentrations in the air at the stands. The air temperature above the canopy at the same stands was simultaneously measured. Similarly to the predictions of the classical day degree models, the period unit model predicts that flowering will take place at a constant accumulation of period units each year, though the calendar time of the flowering varies considerably. This prediction was realised in the tests, so that the period unit model, formulated on the basis of growth chamber experiments, was supported by the observations gathered in these independent field tests (Fig. 3.9).

In addition to Sarvas's (1972) PU curve (Fig. 3.7), relatively few experimentally based air temperature responses have been published for the rate of ontogenetic development in boreal and temperate trees. Some of these responses are discussed in Sect. 8.3.3.2.

3.2.1.4 Introducing the Experimental Results into the Modelling

Introducing the experimentally based air temperature response of the rate of ontogenetic development into model simulations is conceptually straightforward as such: The primary experimental data concerning the developmental time Δt required for growth onset (or any other phenological springtime event modelled) in each of the experimental air temperatures T are first transferred into the corresponding rates of development (Eq. 3.2). Subsequently, a response function f is fitted to the data, so that the rate of ontogenetic development at any given moment, $R_o(t)$, can be calculated as a function of the momentary air temperature T(t):

$$R_o(t) = f(T(t)) \tag{3.4}$$

Under any given air temperature conditions, growth onset is predicted by definition (Eq. 3.2) to occur when the time integral of Eq. 3.4 attains the value of 100. In this ecophysiologically explicit case no arbitrary developmental units, such as day degrees or period units, are used. The unit for the rate variable $R_o(t)$ is % per unit time, so that the integration with respect to time yields the state variable $S_o(t)$ with the unit %.

Historically, however, most modelling studies have been carried out in the framework of arbitrary developmental units, and this also holds for many of the studies reviewed in the present volume (Sarvas 1972, 1974; Cannell and Smith 1983; Hänninen 1990a, b; Kramer 1994a, b; Chuine et al. 1998, 1999). This situation calls for a discussion towards clarifying the relation between the traditional approach, using arbitrary developmental units, and the ecophysiologically explicit approach adopted in the present volume.

Hänninen (1990b) adopted Sarvas's (1972) PU curve (Fig. 3.7) into his generalised modelling framework for growth onset. This was because this response is based on an exceptionally large set of empirical data and because it addresses boreal tree species, which Hänninen (1990b) also studied. Hänninen (1990b) fitted a sigmoidal function to Sarvas's (1972) data. However, as Hänninen (1990b) applied Sarvas's (1972) response with a time step of 1 day instead of the original step of 1 h, he replaced the period unit by a new unit, i.e., the "forcing unit", FU, for the sigmoidal model:

$$R_{FU}(t) = \begin{cases} 0 & T(t) < T_{thr} \\ \left[\dfrac{c}{1+\exp(-a(T(t)-b))}\right] & T(t) \geq T_{thr} \end{cases} \tag{3.5}$$

where $R_{FU}(t)$ = accumulation rate of arbitrary forcing units, and T(t) = air temperature, both at time instant t. In the model fitting, the following values were obtained for the parameters: $a = 0.185\ ^{\circ}C^{-1}$ (steepness of the curve), $b = 18.4\ ^{\circ}C$ (inflexion point on the temperature axis), $c = 28.4$ FU (upper asymptote), and $T_{thr} = 0\ ^{\circ}C$ (threshold air temperature). By definition, the number of accumulated forcing units, $S_{FU}(t)$, is obtained by integrating Eq. 3.5 from the beginning of the simulation to the time instant under study (Eq. 2.4 in Sect. 2.2). Growth onset is predicted to occur when $S_{FU}(t)$ attains a genotype-specific critical forcing unit sum, FU_{crit}:

$$S_{FU}(t) = FU_{crit} \tag{3.6}$$

As pointed out by Kramer (1994b) and Chuine (2000), the parameter c in Eq. 3.5 is redundant. The parameter cancels out, i.e., is replaced with unity when both sides of Eq. 3.5 are divided by it:

$$R_{FU}(t)/c = \begin{cases} 0 & T(t) < T_{thr} \\ \left[\dfrac{1}{1+\exp(-a(T(t)-b))}\right] & T(t) \geq T_{thr} \end{cases} \tag{3.7}$$

By denoting $R_{FU'}(t) = R_{FU}(t)/c$, a modified FU′ scale is introduced, so that the condition expressed by Eq. 3.6 for growth onset to occur is equal to

$$S_{FU'}(t) = \frac{FU_{crit}}{c} \tag{3.8}$$

In Sect. 3.2.1.2, the day degree model was transferred into an ecophysiologically explicit form by Eq. 3.1. Hänninen and Kramer (2007) generalised that equation as follows:

$$R_o(T(t)) = 100\frac{R_{HU}(T(t))}{H_{crit}} \tag{3.9}$$

where t = time, $T(t)$ = air temperature at time instant t, $R_o(T(t))$ = air temperature response of the rate of ontogenetic development, $R_{HU}(T(t))$ = air temperature response of the accumulation rate of any high temperature units, HU, regardless of the form of the air temperature response, and H_{crit} = high temperature requirement of the modelled event. In addition to the piecewise linear responses used in different day degree models with different threshold temperatures covered by Eq. 3.1, all non-linear responses, such as the PU curve (Fig. 3.7), are also included in this generalised model. The response $R_{HU}(T(t))$ needs to be specified by means of a mathematical equation, and once this is done, the high temperature requirement, H_{crit}, is expressed in the high temperature units specific to that particular mathematical equation. It is crucial to notice that for this reason the H_{crit}-values corresponding to different high temperature units are incomparable.[8]

[8] Compare the corresponding case with day degree models in Figs. 3.3 and 3.5.

Following Hänninen and Kramer (2007), the FU′ unit is used in the present volume as the high temperature unit, HU, so that the corresponding Eq. 3.7 is plugged into Eq. 3.9 when the rate of ontogenetic development is calculated; i.e., $R_{HU}(T(t))$ in Eq. 3.9 is calculated according to Eq. 3.7:

$$R_o(t) = \begin{cases} 0 & T(t) < T_{thr} \\ \left[\dfrac{100}{H_{crit}}\right] \cdot \left[\dfrac{1}{1 + \exp(-a(T(t) - b))}\right] & T(t) \geq T_{thr} \end{cases} \tag{3.10}$$

where $R_o(t)$ = rate of ontogenetic development, H_{crit} = genotype-specific high temperature requirement of growth onset in the high temperature units (HU) of the FU′ scale (Eq. 3.8), other symbols as in Eq. 3.7. By definition, the modelled event is predicted to occur when the time integral of Eq. 3.10 attains the value of 100.

To summarise: In the present volume the sigmoidal form of the air temperature response of the rate of ontogenetic development was adopted from Sarvas's (1972) experimental PU curve (Fig. 3.7). However, rather than the time step of 1 h applied in the original model, the time step of 1 day was used in the present volume. The values for the steepness parameter a and the inflexion point parameter b were estimated on the basis of Sarvas's (1972) experimental results (Hänninen 1990b). Finally, following Hänninen and Kramer (2007), the model was converted into an ecophysiologically explicit form. In this way, the sigmoidal form of the response remains unchanged, but the level of the curve is determined in each case by the high temperature requirement, H_{crit} (Fig. 3.10; compare the corresponding case with the day degree models presented in Fig. 3.6a).

Hänninen's (1990b) FU model, applying the forcing units presented in Eq. 3.5, has been used in several studies, which will be discussed in various chapters of the present volume (Hänninen 1990b, 1991, 1996; Hänninen and Hari 1996). When these studies are discussed, the original FU scale is replaced with the FU′ scale obtained by dividing the FU values by the value of c = 28.4 FU (Kramer 1994b; Chuine 2000; Hänninen and Kramer 2007). In this way, the values FU_{crit} = 100 FU and FU_{crit} = 150 FU for instance, typical for central Finnish trees (Hänninen 1990b, 1996), are converted into the values of H_{crit} = 3.5 HU and H_{crit} = 5.3 HU, respectively (Fig. 3.10).

The above discussion shows once again that no arbitrary climatological units, such as day degrees and period units, would actually be needed in ecophysiological research on the spring phenology of trees and other plants if the proper experimentally determined air temperature response of the rate of development (Eq. 3.2) were available in each case. However, in comparison with the vast body of published results for various environmental responses of different ecophysiological processes such as photosynthesis, respiration, and elongation growth (Junttila and Nilsen 1993; Larcher 2003), there are surprisingly few studies where the air temperature response of the rate of development has been determined experimentally (Sects. 3.2.1.3 and 8.3.3.2). Thus, as the proper relative ecophysiological response is not available in most cases, one has to resort to various predefined response

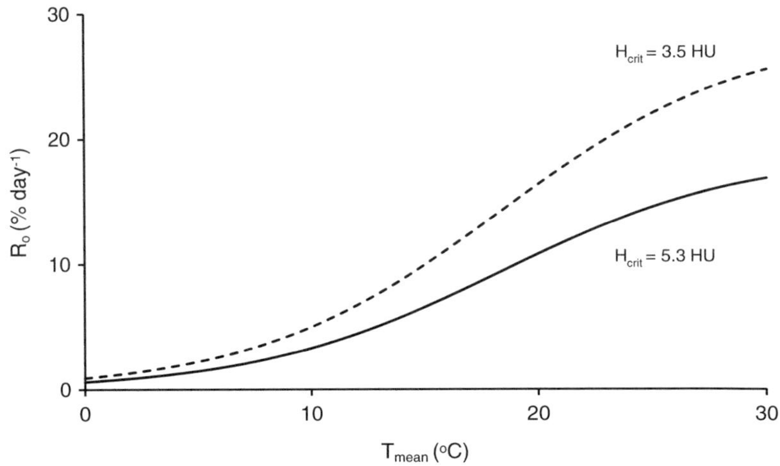

Fig. 3.10 The dependence of the rate of ontogenetic development, R_o, on the daily mean air temperature, T_{mean}, for two typical central Finnish tree genotypes. The shape of the curves drawn according to Eq. 3.10 is similar to that of Sarvas's (1972) PU curve (Fig. 3.7), while the level of the curves depends on the genotype-specific high temperature requirement of growth onset, H_{crit}. *HU* arbitrary high temperature unit specific to the sigmoidal function defined by the right-hand quotient in Eq. 3.10, and to the parameter values of a = 0.185 °C^{-1} and b = 18.4 °C used for drawing the curve (Modified from Hänninen et al. 2007; published with the permission of Oxford University Press)

curves and their particular arbitrary developmental units, especially in field studies carried out in naturally fluctuating temperature conditions (Table 3.1).

There is no need to entirely replace the models using arbitrary developmental units with the ecophysiological percentage approach, but it is nevertheless important to recognise the ecophysiological and methodological background of the models using arbitrary units and even more crucial to recognise the mutual incomparability of models using different arbitrary units (Figs. 3.3 and 3.5). The present volume relies mainly on the ecophysiological percentage scale, emphasising the ecophysiological interpretation of the model variables and the comparison of different tree genotypes (Figs. 3.6a and 3.10). However, the accumulation of high temperature units is also referred to, especially in cases where the discussion is about the properties of the climate rather than the development of trees.

The concepts "high temperature unit" and "high temperature requirement of growth onset" were retained in the present volume, too, as the model of the arbitrary high temperature units (Eq. 3.7) is plugged into the model expressed in the ecophysiological percentage scale (Eq. 3.10). This should not obscure the crucial difference between the traditional approach applying the arbitrary developmental units and the ecophysiologically explicit approach of the present volume. When the model represented by Eq. 3.10 is used, the exact form of the sigmoidal response, i.e., the high temperature units to be applied, is not fixed a priori as it is in the traditional approach. Rather, it is determined when Eq. (3.10) is fitted to experimental data gathered according to the principle presented in Eq. 3.2. In this way the experimental data determine the particular high temperature units to be used (values of parameters a and b) and the number of these units required for growth onset (value of parameter H_{crit}). Furthermore, if necessary, the sigmoidal form of the response function can be replaced with any other form required by the experimental data.

3.2.1.5 Effects of Fluctuating Temperatures

It has been observed in several studies that fewer high temperature units are required for bud burst and other springtime phenological events in fluctuating than in constant temperatures (Campbell and Sugano 1975; Van den Driessche 1975; Hänninen 1990b; Partanen et al. 1998; Sutinen et al. 2009).[9] These findings suggest that fluctuating temperatures are more efficient than constant temperatures in causing ontogenetic development. There are at least four potential explanations for the deviating results obtained in regard to fluctuating versus constant temperatures.[10]

The observed lower requirement of high temperature units at fluctuating as opposed to constant air temperatures may be caused by real physiological phenomena included in the air temperature response of the rate of ontogenetic development. In this case the observation would represent one special case of thermoperiodism, where the temperature fluctuation as such affects the physiological process of the plant. Thermoperiodism is a well-documented phenomenon in plant physiology (Mohr and Schopfer 1995), but its possible role in the ontogenetic development of boreal and temperate trees towards growth onset remains largely unexamined.

However, the different results obtained at constant versus fluctuating temperatures may also be a mathematical artifact caused by the integration error inherent in the numerical integration of a non-linear response function (Smolander 1984). Accordingly, the use of the mean value of an environmental factor averaged over a time period does not result in the mathematically correct, i.e., analytical, value of the state variable, which would theoretically be obtained if the integration were carried out using instantaneous values of the fluctuating environmental factor.

With a concave air temperature response, the averaging of the temperature data over a longer period, say over a day, produces an underestimate of the accumulation of the developmental units. This is readily demonstrated by considering the piecewise linear (and for this reason non-linear) response involved in the day degree models (Fig. 3.3). If the temperature fluctuates during 1 day so that part of the time it is below and part of the time above the threshold T_{thr}, with the daily mean temperature T_{mean} below T_{thr}, then the numerical integration carried out with T_{mean}

[9] However, Myking (1997) found no such difference when studying the bud burst in seedlings of *Betula pubescens*.

[10] Two of the potential explanations are related to the dormancy of the buds and will therefore be discussed in Sects. 3.4.1.2 and 3.4.4.2.

will imply zero accumulation of day degrees for that particular day, even though the air temperature is above T_{thr} part of the time, so that ontogenetic development and corresponding accumulation of developmental units should take place.

In order to remove the integration error from the calculations, the time step used in the averaging of the environmental factor should be so short that the variation in the environmental factor during the time step could be neglected. In the case of air temperature, 1 h is usually a sufficiently short time interval. Thus, if the accumulation of high temperature units is calculated with the time step of 1 h, then it is improbable that any difference observed between the constant and the fluctuating air temperature conditions in the high temperature requirement of growth onset is due to the artifact caused by the integration error. Unfortunately, hourly mean temperatures are often unavailable for natural conditions, so that the calculating of the accumulation of the developmental units has to be based on the daily mean temperatures.

3.2.2 The Unique Role of Air Temperature in Comparison with Other Environmental Factors

In addition to air temperature, discussed in detail in Sect. 3.2.1, several other environmental factors have been found to affect the spring phenology of boreal and temperate trees. The most important of these is night length (or photoperiod; for the terminology, see the discussion in Sect. 1.2); but other factors, too, such as ground frost and soil temperature, concentrations of nutrients in the soil, concentration of atmospheric carbon dioxide, and the intensity of solar radiation have been listed among the environmental factors affecting the spring phenology of boreal and temperate trees (for references, see Sects. 3.3.3 and 3.4.3).

Among all environmental factors, however, the role of springtime air temperature is unique. This is because in all boreal and temperate tree species, the microscopic ontogenetic development (Sect. 3.1.2) leading to the visible phenological events, such as bud burst and onset of height growth, takes place in sufficiently high temperatures only. The high temperature requirement discussed in Sect. 3.2.1 is a universal phenomenon, then, common to all boreal and temperate trees, whereas the effects of the other factors vary among tree species.

The effects of night length on the spring phenology have been generally considered to be indirect, i.e., night length is considered to affect the dormancy status (rest break) of the buds, which then affects the ontogenetic development towards the visible phenological event. That is why the effects of night length and other light- related factors are discussed separately, with reference to the dormancy status of the buds, in Sects. 3.3.3 and 3.4.2. As regards the other environmental factors, such as soil temperature or the concentration of atmospheric carbon dioxide, it is hard to tell whether their effects on ontogenetic development are direct or indirect. Thus there is a brief discussion of these factors presented separately in Sect. 3.4.3.

3.3 Dormancy in the Buds

3.3.1 The Nomenclature for Dormancy

3.3.1.1 Three Categories of Dormancy

In everyday language, the term *dormancy* refers to the opposite of growth, i.e., no visible growth being observed in a plant tissue predisposed to grow. The term is used in this broad sense in the present volume, too. The nomenclature for different types of dormancy is confusing, as different terms have been used for the same type and same terms for different types of dormancy. Reviews of the dormancy phenomena and their nomenclature have been published, among others, by Doorenbos (1953), Samish (1954), Romberger (1963), Weiser (1970), Fuchigami et al. (1982), Sarvas (1974), and Lang et al. (1987). Though different authors use different nomenclatures, they normally subdivide the dormancy phenomena of the buds of boreal and temperate trees similarly into the three categories discussed below. In the present volume, the nomenclature often used in horticultural and forest sciences will be followed (Table 3.3), as that particular nomenclature has been used in studies sharing the modelling approach adopted in the present volume (Fuchigami et al. 1982; Hänninen 1990b; Hänninen and Kramer 2007).

The bud is in the state of *rest* when bud burst and growth are arrested by physiological factors inside the bud itself (Table 3.3).[11] Thus, when the bud is in the state of rest, it will remain dormant no matter how favourable to growth the environmental factors may be. The state of rest remains until some environmental cues have removed the growth-arresting physiological conditions inside the bud. This process is referred to as *rest break* and its endpoint as *rest completion* (Table 3.3). Exposure to chilling temperatures is usually regarded as the main environmental cue causing rest break, but night length is often suggested as another environmental factor regulating rest break (Sects. 3.3.2 and 3.3.3). According to the prevailing view, the buds are usually in the state of rest in autumn and, sometimes, early winter (Worrall and Mergen 1967; Farmer 1968; Erez et al. 1979a, b; Kobayashi and Fuchigami 1983a; Hänninen 1990b; Leinonen 1996a; Hannerz et al. 2003; Søgaard et al. 2008). The physiological mechanisms of rest and rest break still remain enigmatic despite some recent advances in research (Horvath et al. 2003; Arora et al. 2003; Rinne and van der Schoot 2003; Junttila 2007; Rohde and Bhalerao 2007; Ruttink et al. 2007; Cooke et al. 2012; Brunner et al. 2014).

The dormancy category opposite to rest is *quiescence*, during which the dormant condition is caused by environmental factors unfavourable to growth (Table 3.3). Boreal and temperate trees are typically in the state of quiescence from late autumn or early winter to late spring. Thus, in these conditions the unfavourable environmental factor is a low air temperature in most cases. When the air temperature rises,

[11] As the dormant condition is a physiological attribute of the bud in this case, plant physiologists often prefer to restrict the use of the concept 'dormancy' to this case only.

Table 3.3 The nomenclature applied in the present volume to bud dormancy in boreal and temperate trees

Concept	Definition		References	Synonyms
	Visible growth	Growth-arresting factor		
Dormancy	No	Not defined	1, 3, 7,8, 9, 10	–
Rest	No	Physiological condition of bud	3, 4, 6, 8, 9, 10	Winter dormancy[1]
				Endodormancy[7]
				Rest in the narrow sense[2]
				Autumn dormancy[5]
				Dormancy I[5]
Quiescence	No	Environment	2, 3, 4, 6, 8, 9, 10	Imposed dormancy[1]
				Ecodormancy[7]
				Winter dormancy[5]
				Dormancy II[5]

Modified from Hänninen (1990b); published with the permission of the Finnish Forest Research Institute, currently the Natural Resources Institute Finland, and the Finnish Society of Forest Science
[1]Doorenbos (1953)
[2]Samish (1954)
[3]Romberger (1963)
[4]Weiser (1970)
[5]Sarvas (1974)
[6]Fuchigami et al. (1982)
[7]Lang et al. (1987)
[8]Hänninen (1990b)
[9]Kramer (1994a, b)
[10]Hänninen and Kramer (2007)
For Sarvas's (1974) terminology, see also the discussion in Sect. 3.6.1

microscopic ontogenetic development, i.e., cell division and growth, is initiated in the buds (Fig. 3.2). This development, invisible to the naked eye, is the one modelled by the various temperature sum models and the variable state of ontogenetic development, $S_o(t)$ (Sect. 3.2.1). Accordingly, the definition "caused by environmental factors unfavourable to growth" refers here to the low air temperatures that have prevailed for a prolonged period, for the rising of air temperatures to growth-promoting levels for a short period of time does not cause an onset of visible growth at once.

The concepts of rest and quiescence are essential for whole-tree ecophysiology and for studies addressing the effects of climate change; therefore they play a crucial role in the present volume as well. The third dormancy category, *correlative inhibition* (Fuchigami et al. 1982; or paradormancy; Lang et al. 1987), has somewhat less importance in whole-tree ecophysiology, though it is an essential concept in the physiology of deciduous trees. The concept of correlative inhibition refers to the case where the bursting of the bud is arrested by the hormonal regulation

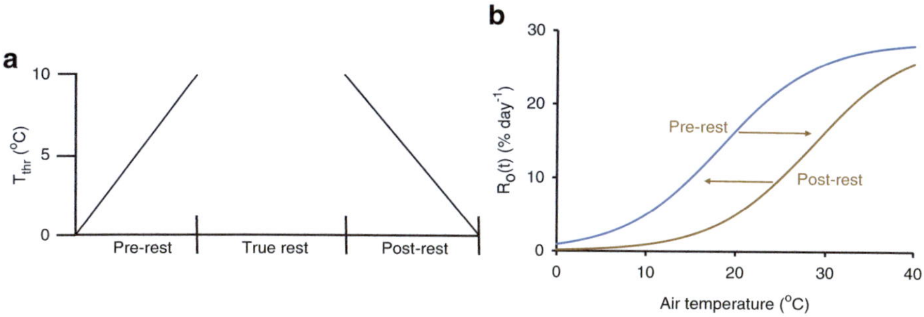

Fig. 3.11 Vegis's (1964) theory of the gradually changing dormancy status of plants, presented for the special case of bud dormancy in boreal and temperate trees. (**a**) A schematic presentation of Vegis's (1964) original theory, addressing the change in the threshold temperature, T_{thr}, for ontogenetic development during pre-rest and post-rest. During true rest, no ontogenetic development towards growth onset takes place (Modified from Hänninen 1990b; published with the permission of the Finnish Forest Research Institute, currently the Natural Resources Institute Finland, and the Finnish Society of Forest Science). (**b**) Extended model, also addressing the change in the rate of ontogenetic development towards growth onset, R_o. The air temperature response of R_o is shown for quiescence (*blue curve*) and for the end of pre-rest and beginning of post-rest (*brown curve*). The change in the response during pre-rest and post-rest is indicated by *arrows*. During true rest, $R_o = 0$ % day^{-1} (not shown in the figure). The phase of pre-rest will not be addressed later in the present volume, so that for brevity, the model will be referred to as the post-rest model

received from other meristems, usually situated in more apical positions than the dormant bud in question. Correlative inhibition typically occurs in the lateral buds relatively late in the growing season (Fuchigami et al. 1982).

3.3.1.2 Vegis's Theory and the Post-Rest Model[12]

In his theory of dormancy phenomena in higher plants, Vegis (1964) also introduced three categories of dormancy. However, his categories are not equivalent to the categories of rest, quiescence, and correlative inhibition discussed above. Rather, his dormancy categories address the gradually changing dormancy status of the plants that is manifested in their air temperature response of growth, so that the growth-promoting range of the air temperature changes as a result of the changing dormancy status.

Among the several special cases discussed by Vegis (1964), the case with changes in the lower limit of the growth-promoting temperature range, i.e., in the threshold temperature, T_{thr}, is relevant for boreal and temperate trees (Fig. 3.11a;

[12] In order to keep the terminology consistent with that used elsewhere in this volume, Vegis's (1964) original concepts predormancy, true dormancy, and postdormancy are replaced with the concepts pre-rest, true rest, and post-rest, respectively.

see also Cooke et al. 2012). In order to introduce this theory into the modelling framework used in the present volume, the theory is extended here by means of the temperature response of the rate of ontogenetic development, R_o (Fig. 3.11b; Hänninen et al. 2013). Vegis's (1964) original theory addresses only the question of whether or not growth occurs at a given temperature (Fig. 3.11a), whereas the extended quantitative model addresses the rate of development towards growth onset (Fig. 3.11b) and, in so doing, also the timing of growth onset. In the following, the extended model is referred to as the post-rest model.

During *pre-rest*, growth is possible, but the growth-promoting air temperature range gets narrower as this phase progresses (Vegis 1964). In the case relevant for boreal and temperate trees, the value of T_{thr} rises during predormancy. At the beginning of this phase, growth – or ontogenetic development in general – occurs at a wide spectrum of air temperatures, but towards the end of pre-rest, growth takes place only in relatively high temperatures (Fig. 3.11). *True rest* is attained when growth is arrested in all environmental conditions (Vegis 1964). With boreal and temperate trees, this implies that there is no more growth, not even in high temperatures.

The phase of true rest is followed by the phase of *post-rest*, which is characterised by a widening of the growth-promoting air temperature range (Vegis 1964). In post-rest, then, the development is the opposite of that in pre-rest. In the case of boreal and temperate trees, this means that the threshold air temperature T_{thr} drops and ontogenetic development towards bud burst is again possible, also in relatively low temperatures (Fig. 3.11a). Thus, in the illustrative example of the post-rest model, there would be very slow ontogenetic development at the temperature of 10 °C at both the end of pre-rest and the beginning of post-rest, i.e., just before and just after true rest (Fig. 3.11b, brown curve), whereas at the end of post-rest, i.e., at the beginning of quiescence, the rate of approximately 5 % day^{-1} would be attained at that temperature (Fig. 3.11b, blue curve).

Despite the different definitions and partially different phenomena addressed, Vegis's (1964) theory of the gradually changing dormancy status in the three dormancy phases may have some relation to the more established categorisation of dormancy into the three types referred to as rest, quiescence, and correlative inhibition in the present volume (Table 3.3). This is obvious in the case of rest and true rest, which, despite the different wording of the two definitions, probably denote the same condition, i.e., growth being arrested no matter how favourable the prevailing conditions might be (true rest; Vegis 1964), and this being due to the internal physiological status of the bud (rest; Fuchigami et al. 1982). As to the remaining four definitions, i.e., pre-rest and post-rest (Vegis 1964) and correlative inhibition and quiescence (Fuchigami et al. 1982), the relationship is less clear.

Vegis's (1964) concept of gradual change in the dormancy status, as manifested in changes in the growth-promoting air temperature range, is physiologically appealing. However, the special case defined for boreal and temperate trees (Fig. 3.11) has received relatively little attention in research, with the exception of the seeds of these trees, where the empirical support of the theory is well documented (Junttila 1970, 1976; Salažs and Ievinsh 2004). Junttila and Hänninen

(2012) tested Vegis's (1964) theory with *Betula pendula* and *B. pubescens* seedlings. They examined the bud burst of the seedlings in various phases of dormancy in the temperature range from 9 to 21 °C. The growth-promoting temperature range was found first to narrow and then to widen, as predicted by Vegis's (1964) original theory (Fig. 3.11a). Junttila and Hänninen (2012) did not examine any potential changes in the rate of ontogenetic development predicted by the extended model introduced in the present volume (Fig. 3.11b).

Vegis's (1964) original theory (Fig. 3.11a) was briefly addressed in some early modelling studies (Hänninen 1987, 1990b; Kramer 1994a), but after that the theory has largely been forgotten in modelling work. The extended model (Fig. 3.11b) has not been explicitly addressed in previous modelling work. In the model of Caffarra et al. (2011a, b), however, a partially similar change in the location of the air temperature response of ontogenetic development (Fig. 3.11b) is included, but Vegis's (1964) theory is not referred to in their study. Caffarra et al.'s (2011a, b) model will be discussed in Sect. 3.4.6.4.

In the present volume, the extended model based on Vegis's (1964) theory (Fig. 3.11b) is discussed as a potential explanation for the paradoxically and drastically different conclusions drawn in studies carried out in controlled conditions on the one hand and in natural conditions on the other (Sect. 3.4.4.2). As discussed in Sect. 8.3.3.4, the phenomenon described by the post-rest model is also critical for studies addressing the effects of climate change on boreal and temperate trees.

Vegis's (1964) theory reveals a problem in the classical dichotomy (Table 3.3) of dormancy being caused either by the physiological status of the bud (rest) or by unfavourable environmental factors (quiescence). This is because according to Vegis's (1964) theory, the concept "unfavourable environmental factor" is not unambiguous. More specifically, in the case relevant for the boreal and temperate trees examined here, a low temperature, i.e., a temperature below a certain threshold, is an "unfavourable environmental factor." If Vegis's (1964) theory is valid, then the threshold temperature changes during the development, implying that the threshold is not unambiguous (Fig. 3.11a). Accordingly, the conclusion on whether the dormant bud is in rest or in quiescence, drawn on the basis of the re-growth test, depends not only on the physiological status of the bud at the beginning of the test but also on the test conditions.

A similar conceptual problem is encountered when the effects of short nights (or long photoperiods) on growth onset are addressed. As discussed in Sect. 3.3.3, short nights often compensate for the lack of chilling in the rest break. However, this is not always the case; accordingly, the strictest definition of rest (or "true physiological dormancy") requires that no growth onset occurs in forcing conditions of high air temperature and continuous light (Myking 1997; Junttila et al. 2003; Junttila and Hänninen 2012).[13]

[13] This definition, equating rest with the existence of an obligatory chilling requirement, is not used in the present volume.

These conceptual problems, caused by interactions with the test conditions, could be avoided if a method for direct measurement of the dormancy condition of the bud were available. But the development of such methods requires considerable advances in our understanding of the physiological mechanisms of rest and rest break (Horvath et al. 2003; Arora et al. 2003; Rinne and van der Schoot 2003; Junttila 2007; Rohde and Bhalerao 2007; Ruttink et al. 2007; Cooke et al. 2012; Brunner et al. 2014); for the present, then, we are restricted to the indirect method based on the re-growth test discussed in Sect. 3.3.2.1. For this reason, the terminology of dormancy requires special explication whenever the phenomenon is addressed. Exact mathematical definitions for the concepts of rest and quiescence are given in Sect. 3.4.1.1.

3.3.2 The Chilling Requirement of Rest Completion

3.3.2.1 Experimental Studies

It has been known since the early work of Coville (1920) that exposure to chilling temperatures, i.e., to temperatures slightly above zero, is a major factor causing rest break, i.e., removal of the growth-arresting physiological conditions in the bud, in natural conditions. Following Coville's (1920) study, numerous studies have addressed the chilling requirement of boreal and temperate trees. The chilling requirement has been an especially central research theme in the case of horticultural fruit trees, where meeting the chilling requirement of the flower buds is essential for fruit production. Studies concerned with the chilling requirement of rest break have been reviewed, among others, by Fuchigami et al. (1982), Saure (1985), Hänninen (1990b), Arora et al. (2003), and Caffarra (2007).

The chilling requirement of rest completion is demonstrated experimentally in a re-growth test by transferring seedlings (or detached twigs) into controlled high temperature conditions, often referred to as forcing conditions, after varying durations of chilling in autumn (Worrall and Mergen 1967; Hänninen 1990b; Myking and Heide 1995). In such a chilling-forcing experiment, the rest status of the buds may be manifested in two potential ways after the transfer into forcing conditions. First, with short durations of chilling, all or part of the buds remain dormant in the forcing conditions. With progressively longer durations of chilling, the bud burst percentage, BB%, observed in subsequent forcing conditions increases (Fig. 3.12a). Second, the duration of the forcing conditions required for bud burst is exceptionally long in non-chilled or partially chilled seedlings. With progressively longer durations of chilling, the mean days to bud burst (mean DBB), i.e., the time required for bud burst in the forcing conditions, decreases exponentially (Fig. 3.12b).

Both the BB% and the DBB measure the rest status of the buds at the time of transfer into the forcing conditions, even though the values of these two indices can be determined only after a prolonged time in the forcing conditions. This indirect method is needed because no direct physiological method for detecting the rest

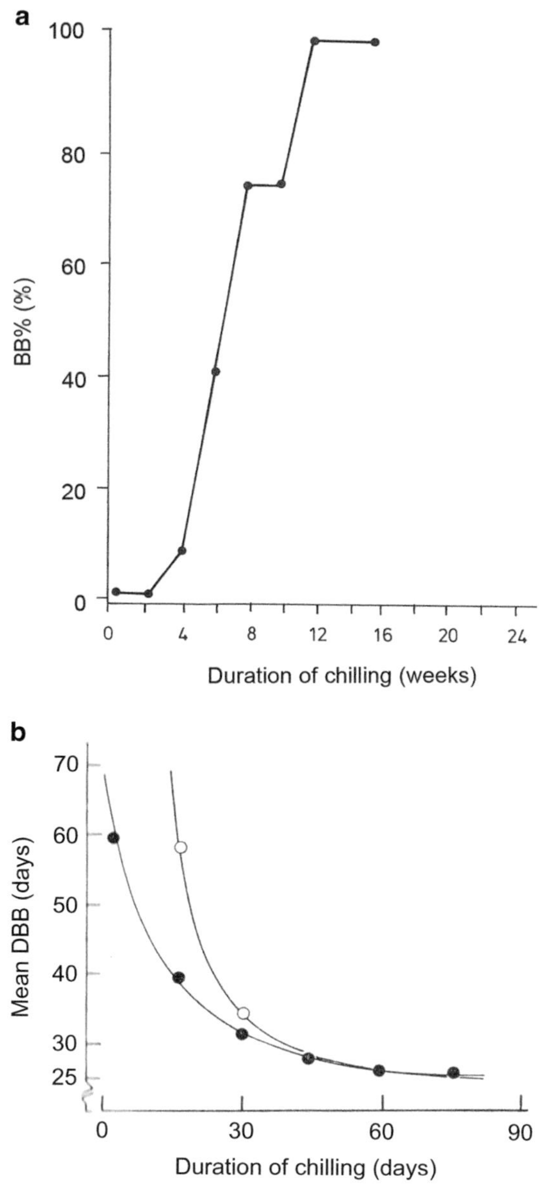

Fig. 3.12 Manifestation of the rest status and rest break of the buds of boreal and temperate trees in a re-growth test with a chilling-forcing experiment. (**a**) Second-year seedlings of *Picea abies* in central Finland. Dependence of the bud burst percentage, BB%, observed in forcing conditions on the duration of previous chilling in a growth chamber (Hänninen 1990b; redrawn with the permission of the Finnish Forest Research Institute, currently the Natural Resources Institute Finland, and the Finnish Society of Forest Science). (**b**) Rooted cuttings of *Picea abies*, a Connecticut clone. The dependence of the mean days to bud burst, mean DBB, observed in forcing conditions with 16-h night length (*white circles*) and 8-h night length (*black circles*), on the duration of previous chilling in a growth chamber (Worrall and Mergen 1967; redrawn with the permission of John Wiley & Sons, Inc.). Depending on the particular study, either the attainment of a given value of BB% or the levelling off of the DBB curve, or both, is required for rest completion to take place. The meeting of both criteria is consistent with the redefinition of rest presented in Sect. 3.4.1.1

status at any given moment is available. The chilling requirement is usually determined by defining a certain bud burst percentage to be attained (Fig. 3.12a; Hänninen 1990b; Erez and Lavee 1971; Junttila and Hänninen 2012) or from the levelling-off of the DBB curve (Fig. 3.12b; Worrall and Mergen 1967; Murray et al. 1989; Myking and Heide 1995). However, due to the gradual decrease of the DBB curve it is hard to determine the exact timing of rest completion (= onset of

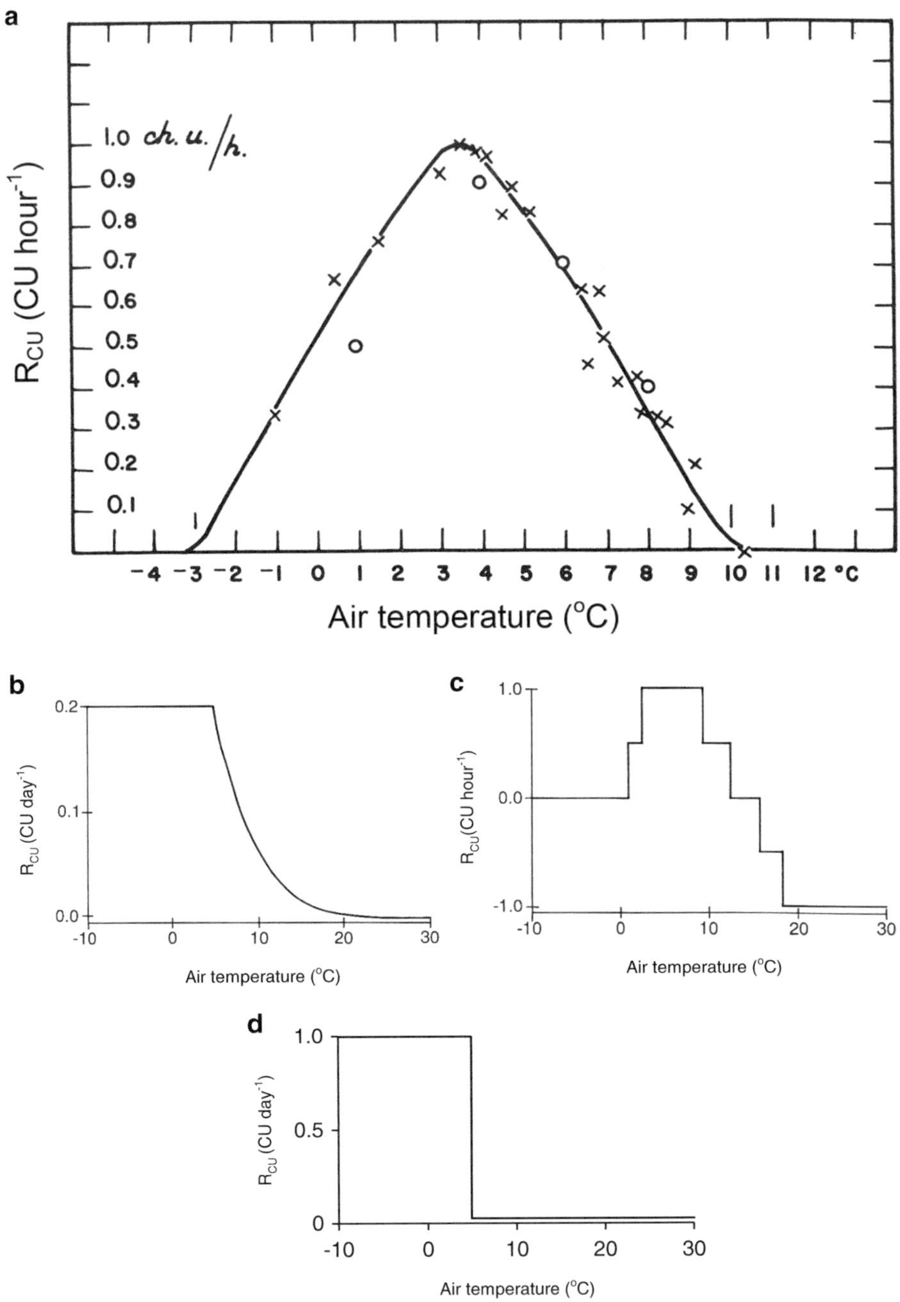

Fig. 3.13 Air temperature responses of the rate of rest break in boreal and temperate trees. (**a**) The air temperature response of the accumulation rate of chilling units, R_{CU}, as determined experimentally by Sarvas (1974) for buds (*circles*) and seeds (x's) of *Betula pubescens*. The air temperature responses of the accumulation rate of chilling units, R_{CU}, in the models of (**b**) Landsberg (1974), (**c**) Richardson et al. (1974) (note the negative values of R_{CU} in high air

quiescence) corresponding to the meeting of the chilling requirement. This is because the chilling temperatures applied in the test often promote slow ontogenetic development, too, and for this reason the DBB curve continues to decrease slowly even after rest completion (Hänninen 1990b).

There is a crucial difference between the curve of the BB% (Fig. 3.12a) and that of the DBB (Fig. 3.12b), though both are indices of rest status. The BB% measures variation among the examined seedlings, twigs, or even individual buds, i.e., in any given test, some buds burst and others do not. Part of the variation among the seedlings or the twigs collected from different trees may be genetic within-population variation. In that case the curve describing the increase of the BB% with progressively longer chilling (Fig. 3.12a) may be taken as an estimate of the cumulative frequency distribution of the chilling requirement in the population (Hänninen 1990b).

In contrast, the declining mean value of the DBB (Fig. 3.12b) measures the increasing rate of ontogenetic development towards bud burst obtained with progressively longer durations of chilling. This rate, too, often evinces within-population variation, but such variation is manifested as deviation from the mean DBB observed at any given duration of chilling rather than in the form of the DBB curve (Myking 1998). This variation is not shown in Fig. 3.12b, where only the mean DBB values are presented.

Rest break is assumed to be a cumulative process driven by exposure to chilling temperatures. Accordingly, rest break is simulated in the whole-tree modelling approach by the accumulation of arbitrary chilling units (CU; Sarvas 1974; Landsberg 1974; Richardson et al. 1974), in a manner basically analogous to the way ontogenetic development is simulated by the accumulation of high temperature units (Sect. 3.2.1).

Sarvas (1974) was the first to present an air temperature response for the accumulation rate of chilling units, R_{CU}, on the basis of direct experimentation (Fig. 3.13a). The seeds of many boreal and temperate trees have a chilling requirement of rest break that is basically analogous to the corresponding requirement of seedlings (Nygren 1987; Dennis 1994). For methodological reasons, Sarvas (1974) therefore used seeds, too, in his experiments though his main objective was to study the rest break of seedlings. The method used for determining the air temperature response of R_{CU}, though more complicated, is basically analogous to the method used for determining the air temperature response of the accumulation rate of period units for ontogenetic development (Table 3.2). The accumulation rate of chilling units, R_{CU}, is calculated as the reciprocal of the developmental time Δt required for rest completion at each temperature, multiplied by the corresponding time required at the reference temperature of 3.5 °C (Sarvas 1974).

Fig. 3.13 (continued) temperatures), and (**d**) Cannell and Smith (1983). Figure (**a**) redrawn from Sarvas (1974), with the permission of the Finnish Forest Research Institute, currently the Natural Resources Institute Finland, and the Finnish Society of Forest Science. Figures (**b**) and (**c**) redrawn from Hänninen (1995a), with the permission of NRC Research Press

The attainment of rest completion in the chilling conditions is manifested by bud burst in the subsequent forcing conditions. Sarvas's (1974) criterion for rest completion was that a half of the buds burst (or seeds germinate) within 21 days in the forcing conditions of 15–19 °C (seedlings) or 17 °C (seeds). He noticed that the duration of chilling required for that response in subsequent forcing conditions, i.e., Δt, decreased with raising the chilling temperature from −1 to +3.5 °C, so that the value of R_{CU} increased in that air temperature range (Fig. 3.13a). With the air temperatures raised further from +3.5 to +10.6 °C, the value of Δt increased and the value of R_{CU} correspondingly decreased, so that in all, a dome-shaped air temperature response for R_{CU} was obtained (Fig. 3.13a).

As in the case of the air temperature response of the rate of ontogenetic development (see Sect. 3.2.1), very few experimentally determined results have been published for the air temperature response of the rate of rest break besides Sarvas's (1974) work. Gilreath and Buchanan (1981) and Erez and Couvillon (1987) determined the air temperature response for *Prunus persica*. Similarly to Sarvas's (1974) results, both of them also found a dome-shaped response, but according to their results, the maximum rate of rest break was attained at +8 °C, i.e., at a temperature about 4° higher than was found by Sarvas (1974) for *Betula pubescens* (Fig. 3.13a). A similar difference of four degrees is observed between the results found for *Prunus persica* (Gilreath and Buchanan 1981; Erez and Couvillon 1987) and those for *Betula pubescens* (Fig. 3.13a) in the upper threshold of the rest-breaking air temperature range.[14]

Campbell and Sugano (1975) examined the rest breaking effects of three temperatures in seedlings of *Pseudotsuga menziesii*. They found that the temperature of 4.4 °C caused rest break more effectively than the two higher temperatures of 7.2 °C and 10.0 °C. This is in accordance with Sarvas's (1974) findings with *Betula pubescens* (Fig. 3.13a). However, contrary to Sarvas's (1974) results, Campbell and Sugano (1975) did not find any major difference between the rest breaking effects of 7.2 °C and 10.0 °C.

Caffarra (2007) found that when seedlings of *Betula pubescens* were chilled for 30–95 days, the mean DBB in the subsequent forcing conditions increased with increasing the chilling temperature from 2 to 10 °C. This finding is in broad agreement with the response that Sarvas (1974) obtained, which suggests that the DBB should increase with increasing the chilling temperature from 3.5 to 10 °C (Fig. 3.13a). After 122 days of chilling, there was no difference in mean DBB among the chilling temperatures (Caffarra 2007). This finding is also in broad agreement with the response Sarvas (1974) obtained, as with such a long period of chilling, rest may be completed even at the high chilling temperature of 10 °C. With *Tilia cordata*, however, Caffarra (2007) found that the mean DBB observed during subsequent forcing did not vary at all among chilling temperatures ranging from 2 to 10 °C. This finding suggests that rather than the dome-shaped response

[14] For copyright reasons, the responses published by Gilreath and Buchanan (1981) and Erez and Couvillon (1987) are not redrawn here.

observed earlier for *Betula pendula* (Fig. 3.13a) and *Prunus persica* (Gilreath and Buchanan 1981; Erez and Couvillon 1987), the response of *Tilia cordata* is a plateau in that temperature range.

Recently, Jones et al. (2015) carried out an experimental study on the effects of different chilling temperatures on rest break in *Ribes nigrum*. They found that temperatures below 0 °C also cause rest break in that species. Furthermore, they found differences among various *Ribes* cultivars in the air temperature response of the rate of rest break. Once again, this finding shows that in order to develop models of high realism for the annual cycle of woody plants, one needs to carry out experimental studies of their features not only species-specifically but also provenance- and cultivar-specifically (Hänninen and Tanino 2011).

It has been observed in experimental studies that intermittent exposure to high temperatures during the chilling period may counteract the previous chilling, so that a higher total accumulation of chilling is required for rest completion than would be with continuous chilling (Overcash and Campbell 1955; Erez and Lavee 1971; Erez et al. 1979a, b). However, it has also been found that intermittent high temperature exposure may enhance the effects of chilling (Erez and Lavee 1971; Erez et al. 1979a) and that the effect of intermittent high temperature exposure may depend on its timing (Hänninen and Pelkonen 1989; Young 1992; Granhus et al. 2009) A review of studies addressing these complicated dynamics is outside the scope of the present volume, but the chilling negation observed in some of these experimental studies shows that unlike ontogenetic development, rest break is at least partially reversible. This has important implications for the modelling of rest break, as is shown in the next section.

3.3.2.2 Introducing the Experimental Results into the Modelling

Unlike in Sarvas's (1972, 1974) model, the air temperature responses of the accumulation rates of chilling units applied in other dynamic models are usually based on various indirect evidence and educated guesses. Cannell and Smith (1983), for example, took a simple approach by calculating the state of chilling, S_{CU}, simply as the number of days when the daily mean air temperature is below the threshold of +5 °C. In their model, then, the value of R_{CU} is constant at any air temperature below the threshold of +5 °C and zero at temperatures above it (Fig. 3.13d). In this case, freezing temperatures are assumed to cause rest break similarly to the chilling temperatures below the threshold.[15] That is also the case in Landsberg's (1974) model, where, however, temperatures above the threshold also cause rest break but at a rate decreasing exponentially with rising air temperature (Fig. 3.13b). In Richardson et al.'s (1974) model, the chilling negation occurring at high air temperatures is addressed with negative values of R_{CU} (Fig. 3.13c; see previous paragraph).

[15] Cannell and Smith's (1983) model is discussed further in Sect. 3.4.6.5.

Similarly to the day degree (Fig. 3.6) or period units (Fig. 3.7) used for modelling the ontogenetic development towards bud burst, the chilling units used for modelling the rest break (Fig. 3.13) are arbitrary, and for this reason not mutually comparable.

Thus, in order to attain the ecophysiological percentage scale, the accumulation rate of chilling units has to be divided by the genotype-specific parameter *chilling requirement of rest completion*, C_{crit}, and multiplied by 100 (Hänninen 1995a; Hänninen and Kramer 2007; compare with Eq. 3.9 in Sect. 3.2.1.4):

$$R_r(T(t)) = 100 \cdot \frac{R_{CU}(T(t))}{C_{crit}} \tag{3.11}$$

where T(t) = air temperature at time instant t, $R_r(T(t))$ = air temperature response of the rate of rest break at time instant t, and $R_{CU}(T(t))$ = air temperature response of the accumulation rate of chilling units at time instant t. The chilling requirement of rest completion, C_{crit}, is expressed in the chilling units (CU) specific to the equation used for calculating the value of $R_{CU}(T(t))$.

In the present volume, the air temperature response obtained by Sarvas (1974) (Fig. 3.13a), as reformulated by Hänninen and Kramer (2007), is used for the rate of rest break, $R_r(t)$ (Fig. 3.14a):

$$R_r(t) = \begin{cases} 0 & T(t) < T_1 \\ \left[\dfrac{100}{C_{crit}}\right] \cdot \left[\dfrac{T(t) - T_1}{T_2 - T_1}\right] & T_1 \le T(t) \le T_2 \\ \left[\dfrac{100}{C_{crit}}\right] \cdot \left[\dfrac{T(t) - T_3}{T_2 - T_3}\right] & T_2 < T(t) \le T_3 \\ 0 & T(t) > T_3 \end{cases} \tag{3.12}$$

where $T_1 = -3.4$ °C, $T_2 = +3.5$ °C, $T_3 = +10.4$ °C, and C_{crit} = chilling requirement of rest completion in chilling units (CU) specific to their accumulation rate determined by the right-hand quotients in Eq. 3.12 corresponding to the curve in Fig. 3.13a.

Similarly to the way the transformation involved in Eq. 3.9 facilitates the comparison of ontogenetic development in different genotypes, the transformation in Eq. 3.11 facilitates the comparison of rest break and rest completion in different genotypes (Fig. 3.14). In the case of the genotype $C_{crit} = 20$ CU, the predicted rest completion takes place on 13 October, autumn 1970, at Jyväskylä, whereas with the genotype $C_{crit} = 40$ CU, the rest completion is postponed until 18 December (Fig. 3.14b).

The illustrative example of Fig. 3.14b also reveals that with the dome-shaped air temperature response of the accumulation rate of chilling units, R_{CU}, (Figs. 3.13a and 3.14a), the accumulation of chilling is restricted not only by non-chilling high temperatures ($T_{mean} > +10.4$ °C) in early autumn but also by non-chilling low temperatures ($T_{mean} < -3.4$ °C) in late autumn and winter. Thus, for instance, the assumed value of $C_{crit} = 40$ CU in this simulation was near the total accumulation

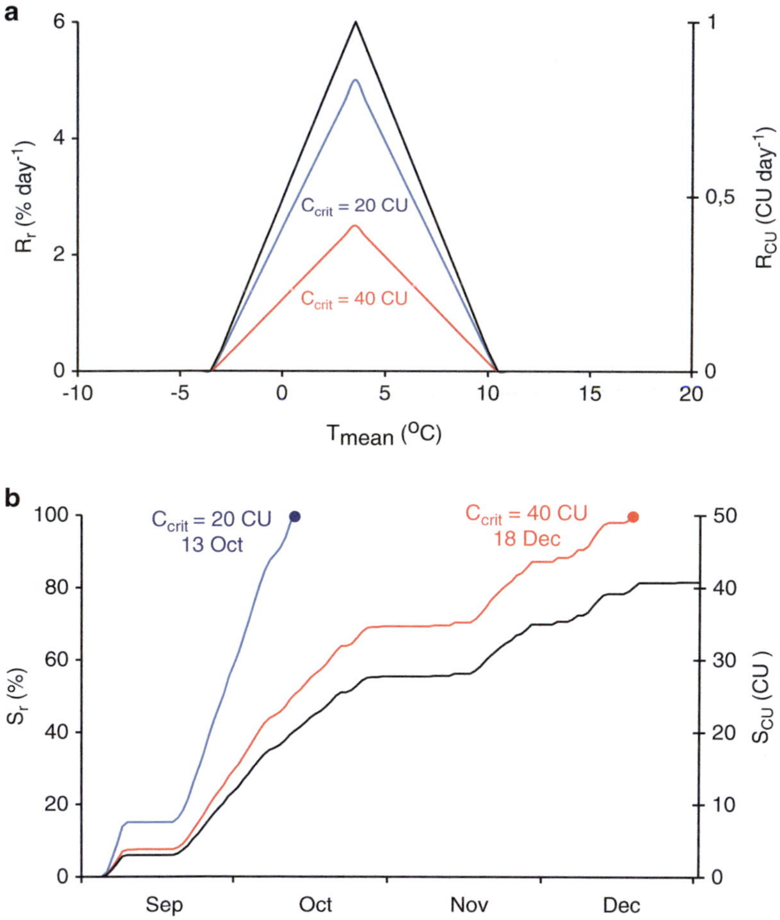

Fig. 3.14 An ecophysiological interpretation of modelling the rest break of buds in boreal and temperate trees by accumulating chilling units. In both (**a**) and (**b**) the *black line* and the *vertical axis on the right* denote the meteorological accumulation of chilling units, while the *coloured lines* and the *vertical axis on the left* denote the simulated rest break of the buds of two tree genotypes characterised by their corresponding values for the chilling requirement of rest completion, C_{crit}. (**a**) The air temperature response of the accumulation rate of arbitrary chilling units (CU), R_{CU}, and the corresponding air temperature responses of the rate of rest break, R_r, for the two tree genotypes. T_{mean} daily mean temperature. (**b**) The accumulation of chilling units, S_{CU}, and the predicted state of rest break, S_r, for the two tree genotypes. Starting on 1 September 1970, the calculations were carried out with daily mean air temperature data from Jyväskylä, central Finland as input. The predicted date of rest completion for each genotype is indicated at the end of the respective lines of S_r

of chilling units in September–December 1970 at Jyväskylä. With a higher value of C_{crit}, the predicted rest completion would have been postponed until mid-winter or early spring, depending on when temperatures in the assumed rest-breaking range ($-3.4\ ^{\circ}C < T_{mean} < +10.4\ ^{\circ}C$) would have occurred again. It goes without saying that this is not the case with those air temperature responses where frost temperatures are assumed to contribute to the accumulation of chilling equally to temperatures slightly above zero (Fig. 3.13b, d). With these responses, the time window for the accumulation of chilling is restricted only by the occurrence of high non-chilling air temperatures in early autumn and, in mild and maritime climates, sometimes also in winter.

Similarly to the case of ontogenetic development, no arbitrary chilling units would actually be needed in the modelling of the rest break, either, if the primary ecophysiological air temperature response (= air temperature response of the reciprocal of Δt) were available for each species and genotype. However, not only because this is not the case but also because of historical and conventional reasons, there is no need to entirely replace the models using arbitrary chilling units with the percentual ecophysiological approach. As in the case of the high temperature units used for simulating ontogenetic development, the restrictions concerning the mutual incomparability of the different chilling units entailed by different air temperature responses (Fig. 3.13) should be recognised (see the corresponding discussion for models of ontogenetic development in Sect. 3.2.1.4). Emphasizing the ecophysiological interpretation of the model variables and the comparison of different tree genotypes, the present volume mainly applies the ecophysiological percentage scale. However, the accumulation of chilling units is also referred to, especially in cases where, rather than the development of trees, the properties of the climate are discussed.

3.3.3 Effects of Night Length on Rest Break

It has been known for a long time that in many boreal and temperate tree species long days, or short nights, compensate for lack of chilling (e.g., Nienstaedt 1967; Farmer 1968; Myking and Heide 1995; Caffarra and Donnelly 2011). This can also be seen in Fig. 3.12b, where short nights decreased the value of the mean DBB after short durations of chilling, but after long durations of chilling, i.e., when the chilling requirement of rest completion was met (as indicated by the levelling off of the DBB-curve), there was no difference in the mean DBB between the two night lengths. Worrall and Mergen (1967) concluded that despite this clear effect in experimental conditions, night length would not affect the timing of bud burst in natural conditions, where the chilling requirement is always met well before the occurrence of short nights potentially compensates for lack of chilling in spring. However, with some species and climates this notion may not be valid, i.e., the accumulation of chilling may not meet the chilling requirement in all years. In those cases, rest break is regulated by an interaction of the accumulation of chilling and the decreasing night length (Campbell 1978).

In evolutionary terms, the compensating effect of short nights may be interpreted as a safety mechanism in conditions of a mild climate (Kramer 1994a). If there is exceptionally little accumulation of chilling during a given dormant period of the tree, then the short nights in late spring cause rest completion, and the tree resumes growth in due course at the beginning of the growing season. In the case of the dome-shaped air temperature response of the rate of rest break (Fig. 3.14a), night length may have the same effect in an extremely continental climate, where the time window for the accumulation of chilling is short because air temperatures in autumn drop rapidly from non-chilling high values to non-chilling low values.

For a long time, it has been suggested that night length and other light-related factors may have a more profound effect on rest break in many boreal and temperate trees species than just compensating for the lack of chilling in experimental conditions. Even so, the evidence for this notion remains unclear and partly even controversial (Worrall and Mergen 1967; Nienstaedt 1966, 1967; Farmer 1968; Campbell 1978; Nizinski and Saugier 1988; Hänninen 1995a; Linkosalo and Lechowicz 2006; Partanen et al. 1998; Körner and Basler 2010a,b; Chuine et al. 2010; Caffarra and Donnelly 2011; Caffarra et al. 2011a, b; Basler and Körner 2012; Blümel and Chmielewski 2012). To identify the species and the environmental conditions in which the effects of night length on rest break need to be addressed in modelling studies, a thorough and critical literature review of the experimental studies is needed, but such a review is outside the scope of the present volume. The modelling of the effects of night length on rest break is discussed in Sects. 3.4.2 and 3.4.6.4.

The role of night length in the rest break of *Fagus sylvatica* has been studied for a long time, but despite the large number of studies conducted, its effect remains controversial even in this tree species (Wareing 1953; Falusi and Calamassi 1990; Heide 1993b; Caffarra and Donnelly 2011). Recently, Vitasse and Basler (2013) presented a thorough review of studies addressing bud burst phenology in *Fagus sylvatica* both in natural and in controlled conditions. They introduced a reconciling conceptual model to explain the earlier, seemingly contradictory results. The exceptionally high chilling requirement of *Fagus sylvatica* is an essential cornerstone in Vitasse and Basler's (2013) conceptual model. Accordingly, night length often affects rest break in *Fagus sylvatica* because the high chilling requirement of this species is not always met in natural conditions.

The conceptual model was able to explain the spatial and the temporal variation in European populations of *Fagus sylvatica*. In this way, Vitasse and Basler's (2013) study provided new insights to the environmental regulation of bud burst timing in *Fagus sylvatica*, thus also facilitating its dynamic modelling in forthcoming studies of this species. In their conceptual model Vitasse and Basler (2013) suggested a type of night length effect similar to that included in the DORMPHOT model presented for *Betula pubescens* by Caffarra et al. (2011a, b). This model is briefly discussed in Sect. 3.4.6.4.

3.4 Overall Environmental Regulation of Spring Phenology

3.4.1 Synthesizing the Effects of Chilling and Those of High Temperatures

3.4.1.1 A Generalised Model for Growth Onset

Models addressing both the effects of the chilling requirement of rest completion and the high temperature requirement of growth onset on the spring phenology of boreal and temperate trees have been available since the 1970s (for reviews, see Cannell 1985, 1989; Hänninen 1987; Kramer 1994a, b; Chuine et al. 1998, 1999, 2013; Chuine 2000; Hänninen and Kramer 2007).[16] On the basis of his literature review, Hänninen (1990b) concluded that these models involve three different ecophysiological aspects. In addition to the chilling requirement of rest completion and the high temperature requirement of growth onset, a third one is needed to link these two.

On the basis of this analysis, Hänninen (1990b) presented a generalised formulation, where an overall model is constructed with the aid of three sub-models, one for each ecophysiological aspect addressed (Fig. 3.15). By specifying each sub-model, any previous model can be formulated as a special case of the generalised formulation. Furthermore, the generalised formulation facilitates the consideration of new models, obtained either by novel combinations of previous sub-models or combinations of novel sub-models. Using this technique, Hänninen (1995a) developed a total of 96 models of growth onset and tested them against data gathered with *Pinus sylvestris* saplings in a whole-tree chamber experiment.[17] Hänninen's (1990b) generalised formulation was later adopted and applied to sets of long-term phenological data by Kramer (1994a, b).

Landsberg (1977) introduced the concept "competence" to refer to the ability of a plant to react to environmental factors. This general concept was adopted by Hänninen (1990b) to mediate the effects of rest to the ontogenetic development towards growth onset. He then introduced the concept of *ontogenetic competence*, which facilitates the exact definition of the concepts of rest and quiescence (Table 3.4; Fig. 3.16).[18] These new definitions do not facilitate the direct experimental determination of the rest status of the buds, but as shown in Sect. 3.4.1.2, they facilitate the deducing of the predictions of different dormancy models for

---

[16] The models address various springtime developmental events, such as bud burst, leaf unfolding, the growth onset of vegetative buds, and the flowering of generative buds, but for the sake of brevity, only growth onset will be referred to in the description of the generalised model.

[17] The experimental results of Hänninen (1995a) are discussed in Sects. 3.4.6.1 and 8.3.2.2.

[18] The concept originally used by Hänninen (1990b) was "growth competence", which was then replaced with the more informative concept "ontogenetic competence" in Hänninen and Kramer's (2007) review.

Fig. 3.15 A generalised model of the timing of growth onset (including bud burst and leaf unfolding) and flowering in boreal and temperate trees (Hänninen 1990b, 1995a; Kramer 1994a, b; Hänninen and Kramer 2007). The generalised model is modular, consisting of three sub models. Sub-model I describes the effects of chilling on rest break, Sub-model II concerns the effects of high temperatures on the potential rate of ontogenetic development, and Sub-model III mediates the effects of the rest status of the buds on the (actual) ontogenetic development towards growth onset or flowering. Accordingly, the *arrows* on the *left-hand* and the *right-hand* side describe the chilling requirement of rest completion (Sect. 3.3.2) and the high temperature requirement of growth onset (Sect. 3.2.1), respectively. The text in *red* identifies the constant structural components of the generalised model, and the text in *blue* indicates the optional functions that were selected for use as the Sub-models in the simulations of the present volume. The generalised model can also be used for describing the effects of other environmental factors than chilling on rest break and ontogenetic development, though that is not shown in the figure

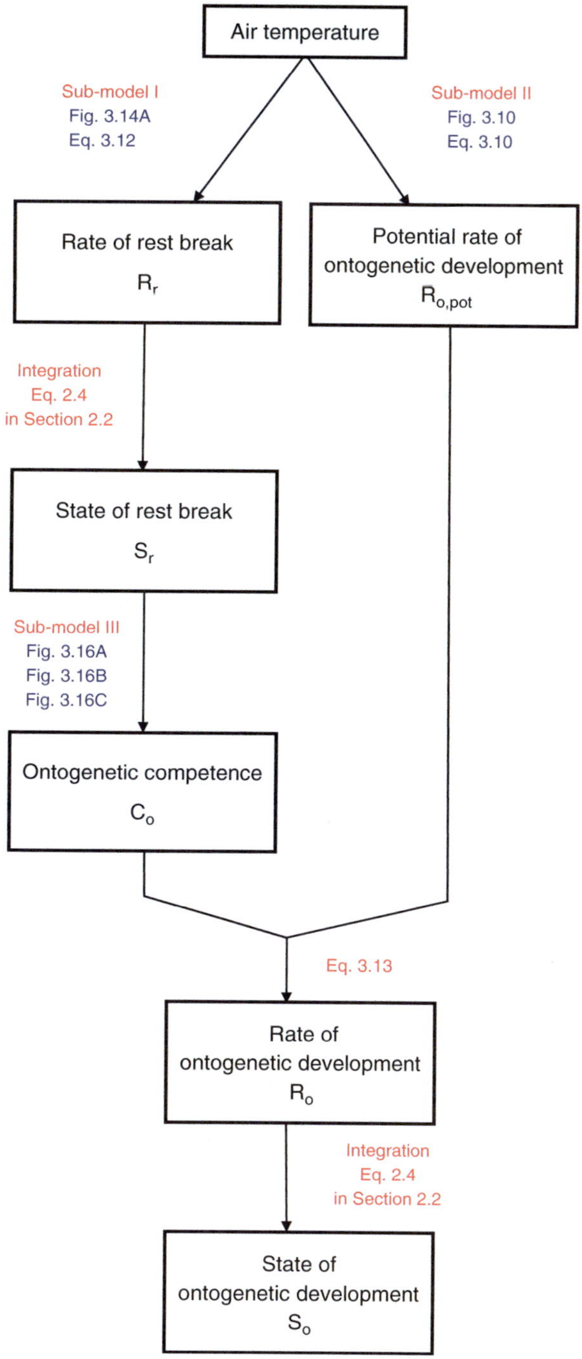

Table 3.4 A redefinition of the dormancy concepts for the modelling framework

Concept	Classical definition in empirical studies		Redefinition in modelling studies
	Visible growth	Growth-arresting factor	Ontogenetic competence
Dormancy	No	Not defined	Not defined
Rest	No	Physiological condition of bud	None or reduced
Quiescence	No	Environment	Full

(Hänninen 1990b, 1995a; Kramer 1994a; Hänninen and Kramer 2007)
Modified from Hänninen (1990b); published with the permission of the Finnish Forest Research Institute, currently the Natural Resources Institute Finland, and the Finnish Society of Forest Science

experimental testing. In this way the new definitions, given within the hypothetico-deductive modelling approach adopted in the present volume, help to overcome the problems associated with the classical definitions of rest and quiescence, discussed in Sect. 3.3.1.

When ontogenetic development is fully arrested by the rest condition, the bud has no ontogenetic competence, i.e., competence for ontogenetic development (Hänninen and Kramer 2007). However, the arresting of ontogenetic development may also be partial or quantitative. Therefore, following the idea put forward by Hari et al. (1970) and Hari (1972) to model the rate of height growth in boreal trees, the variable of ontogenetic competence, C_o, was introduced as a [0, 1] multiplier for calculating the rate of ontogenetic development (Figs. 3.16 and 3.17; Hänninen 1990b; Hänninen and Kramer 2007):

$$R_o(t) = C_o(S_r(t)) \cdot R_{o,pot}(t) \tag{3.13}$$

where $R_o(t)$ = rate of ontogenetic development at time instant t, $R_{o,pot}(t)$ = potential rate of ontogenetic development at time instant t, and $C_o(S_r(t))$ = ontogenetic competence at time instant t, as determined by the state of rest break at time instant t, $S_r(t)$ (Fig. 3.16). The potential rate identifies the unrestricted rate during quiescence, when the growth-arresting physiological conditions have been entirely removed from the bud. In this case $C_o(t) = 1$, so that the prevailing air temperature alone regulates the rate of ontogenetic development towards growth onset. Equation 3.13 also provides an illustrative example of the general principle of indirect

Fig. 3.16 (continued) development during the phases of rest and quiescence, respectively. (**a**) The sequential model, (**b**) the parallel model, and (**c**) an intermediate model (Hänninen 1990b, 1995a). Regardless of the model applied, the buds have full ontogenetic competence by definition during quiescence (Table 3.4). (**a** and **b** modified from Hänninen 1990b; published with the permission of the Finnish Forest Research Institute, currently the Natural Resources Institute Finland, and the Finnish Society of Forest Science; **c** modified from Hänninen 1995a; published with the permission of NRC Research Press)

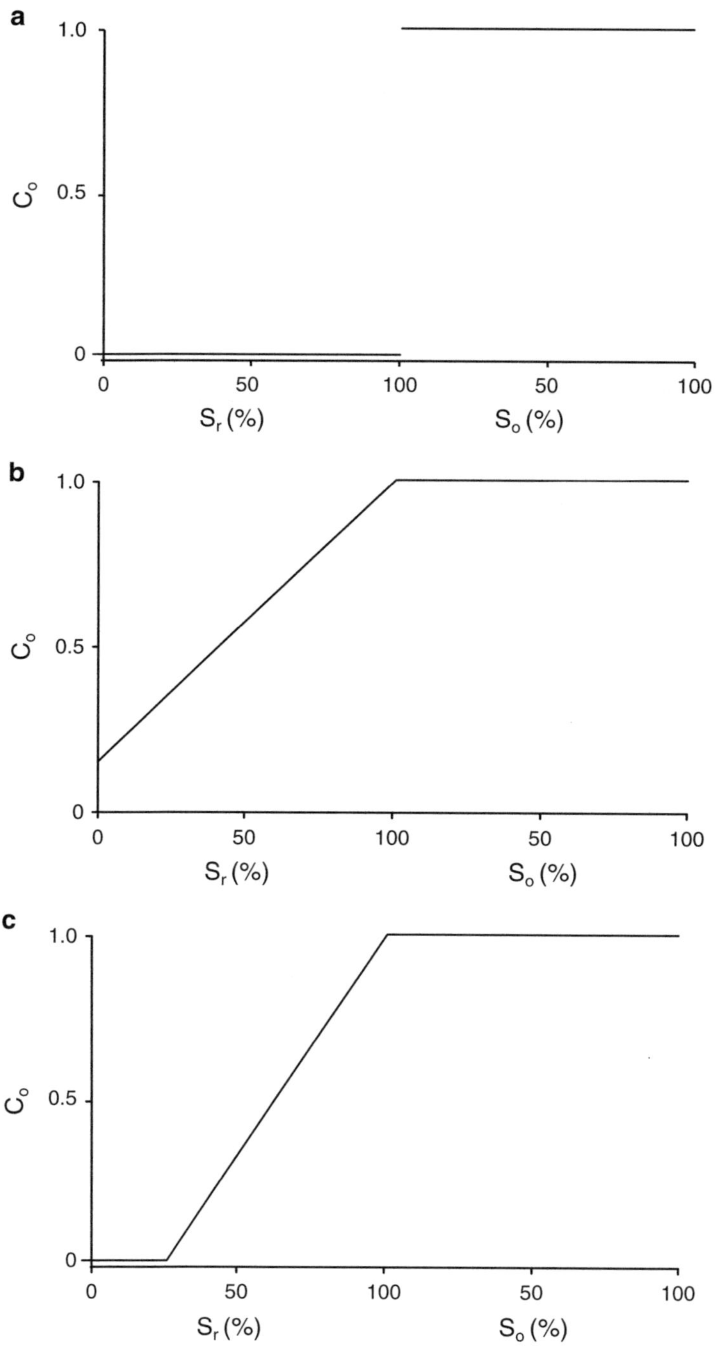

Fig. 3.16 The dependence of the ontogenetic competence, C_o, of buds in boreal and temperate trees on the state of development during dormancy. The state of rest break, S_r, and the state of ontogenetic development, S_o, both represented by the horizontal axes, quantify the state of

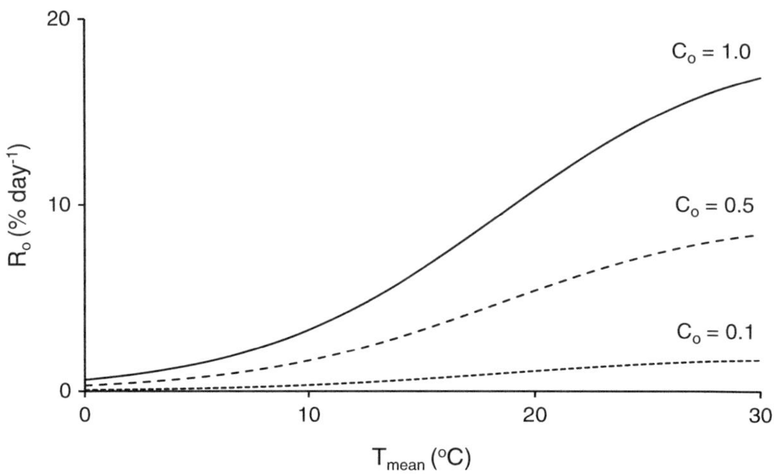

Fig. 3.17 The effect of ontogenetic competence, C_o, on the air temperature response of the rate of ontogenetic development, R_o (Eq. 3.13); T_{mean} = daily mean air temperature. With zero ontogenetic competence ($C_o = 0$) the rate of ontogenetic development is zero ($R_o = 0$ % day^{-1}) irrespective of T_{mean} (not shown in the figure). With full ontogenetic competence ($C_o = 1$), R_o attains its potential value indicated by $R_{o,pot}$ (Eq. 3.14). This occurs during quiescence irrespective of the model assumed for ontogenetic competence (Fig. 3.16). The parallel model involves a gradual increase of C_o during rest (Fig. 3.16b), thus implying the gradual increase in the level of R_o indicated in the figure. The sequential model involves an abrupt increase of C_o from zero to unity at the beginning of quiescence, thus implying an abrupt change from $R_o = 0$, irrespective of air temperature, to the air temperature response obtained with full ontogenetic competence ($C_o = 1$). See also Campbell (1978)

regulation introduced in Eq. 2.3 in Sect. 2.2, i.e., that the state of development of one attribute of the annual cycle (rest break) affects the rate of development of another attribute (ontogenetic development towards growth onset) (Figs. 3.16 and 3.17).

For the air temperature responses of the rate of rest break and the potential rate of ontogenetic development, Hänninen (1990b) selected the ones presented by Sarvas (1972, 1974). That was because these two responses are among the few that have been determined on direct experimental basis and because they concern boreal tree species, which is also what Hänninen (1990b) studied. In the present volume the responses are used as reformulated by Hänninen and Kramer (2007). Accordingly, Eq. 3.12 (Fig. 3.14a) is used for simulating the rate of rest break, $R_r(t)$. For the potential rate of ontogenetic development, $R_{o,pot}(t)$, the sigmoidal equation based on Sarvas's (1972) PU curve (Fig. 3.7 in Sect. 3.2.1.3) is used:

$$R_{o,pot}(t) = \begin{cases} 0 & T(t) < T_{thr} \\ \left[\dfrac{100}{H_{crit}}\right] \cdot \left[\dfrac{1}{1 + \exp(-a(T(t) - b))}\right] & T(t) \geq T_{thr} \end{cases} \qquad (3.14)$$

For the explanation of the symbols, see the corresponding Eq. 3.10 in Sect. 3.2.1.4.

Following Campbell's (1978) presentation, the ecophysiological interpretation of Eq. 3.13 is best illustrated by a graphical examination of the curves for the air temperature response of the rate of ontogenetic development, R_o, implied by the different values of ontogenetic competence, C_o (Fig. 3.17). When the value of C_o is zero, then no ontogenetic development takes place, i.e., regardless of the prevailing air temperature, T(t), the rate of ontogenetic development, $R_o(t) = 0$ % day^{-1} (not shown in Fig. 3.17). This is the state of rest, or 'true rest', in the strictest sense of the word. Such strict rest is assumed in Sarvas's (1972, 1974) and Richardson et al.'s (1974) models, as in both of them the simulated ontogenetic development is initiated only after the simulated rest completion has taken place. In his review, Hänninen (1987) referred to this type of model by the term *sequential model*, because in these models the arbitrary developmental units are accumulated sequentially, i.e., the chilling units are accumulated first until the chilling requirement of rest completion, C_{crit}, is met, and the accumulation of high temperature units is initiated only after that. This condition is simulated with ontogenetic competence, so that $C_o = 0$ during rest, i.e., when $0 < S_r(t) < 100$, and $C_o = 1$ during quiescence, i.e., when $S_r(t)$ has attained the value 100 (Fig. 3.16a).

With intermediate values of ontogenetic competence, i.e., with $0 < C_o < 1$, ontogenetic development takes place, but the rate is reduced in comparison with the maximal rate during quiescence, when $C_o = 1$ (Fig. 3.17). This case can be used to model the gradual rest break assumed in Landsberg's (1974) and Cannell and Smith's (1983) models. In these models, ontogenetic competence increases gradually, rather than abruptly, with increasing values of the state of rest break, S_r (Fig. 3.16b). Thus, the high temperature units are already accumulated before rest completion, i.e., simultaneously (or parallelly) with the accumulation of the chilling units. Accordingly, this type of model was referred to as the *parallel model* in Hänninen's (1987) review.

All other things being equal, the parallel model predicts an earlier growth onset in natural conditions than the sequential model does (Hänninen 1990b). This is because in the parallel model the predicted ontogenetic development is initiated during the rest period already, since the buds already have a non-zero ontogenetic competence then (Fig. 3.16b). In 1904, the beginning of September was warm in Jyväskylä, central Finland, so that the parallel model predicted ontogenetic development to occur even before the simulated process of rest break started in mid-September (Fig. 3.18). The sequential model, in contrast, predicted ontogenetic development to start only on 1 November, after rest was completed (Fig. 3.18). Thus, the sequential and the parallel model predicted that roughly 10 % and 50 %, respectively, of the ontogenetic development towards growth onset would take place in autumn and winter. In April, then, there was a large difference between the states of ontogenetic development, S_o, predicted by the two models. However, due to a rapid rise of the air temperature in May, there was only a 16-day difference between the two models in the predicted time of growth onset (Fig. 3.18).

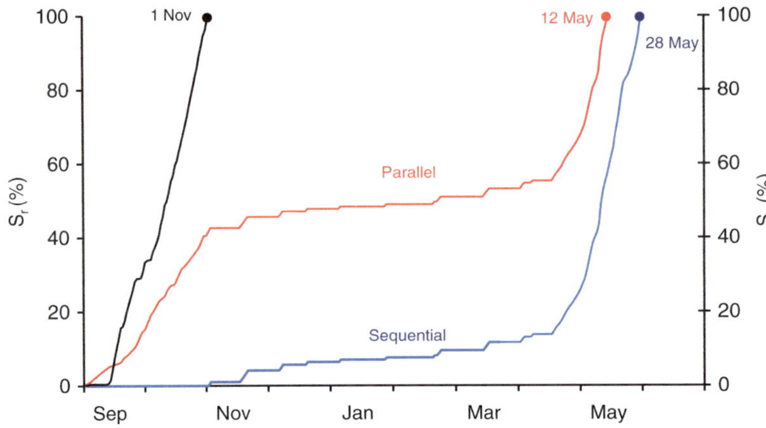

Fig. 3.18 Simulated rest break in buds and simulated ontogenetic development towards growth onset in a generalised central Finnish tree species in Jyväskylä in 1904–1905. S_r denotes the state of rest break and S_o the state of ontogenetic development. Rest break was simulated with Eq. 3.12 (Sect. 3.3.2.2) with $CU_{crit} = 30$ CU (*black curve*, predicted rest completion on 1 November). Ontogenetic development was simulated with Eqs. 3.13 and 3.14 with $H_{crit} = 5.3$ HU, using both the sequential (*blue curve*) and the parallel (*red curve*) sub-model for ontogenetic competence (Fig. 3.16). The models predicted bud burst to occur on 28 and 12 May, respectively

3.4.1.2 Predictions for Model Testing

Using the hypothetico-deductive approach introduced in Chap. 2, Hänninen (1990b) deduced the predictions of the different types of model for the outcomes of chilling-forcing experiments (Fig. 3.19). The parallel model predicts a bud burst percentage of 100 for any duration of chilling (Fig. 3.19c), as the ontogenetic competence, C_o, is above zero even with non-chilled seedlings (or twigs), but due to the increasing C_o with increased duration of chilling (Fig. 3.16b), the model predicts an exponentially declining value of days to bud burst, DBB (Fig. 3.19d).

The sequential model, in contrast, predicts an increasing bud burst percentage with increased duration of chilling, provided that there is within-population variation in the chilling requirement (Fig. 3.19a). However, as a non-zero ontogenetic competence always has the value of $C_o = 1$ in the sequential model (Fig. 3.16a), that model predicts an almost constant value of the DBB for all durations of chilling that are long enough to cause bud burst in at least part of the seedlings or twigs under forcing conditions. Due to the slow ontogenetic development predicted to take place in the chilling conditions already, the value of the DBB actually decreases slightly, but not exponentially, with an increasing duration of chilling (Fig. 3.19b). Thus, as rest completion (= onset of quiescence) is defined in this volume as the time instant when full ontogenetic competence is attained (Table 3.4), it is marked by the levelling off of the DBB curve in Fig. 3.19d.

In his literature review, Hänninen (1990b) found that most experimental studies supported a model intermediate between the sequential and the parallel model. This

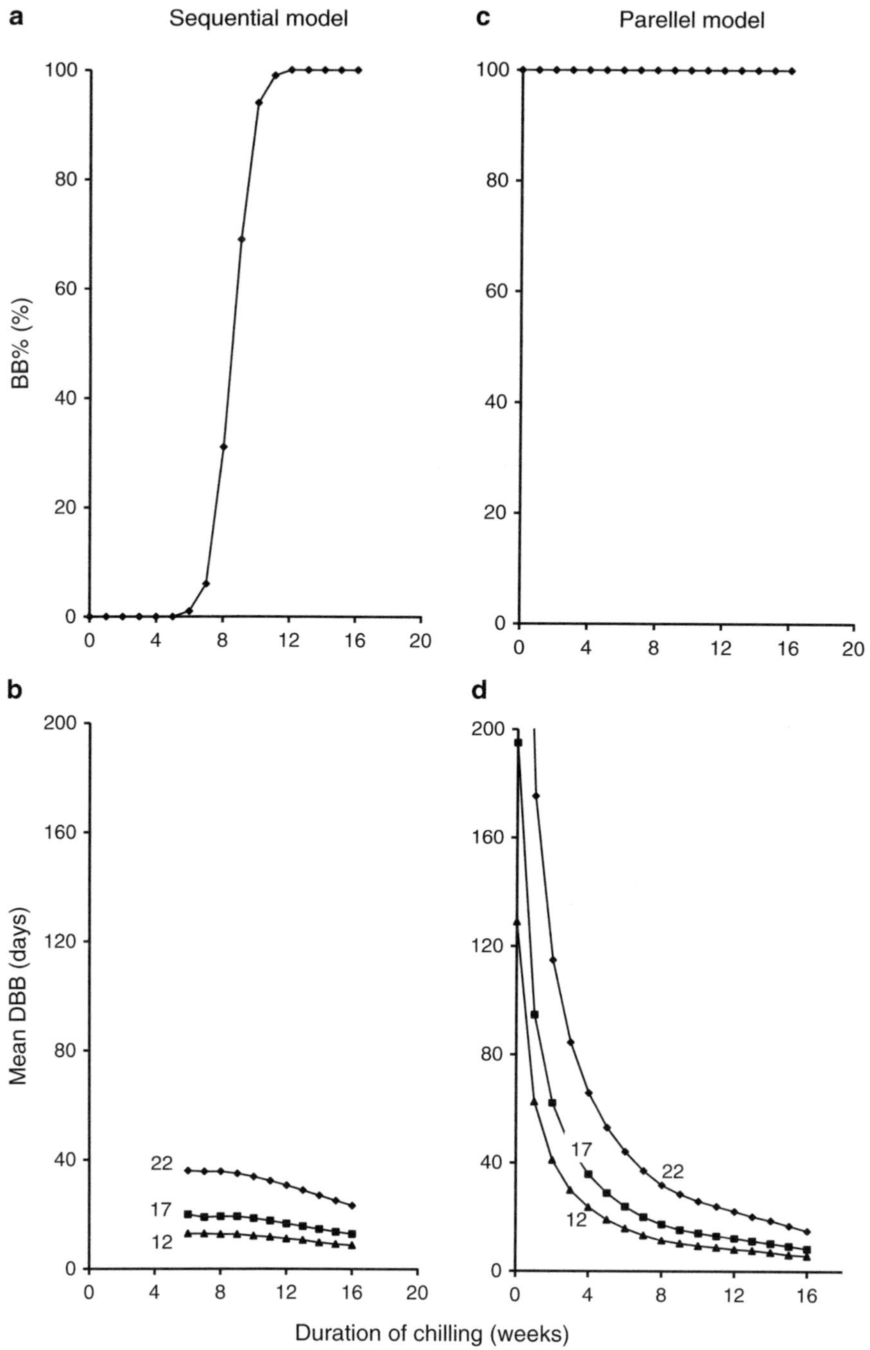

Fig. 3.19 Prediction of (**a**, **b**) the sequential and (**c**, **d**) the parallel model for the outcomes of a re-growth test in a chilling-forcing experiment. BB% bud burst percentage, DBB days to bud burst in the forcing conditions. Predictions for three forcing temperatures are indicated for DBB, but for BB% the predictions are identical in all three forcing temperatures (Hänninen 1990b; redrawn with the permission of the Finnish Forest Research Institute, currently the Natural Resources Institute Finland, and the Finnish Society of Forest Science)

intermediate case can be readily addressed by means of the concept of ontogenetic competence (Fig. 3.16c): with the intermediate model, the value of C_o remains at zero with short durations of chilling, as in the sequential model. After the lower chilling requirement has been attained, the value of C_o starts to increase with increasing chilling, as in the parallel model. After the upper chilling requirement has been attained, the state of quiescence begins, so that the intermediate model, like both the original models (Figs. 3.16a, b), predicts full ontogenetic competence (Fig. 3.16c).

Myking (1997) suggested that interactions with the chilling requirement may partially explain the findings where fluctuating temperatures have been noted to be more efficient than constant temperatures in fulfilling the high temperature requirement, i.e., in causing ontogenetic development towards bud burst (see Sect. 3.2.1.5). Myking (1997) addressed the case where the high temperature requirement is determined in a forcing experiment before rest completion, i.e., before the chilling requirement is met. According to the predictions of the parallel and the intermediate model (Fig. 3.16b, c), such partial chilling will result in higher values of mean DBB than are observed in forcing after the chilling requirement has been met (Fig. 3.19d). However, when the forcing is carried out in fluctuating temperatures, the lowest temperatures involved in the fluctuation may have a chilling effect, thus causing further rest break in the partially chilled seedlings after the transfer to the forcing conditions. Accordingly, the ontogenetic competence of these seedlings increases, so that the mean DBB decreases. This reasoning is valid only in the case of partially chilled seedlings (Myking 1997), so that the hypothesis can be readily tested by forcing both partially and fully chilled seedlings in both constant and fluctuating air temperatures.

3.4.1.3 Comparison of the Parallel and the Post-Rest Model

Though both the parallel model (Landsberg 1974) and the post-rest model based on Vegis's (1964) theory assume a gradual rest break, there is a strict conceptual difference between the two. The parallel model assumes a gradually increasing ontogenetic competence as the phase of rest progresses (Fig. 3.16b), resulting in an increasing rate of ontogenetic development at any temperature within the range that promotes ontogenetic development. Accordingly, the level of the air temperature response increases as a result of rest break (Fig. 3.17), but there is no interaction between the rest phase and the prevailing temperature. In contrast, the post-rest model assumes such an interaction, so that there is a shift in the air temperature response towards lower air temperatures as the phase of post-rest progresses but no change in the shape of the curve (Fig. 3.11b in Sect. 3.3.1.2). Despite this difference, the two models have partly similar implications, and it may sometimes be hard to infer which of the two is manifested in experimental results (Campbell 1978; Cannell 1989).

First, when only one forcing temperature, or a narrow temperature range, is used in the experiment, then the post-rest model may predict results similar to those predicted by the parallel model, i.e., the rate of development in a given forcing temperature, say 15 °C, is predicted to be higher after a long than after a short duration of chilling (Figs. 3.11b and 3.17). Second, according to the prediction of the parallel model, chilling affects the days to bud burst, DBB, in forcing conditions differently, depending on the forcing temperature used (Fig. 3.19d). In absolute terms, chilling lowers the value of the DBB more when the forcing is done at a low (+12 °C) rather than a high (+22 °C) air temperature. Furthermore, for the levelling off of the DBB curve, often used to determine the chilling requirement (Fig. 3.12b), less chilling is needed at a high (22 °C) than a low (+12 °C) forcing temperature (Fig. 3.19d).

These apparent interactions between the duration of chilling and the forcing temperature may be interpreted to be manifestations of Vegis's (1964) post-rest theory (Campbell 1978; Cannell 1989). However, the deductions documented in Fig. 3.19d show that these results are also implied by the parallel model. In that model, chilling increases the rate of ontogenetic development proportionally at any temperature within the range that promotes ontogenetic development (Fig. 3.17), which is contrary to the assumption of the post-rest model based on Vegis's (1964) theory (Fig. 3.11b).

Thus, though formally different, these two models have much in common descriptively. It is also possible that a synthesis of the two models is manifested in real trees, i.e., that both the level of the curve (Fig. 3.17) and its location (Fig. 3.11b) change during rest break. In all cases, the ecophysiological percentage scale, obtained by the reciprocal of the developmental time required for growth onset, facilitates the explication and comparison of the different air temperature responses and their changes during the rest period (Campbell 1978; Cannell 1989).

3.4.2 *Effects of Night Length*

As discussed in Sect. 3.3.3, night length is evidently included in the environmental regulation of rest break in many boreal and temperate tree species. In modelling studies, however, the rest-breaking effects of night length have been addressed relatively rarely. It has been assumed in some modelling studies that rest is completed simply on a fixed calendar day. This assumption may be interpreted so that rest completion is caused by the attainment of a critical night length corresponding to the fixed calendar day (Häkkinen et al. 1998; Linkosalo et al. 2000; Sect. 3.4.4). Kramer (1994a) took a more complicated approach by considering a potential additive effect of chilling and night length on rest break in *Fagus sylvatica*. On the basis of Campbell and Sugano's (1975) experimental results, Campbell (1978) suggested that similarly to chilling temperatures (Fig. 3.17), night length affects the form of the sigmoidal air temperature response of the rate of ontogenetic development in *Pseudotsuga menziesii*. Nizinski and Saugier (1988) adopted Cannell and Smith's (1983) alternating model for modelling the effects of night length on rest break.[19]

Following the approach taken in the present volume, the rest-breaking effects of night length can be modelled with the generalised model (Fig. 3.15) used for modelling the effects of chilling in Sects. 3.4.1.1 and 3.4.1.2. The key concept in the generalised model is the ontogenetic competence, C_o. It can be used for mediating the rest-breaking effects of any environmental factor to the ontogenetic development towards growth onset. If there is an absolute night length requirement of rest completion corresponding to the sequential model (Fig. 3.16a) and it is not met, then $C_o = 0$ and in this way the rate of ontogenetic development, $R_o(t), = 0$ (Eq. 3.13). If there is a quantitative night length requirement corresponding to the parallel model (Fig. 3.16b) and it is partially met, then $0 < C_o < 1$ and the ontogenetic development takes place at a reduced rate (Fig. 3.17). This approach was taken by Hänninen (1995a) in testing various models of rest break, ontogenetic competence, and potential rate of ontogenetic development for *Pinus sylvestris* with a whole-tree chamber experiment.

The approach based on the concept of ontogenetic competence (Hänninen 1995a) is quite close to Campbell's (1978) approach discussed in the first paragraph of this section. This is because in both cases the effects of night length are ultimately considered in the air temperature response of the rate of ontogenetic development. However, without further development, neither of these approaches facilitates modelling the simultaneous, let alone interactive, effects of chilling and night length. A novel approach for this was introduced by Caffarra et al. (2011a, b). Their model is discussed in Sect. 3.4.6.4.

3.4.3 Effects of Other Environmental Factors

In addition to air temperature and night length, the role of the following environmental factors in the regulation of the spring phenology of boreal and temperate trees has been discussed in the literature: ground frost and soil temperature (Huikari and Paarlahti 1967; Lavender et al. 1973; Harding et al. 1976; Emmingham 1977; Sorensen and Campbell 1978; Hammond and Seeley 1978; Repo et al. 2005, 2008); concentrations of nutrients in the soil (Murray et al. 1994); concentration of atmospheric carbon dioxide (Murray et al. 1994; Hänninen 1995a; Repo et al. 1996; Kilpeläinen et al. 2006; Slaney et al. 2007); and the intensity of solar radiation (Höyhtyä and Hänninen 1991; Caffarra and Donnelly 2011).

For each of these factors, the empirical evidence remains insufficient or even contradictory, but on the basis of the limited data available, the potential effects of any of them on the spring phenology of some boreal and temperate tree species

[19] The alternating model is discussed in Sect. 3.4.6.5.

cannot be ruled out. Introducing these environmental factors into the ecophysiological framework adopted in the present volume (Fig. 3.15) is difficult because whenever an effect on spring phenology is documented, it is still hard to tell whether the environmental factor affects the ontogenetic development directly or indirectly, by affecting the process of rest break. Given the universal role of air temperature in the environmental regulation of spring phenology (Sect. 3.2.2), it may be pragmatic to model the effects of the environmental factors discussed in this section by altering the air temperature response according to the empirical results on the effect of the other factor(s). This approach was taken by Murray et al. (1994) in their modelling of the effects of concentration of atmospheric carbon dioxide on the bud burst phenology in *Picea sitchensis* (Sect. 8.3.1.3).

In their analysis of long-term phenological records, Fu et al. (2015) found that the high temperature requirement of growth onset in several tree species had risen between 1980 and 2012 in both Europe and North America. They attributed the rise to increased precipitation and decreased incoming solar radiation. They suggested that as the tissue temperature may rise considerably above the air temperature due to incoming solar radiation (Grace et al. 1989), the finding is an artifact caused by the use of the air temperature rather than the tissue temperature in the calculation of the high temperature requirement. Fu et al.'s (2015) finding shows that environmental factors may also have indirect effects on the spring phenology of trees. The finding does not directly affect the ecophysiological temperature responses discussed in this volume; rather, it calls for developing separate models of the effects of climatic factors on tissue temperatures so as to find a tissue temperature usable as input in the ecophysiological models.

3.4.4 The Paradox of Rest Completion

3.4.4.1 Contradictory Findings Concerning Rest Completion

According to the traditional conception cited in most of the literature, the accumulation of chilling causes rest break in the buds of boreal and temperate trees. Furthermore, the experimentally determined chilling requirements are so small in most cases that rest completion, in the sense of attaining full ontogenetic competence (Table 3.4; Fig. 3.16), takes place in early to late autumn (Fig. 3.18; Table 3.5).

However, since the mid-1990s, contradictory results have been obtained in studies examining the bud burst, growth onset, or flowering of mature trees growing in natural conditions (Table 3.5; Alternative conception). Many of these studies are based on modelling analyses of sets of long-term historical data on the timing of springtime phenological events, combined with air temperature data gathered in standard meteorological stations located near the observed trees. Using this approach, Häkkinen et al. (1998) and Häkkinen (1999b) compared ecophysiologically explicit models addressing both the chilling requirement of rest

Table 3.5 Two contradictory conceptions of rest completion in boreal and temperate trees

	Traditional conception	Alternative conception
Time of rest completion	Early to late autumn	Winter to spring
Environmental factor(s)	Chilling	Chilling, night length, light factors
References	1–9	10–19

[1]Worrall and Mergen (1967)
[2]Farmer (1968)
[3]Erez et al. (1979a)
[4]Erez et al. (1979b)
[5]Kobayashi and Fuchigami (1983a)
[6]Hänninen (1990b)
[7]Leinonen (1996a)
[8]Hannerz et al. (2003)
[9]Søgaard et al. (2008)
[10]Hänninen (1995a)
[11]Häkkinen et al. (1998)
[12]Hannerz (1999)
[13]Häkkinen (1999b)
[14]Linkosalo (2000)
[15]Linkosalo et al. (2000)
[16]Linkosalo et al. (2008)
[17]Partanen et al. (2005)
[18]Hänninen et al. (2007)
[19]Fu et al. (2012)

completion, C_{crit}, and the high temperature requirement of growth onset, H_{crit}, with a simplified model where rest is assumed to be completed on a fixed calendar day, so that only the accumulation of high temperature units and the ontogenetic development are explicitly modelled. For rest break caused by chilling and ontogenetic development, they used the air temperature responses published by Sarvas (1972, 1974) (Figs. 3.7 and 3.13a). In this way, two parameters were estimated for each model, i.e., H_{crit} for all models and, depending on the model, either C_{crit} (for chilling models) or the constant calendar day of rest completion (for the simplified model).

Using a 60-year phenological time series for *Betula pendula* in central Finland, Häkkinen et al. (1998) and Häkkinen (1999b) found that the simplified model predicted the timing of bud burst more accurately than the more complicated models, which also addressed the chilling requirement of rest completion. Furthermore, the estimated date of rest completion was March 21, i.e., several months later than the dates generally observed in experimental studies addressing the chilling requirement (Figs. 3.12 and 3.18). The conclusions were confirmed for a wider spectrum of tree species and springtime phenological events later on by Linkosalo (2000), Linkosalo et al. (2000, 2008), and Fu et al. (2012).

Besides the studies with long-term historical time series, similar results have also been found in other types of studies carried out with mature trees. Hannerz (1999) tested various models with bud burst data from a field trial with 17 clones of *Picea abies*. Similarly to the studies with long-term historical data sets, Hannerz (1999) found that simple temperature sum models that assumed rest completion on

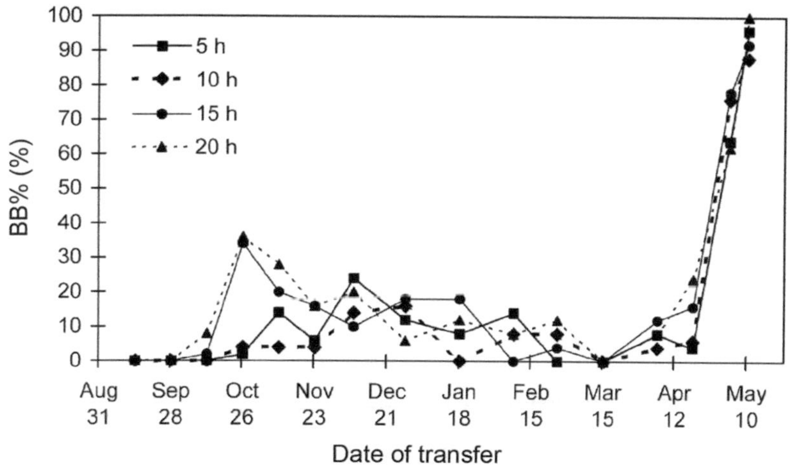

Fig. 3.20 Bud burst percentage, BB%, observed in detached twigs of 56-year-old eastern Finnish *Picea abies* trees after transferring them into forcing at 20 °C in a greenhouse between autumn 1992 and spring 1993. Four separate greenhouse departments with four respective photoperiods were used (Partanen et al. 2005; redrawn with the permission of Springer Science+Business Media)

a fixed calendar day predicted the timing of bud burst more accurately than more complicated models that also addressed the effects of chilling. Partanen et al. (2005) studied the rest completion of naturally growing 56-year-old *Picea abies* trees by transferring twigs detached between September and May into forcing conditions in a greenhouse. Regardless of the night length applied in the forcing conditions, bud burst consistently occurred only in transfers made after mid-April (Fig. 3.20). Finally, studies carried out under elevated temperature conditions in whole-tree chambers (WTCs) have also yielded results suggesting that rest completion takes place relatively late in spring. These studies will be discussed in detail in Sect. 8.3.2.2.

3.4.4.2 Attempts to Resolve the Paradox

The night length prevailing on a given calendar day is constant at each growing site. For this reason, the result obtained in several modelling studies with long-term phenological data sets, suggesting that rest completion takes place on a fixed calendar day during the spring, has often been interpreted to provide indirect support for the hypothesis that rest completion is caused by the attainment of a critical night length during the spring (Häkkinen et al. 1998; Linkosalo 2000; Linkosalo et al. 2000, 2008). According to this interpretation, the fixed calendar day in the modelling analysis represents the critical night length that is the causal factor inducing rest completion. However, as stated in Sect. 3.3.3, the empirical evidence for the effects of night length in the rest break of boreal and temperate trees is contradictory. Thus, while the role of night length in the rest break of boreal and temperate trees cannot be ruled out, it is not likely that the paradox of rest completion (Table 3.5) could be entirely explained by the rest-breaking effects of night length.

Linkosalo et al. (2006) tried to resolve the paradox by introducing a new assumption into the sequential model. They hypothesised (i) that rest completion in autumn takes place as a result of accumulated chilling, as assumed in the prevailing models, but (ii) that, contrary to the prediction of these models, the meteorological accumulation of high temperature units in late autumn and winter is not physiologically effective, i.e., it does not cause ontogenetic development. On that basis they posited that the predictions of the parallel model and, to a lesser extent, those of the sequential model, too, are erroneous, i.e., the increase in the state of ontogenetic development, S_o, should commence in spring, not in autumn as implied by the prevailing models (Fig. 3.18).

Linkosalo et al. (2006) further hypothesised that the physiological effect of the meteorological accumulation of high temperature units in mild winter periods is nullified by subsequent cold periods. Ontogenetic development as such is by definition irreversible, so that it cannot be nullified, but the accumulation of high temperature units probably also simulates some related physiological phenomena that may be reversible. On the basis of such reasoning, Linkosalo et al. (2006) formulated two novel models, where different types of nullifying, or reversing, the accumulation of high temperature units is assumed. These models provided more accurate predictions than the sequential model, but again, all the models that explicitly addressed chilling, too, were outperformed by the simplified model that assumed that the onset of the accumulation of high temperature units and the corresponding ontogenetic development occur on a given date in spring. Thus Linkosalo et al.'s (2006) hypothesis does not provide an adequate solution to the paradox (Table 3.5).

Partanen et al. (2005) found that the pattern of bud burst percentage, BB%, obtained with 15-year-old Norway spruce trees was different from both the typical pattern observed earlier with seedlings and the other pattern observed by Partanen et al. (2005) with older trees. The BB% of the 15-year-old trees first increased in autumn, basically similarly to that observed with seedlings (Fig. 3.12a), but before the end of year it dropped to near zero, then increased again at approximately the same time as with the older trees (Fig. 3.20). These deviant findings obtained with differentially aged trees led Partanen et al. (2005) to hypothesize that the environmental regulation of rest break and bud burst changes as the trees get older. This hypothesis is physiologically motivated, since several aspects of the seasonality of trees change with ageing (Ununger et al. 1988; Vitasse 2013). Furthermore, many of the studies supporting the traditional conception (Table 3.5) have been carried out with seedlings. That is not the case, however, in all the studies supporting the conception (Worrall and Mergen 1967; Erez and Lavee 1971; Erez et al. 1979b). Therefore the hypothesis put forward by Partanen et al. (2005) can provide only a partial explanation for the paradoxical results.

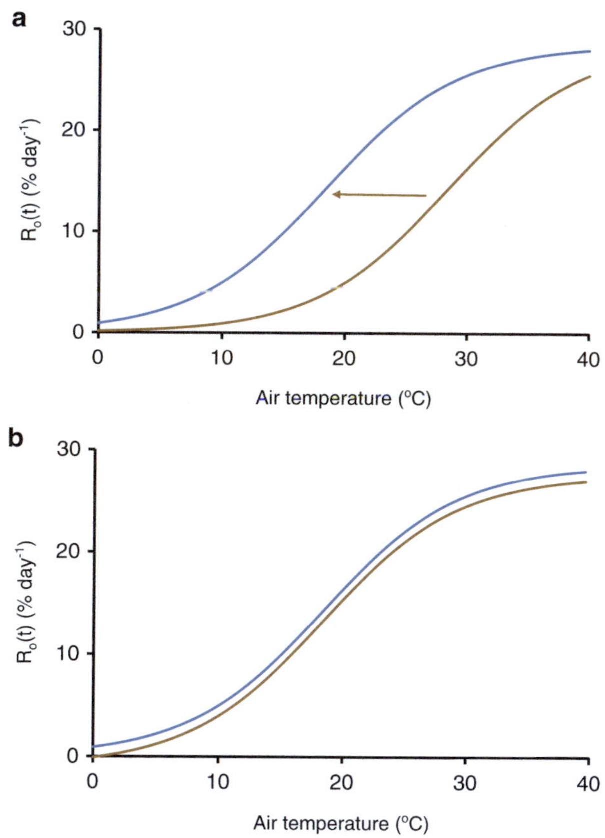

Fig. 3.21 A hypothesis explaining apparently contradictory findings for rest completion in boreal and temperate trees (Table 3.5). It is assumed in the hypothesis that the post-rest model (Hänninen et al. 2013) introduced in Sect. 3.3.1.2 on the basis of Vegis's (1964) theory is realised in trees. The *curves* represent the air temperature response of the rate of ontogenetic development, $R_o(t)$, in (**a**) late autumn and winter and (**b**) in spring, both in the prevailing models (*blue curves*) and in the post-rest model (*brown curves*). The *arrow* in (**a**) indicates the change in the response taking place in winter according to the post-rest model. See the text for details

In the present volume, the post-rest model (Hänninen et al. 2013) introduced in Sect. 3.3.1.2 on the basis of Vegis's (1964) theory is suggested as a hypothetical explanation for the earlier paradoxical results (Table 3.5). To recapitulate, the paradox arises from two types of observations: studies carried out mainly with adult trees in natural conditions and in whole-tree chambers suggest that the meteorological accumulation of high temperature units is not physiologically effective from late autumn to early spring, i.e., the accumulation does not cause any ontogenetic development towards bud burst, whereas experiments carried out by interrupting the overwintering by artificial forcing show that ontogenetic development leading to bud burst is possible during that time. The post-rest model may explain this contradiction as follows (Fig. 3.21):

In early autumn, the buds are in true rest, so that no ontogenetic development will take place and no bud burst will be observed when the seedlings are transferred into experimental forcing at that time. Later in the autumn, chilling has accumulated to the extent that the tree progresses from true rest to post-rest. Thus the buds have regained their ontogenetic competence, but only partially, so that ontogenetic development is possible only at relatively high air temperatures (e.g., $T > 10$ °C;

Fig. 3.21a, brown curve). Such high air temperatures are generally used in experimental forcing, so that bud burst will take place after the transfer into forcing conditions in late autumn, as implied in the traditional conception (Table 3.5). In natural conditions, however, the air temperature rarely rises above 10 °C at that time of the year even during mild spells, so that little or no ontogenetic development will take place (Fig. 3.21a, brown curve), as implied in the alternative conception (Table 3.5).

The phase of post-rest lasts until relatively late in spring, when the accumulation of chilling, and perhaps also other environmental cues (Caffarra et al. 2011a, b), causes a release from post-rest and the attainment of the phase of quiescence. During this development, lasting several months, the air temperature response of the rate of ontogenetic development, $R_o(T(t))$, shifts gradually towards lower air temperatures (Fig. 3.21a, brown arrow). Thus, progressively lower air temperatures below 10 °C start to cause ontogenetic development, and in spring, the air temperature response of $R_o(T(t))$ corresponding to the prevailing models (Fig. 3.21b, blue curve) is finally realised in the trees (Fig. 3.21b, brown curve).

According to this reasoning, which assumes the existence of the post-rest phase in real trees, the alternative conception obtained in analyses of long-term phenological data gathered in natural conditions (Table 3.5) is based on a mathematical artifact in the analysis of the data. The prevailing models predict considerable ontogenetic development to take place whenever the air temperature rises above 0 °C in late autumn and winter (Fig. 3.21a, blue curve). According to the post-rest hypothesis discussed here, this modelled ontogenetic development does not have any counterpart in real trees (Fig. 3.21a, brown curve). In order to eliminate this unrealistically modelled wintertime ontogenetic development, the fitting algorithm minimises the difference between the predicted and the observed time of growth onset by postponing rest completion till spring (Table 3.5), when ontogenetic development takes place at temperatures of 0–10 °C in real trees, too, which are here assumed to follow the post-rest model (Fig. 3.21b).

The above reasoning may also explain why higher values of the high temperature requirement of growth onset, H_{crit}, have often been obtained in greenhouse forcing at a constant air temperature than have been estimated on the basis of studies carried out in fluctuating temperatures under natural conditions (there may also be additional explanations for this; see Sects. 3.2.1.5 and 3.4.1.2). According to this explanation, the seedlings are in post-rest when brought into forcing conditions in the autumn. Thus, at a typical forcing temperature of 15 °C, for instance, ontogenetic development towards growth onset takes place (Fig. 3.21a, brown curve), but it does so at a lower rate than anticipated in the prevailing models (Fig. 3.21a, blue curve). Thus a larger number of high-temperature units, as calculated in the prevailing models, will be needed under forcing conditions in the autumn (Fig. 3.21a) than under natural conditions in the spring, when the prevailing models do not overestimate the rate of development any longer (Fig. 3.21b). According to this explanation, there is a difference in the outcomes between forcing in a fluctuating air temperature and in a constant one when the forcing experiment is carried out in the autumn, but the difference vanishes when the experiment is postponed until the spring.

In conclusion, the post-rest model based on Vegis's (1964) theory provides a promising candidate to explain the paradoxical results in most of those cases where growth onset was observed in experimental forcing conditions in autumn but yet other studies carried out in the field or in whole-tree chamber conditions suggested that no ontogenetic development is possible before late spring. However, the post-rest model does not explain the case where very little bud burst was observed when detached twigs of Norway spruce were forced at the temperature of 20 °C in autumn (Fig. 3.20). Despite this, the model deserves to be tested thoroughly in future work, both experimentally and with long-term phenological time series. Quite possibly, the research practice of sticking to the fixed location of the air temperature response of the rate of ontogenetic development on the temperature axis (Fig. 3.17), thus not allowing the shifting of the response postulated in the post-rest model (Fig. 3.21a), has formed a paradigm that hinders the progress of the research work. Consideration of the post-rest model may also be essential for assessing the implications of the projected climate change, as discussed in detail in Sect. 8.3.3.4.

For any attempts to resolve the paradox (Table 3.5), it would be helpful to have additional point events for testing the different models of rest break and bud burst. Most importantly, it would seem necessary to identify rest completion and the onset of ontogenetic development in field conditions. To this end, physiological and microscopic studies are needed (Fig. 3.2; Sutinen et al. 2009, 2012; Viherä-Aarnio et al. 2014).

3.4.5 Quantitative Dormancy Induction

According to the prevailing view, rest is induced in many boreal and temperate trees by exposure to long nights in early autumn (Junttila 2007; Hänninen and Tanino 2011).[20] However, there is increasing evidence for the notion that the air temperature prevailing during the long-night dormancy induction affects the deepness of the dormancy, as manifested in either the chilling requirement of rest completion, C_{crit}, or the high temperature requirement of growth onset, H_{crit}, or both. In general, it has been observed that relatively high air temperatures during the long-night dormancy induction increase the deepness of dormancy, i.e., the value of C_{crit} and/or H_{crit} is increased by the high air temperatures (Dormling 1989; Westergaard and Eriksen 1997; Heide 2003; Junttila et al. 2003; Kalcsits et al. 2009). In the present volume, this phenomenon is referred to with the term

[20] According to Junttila (2007) there is actually very little, if any, direct experimental evidence for this notion. However, as height growth cessation is needed for rest induction, this notion is plausible for species in which long nights cause height growth session (Sect. 3.5.2).

quantitative dormancy induction.[21] The effect of the air temperature prevailing during the long-night dormancy induction on the timing of bud burst can be substantial. Søgaard et al. (2008), for instance, found that the time required for bud burst in *Picea abies* seedlings increased from approximately 8 to 30 days when the dormancy induction temperature rose from 9 to 21 °C.

The chilling requirement of rest completion, C_{crit}, and the high temperature requirement of growth onset, H_{crit}, are assumed to be genotype-specific constants in all prevailing dynamic models of dormancy release and growth onset (Hänninen and Kramer 2007; Chuine et al. 2013). In other words, at the beginning of the rest phase, a given tree genotype is assumed to always be in a similar physiological condition with respect to phenomena related to dormancy, i.e., rest and quiescence, regardless of the environmental factors prevailing during the dormancy induction in late summer or early autumn. However, the experimental findings supporting the concept of quantitative dormancy induction call for a revision of the models.

This is technically straightforward enough, as the values of C_{crit} and H_{crit} can be modelled to depend on the air temperatures prevailing in autumn. However, in order to attain quantitative data for this, quite complicated hierarchical experimental set-ups, such as were used, e.g., by Junttila et al. (2003), need to be applied, so that the amount of experimental work required is substantial. This notion is further emphasised by the findings of Kalcsits et al. (2009). Working with clones of *Populus* x spp., they found that the effects of night temperatures were largely opposite to those of day temperatures and that the different clones responded differently to the air temperatures.

However, despite the great amount of experimental work required in order to develop realistic models for quantitative dormancy induction, it should be addressed in modelling the annual phenological cycle of boreal and temperate trees because its effect on the timing of growth onset is considerable. Even without any simulation studies, it is obvious in qualitative terms that higher-than-average air temperatures during dormancy induction may delay growth onset in the next spring, and this phenomenon may be essential for the implications of climatic warming as well (Heide 2003). This issue is discussed further in Sect. 8.3.3.6.

3.4.6 Alternative Modelling Approaches

3.4.6.1 Ecophysiological, Phenological, and Intermediate Approach

Hänninen et al. (2007) divided the methodological approaches to the modelling of the rest break and growth onset of boreal and temperate trees into three categories,

[21] Following the nomenclature adopted for the present volume, the concept of dormancy refers here to both rest and quiescence, so that the deepness of the dormancy can be manifested in the values of both C_{crit} and H_{crit}. In some of the studies referred to in this section, the concept of dormancy refers to rest only.

i.e., the ecophysiological, the intermediate, and the phenological. In the *ecophysiological approach*, as many as possible of the ecophysiological aspects discussed in the previous sections of this chapter are addressed for the tree species and provenance in question by means of specific experiments, with each experiment designed for one particular aspect (e.g., Figs. 3.7, 3.12, and 3.19). Thus, models developed and tested in this way have sound empirical support.

It goes without saying, though, that due to practical restrictions, it is not always possible to adopt this scientifically preferable approach. The *intermediate approach* denotes those cases where the air temperature responses of the accumulation rates of chilling units and high temperature units are fixed on the basis of a priori information but the chilling requirement of rest completion, C_{crit}, the high temperature requirement of growth onset, H_{crit}, and the model for ontogenetic competence (Fig. 3.16) are determined by fitting the model to the data in question.[22] This approach has been taken especially in testing the models with sets of long-term historical phenological data. Relying on the exceptionally strong empirical support of the air temperature responses published by Sarvas (1972, 1974) (Figs. 3.7 and 3.12a), Finnish researches have frequently used these fixed responses when taking the intermediate approach to modelling studies with sets of long-term historical phenological data (Häkkinen et al. 1998; Linkosalo 2000; Linkosalo et al. 2000).

The intermediate approach was also adopted by Hänninen (1995a) in testing 96 models of growth onset, which were generated by combining a number of previous sub-models in various ways with growth onset data on *Pinus sylvestris* gathered in a whole-tree chamber (WTC) experiment. He found that ecophysiologically different models had almost equal predictive power in the test.[23] These findings revealed the importance of the concepts of model accuracy and model realism discussed in Chap. 2 (Hänninen 1995b). Many of the models tested have high accuracy, but due to the different ecophysiological assumptions included in them, only one of them can have high realism as well, i.e., the assumptions of only one of the conflicting models can be realised in real trees. The practically identical accuracy of the different models did not enable any inferences about their realism, nor was there any a priori information available for such inferences.

On the basis of his findings, Hänninen (1995a) concluded that observations of the growth onset in natural conditions, or even in the elevated temperature conditions of the WTCs, were not sufficient by themselves for testing the models. This is because each model is actually a combination of three sub-models, with each representing a particular ecophysiological aspect, and because the models describe developmental phenomena occurring in the trees during several months before the observed growth onset (Fig. 3.18). The information contained in the data, i.e., the date of growth onset observed in each case after overwintering in the specific

---

[22] In the modelling of the effects of night length on rest completion, it is the value of the critical night length (or the corresponding calendar day) that is estimated instead of the value of the chilling requirement of rest completion (Häkkinen et al. 1998).

[23] The results are discussed in detail in Sect. 8.3.2.2.

climatic conditions of the year (or of the WTC) is not sufficient to identify the most appropriate model from among several competing ones simulating the ecophysiological processes preceding the growth onset. This reasoning calls for more thorough experimental testing of the various ecophysiological aspects represented by the different sub-models, one at a time, with each experiment designed for one particular aspect (Hänninen 1995a). In other words, this reasoning calls for the ecophysiological approach (Hänninen et al. 2007).

Hunter and Lechowicz (1992) generated sets of artificial data for the timing of bud burst by calculating the predicted timing with different models and using real air temperature data as input in their calculations. Examining the predictive power of the different phenological models, they found that several other models besides the one used in generating the data were equally accurate in predicting the generated observations. These findings provide perhaps the clearest demonstration of the methodological pitfalls hampering the efforts to test phenological models with data gathered in natural conditions. In order to develop realistic models, which account for the environmental responses of real trees, approaches using field data need to be augmented with ecophysiological approaches, where various aspects of rest break and ontogenetic development are examined experimentally.

In the *phenological approach* all parameters of the model are estimated by fitting the model to phenological and air temperature data (Kramer 1994a, b; Linkosalo et al. 2008; Morin et al. 2009). In this approach the problem of insufficient information in the data is magnified (Hänninen et al. 2007; Hänninen and Kramer 2007). For instance, Kramer (1994b) fitted the sequential model into sets of historical data on several species by estimating the values of all the parameters related to rest break (Eq. 3.12) and ontogenetic development (Eq. 3.10). For the air temperature response of the rate of rest break in *Fraxinus excelsior*, the model-fitting resulted in the values of $T_1 = -20.4$ °C and $T_3 = 165.8$ °C for the lower and the higher thresholds of the rest-breaking temperature range, respectively (Eq. 3.12). It goes without saying that air temperatures above 100 °C are lethal and thus cannot cause rest-breaking chilling effects, so that in one way or another, the fitting result is anomalous.

Chuine et al. (1998) attributed the problems met in testing and comparing various phenological models with field data to shortcomings in the computer algorithms used for the model-fitting. Due to the large number of parameters involved, the algorithms are not able to examine all the potential parameter combinations. Thus it may happen that rather than the global minimum of error variance, the algorithm finds a local minimum and thus also estimates erroneous values for the parameters. So, Chuine et al. (1998) introduced a new algorithm, i.e., the simulated annealing, into the studies of modelling bud burst in boreal and temperate trees and suggested that this new algorithm would facilitate the finding of the global minimum and thus facilitate the identification of ecophysiologically meaningful models. By applying the new algorithm to sets of large-scale data representing several tree species, they found support for this notion (Chuine et al. 1998, 1999).

Unfortunately, however, the improved algorithm does not solve the methodological problem entirely. This was demonstrated by Hänninen et al. (2007), who applied the phenological approach to bud burst data gathered on *Picea abies* in WTCs.[24] The models used by Hänninen et al. (2007) were so simple that it was possible to examine all the parameter combinations in the model-fitting. Thus the fitting would, in all cases, find the combination of parameter values that implied the global minimum of error variance. Despite this, Hänninen et al. (2007) were unable to identify the most appropriate model, since ecophysiologically different models had practically identical accuracy.

This shortcoming cannot be attributed to difficulties with the fitting algorithm, for all combinations in the parameter space were examined, i.e., there is no way to improve the performance of the algorithm. Rather, the result reveals, once again, the problem inherent in the phenological approach, i.e., that the information contained in phenological data is simply insufficient for judging among different competing ecophysiological models. This conclusion can be easily understood by reconsidering what happens in the model simulations and the model fitting. During one annual cycle, the values of two state variables, i.e., S_r and S_o, are calculated for approximately 250 days (Fig. 3.18), but the only test of all these simulated dynamics takes place at the endpoint of the simulation, in the comparison of the predicted date of bud burst when $S_o(t) = 100$ with the observed one.[25]

Despite these problems and criticisms, the phenological approach should not be altogether abandoned. Rather, when applied carefully, it can provide valuable information, also for studies addressing the climatic change, especially since more reliable results obtained with the ecophysiological approach are not available. This notion is discussed in more detail in Chap. 8.

3.4.6.2 Mechanistic Models

The physiological mechanisms of rest break are still only partially understood (Horvath et al. 2003; Arora et al. 2003; Rinne and van der Schoot 2003; Junttila 2007; Rohde and Bhalerao 2007; Ruttink et al. 2007; Cooke et al. 2012; Brunner et al. 2014). Despite this, hypothetical models addressing the rest break at the physiological level have also been suggested. This was first done by Fishman et al. (1987a, b), who presented a mechanistic model of rest break based on the production of theoretical rest-breaking factors. Even though the model was based on hypothetical assumptions, it was able to explain several complicated patterns of the effects of air temperature on rest break in *Prunus persica*, including the negation

[24] The results are discussed in detail in Sect. 8.3.2.2.

[25] Recently, Clark et al. (2014) addressed this problem and introduced a novel approach, the continuous development model (CDM), for phenological studies. While opening new avenues for phenological modelling, Clark et al.'s (2014) approach is substantially different from the one adopted in the present volume. Therefore the reader is referred to Clark et al.'s (2014) original study rather than discussing the CDM model here.

of chilling (for a further discussion of the model, see also Cannell 1989; Pitacco 1990). More recently, Schaber and Badeck (2003) have presented a mechanistic model based on the dynamics of theoretical growth promoters and inhibitors. They found that in predicting the timing of bud burst with several tree species in Germany, their novel model was more accurate than the classical models in several cases.

The mechanistic models may be helpful in efforts to create a new synthesis between the whole-tree approach adopted in the present volume and the physiological and molecular approaches (Horvath et al. 2003; Arora et al. 2003; Rohde and Bhalerao 2007; Brunner et al. 2014) to the study of dormancy in boreal and temperate trees. As discussed in Hänninen and Kramer (2007), the concept of ontogenetic competence may also be useful in efforts towards such a synthesis. The concept facilitates the mediating of the effects of the rest status to the rate of ontogenetic development in all cases where the rest status can be expressed quantitatively by means of the variable state of rest, S_r, (Figs. 3.15 and 3.16), regardless of the complexity of the processes addressed in the calculating of the value of S_r.

When mechanistic models are used as part of the phenological approach, the methodological problems inherent in that approach are often amplified. As discussed above, the information contained in phenological data is insufficient for comparing different classical ecophysiological models addressing the chilling requirement of rest completion and the high temperature requirement of growth onset. It is therefore quite clear that this problem is amplified when the physiological complexity of the model is increased. Due to their overparameterisation, mechanistic models may easily be fitted not only to the variation reflecting real year-to-year variation in the timing of phenological events but also to the noise contained in long-term phenological data (Linkosalo et al. 2008).

3.4.6.3 The Unified Model of Growth Onset

The main idea in introducing the generalised formulation for models of rest break and growth onset (Fig. 3.15) was to provide a unifying framework for addressing different models originally formulated in different ways and a framework for formulating novel models (Hänninen 1990a, b, 1995a; Kramer 1994a, b; Hänninen and Kramer 2007). This line of reasoning was developed further by Chuine (2000), who introduced a unified model that maximised model generality. The sophisticated equations of Chuine's (2000) model are so flexible that with the proper choice of parameter values, they will reproduce any particular model as a special case. Rather than using any a priori information, all aspects of Chuine's (2000) model are determined when fitting the model to empirical data. That is not the case in the less generalised formulation applied in the present volume (Fig. 3.15), where the choice among the three sub-models is restricted by the use of a priori information. It is possible, for instance, to address both the sequential and the parallel model (Figs. 3.16a, b), but a conscious choice between the two needs to be made. Chuine's

(2000) unified model, in contrast, is flexible in that respect, too, so that the model fitting also takes care of the choice between the sequential and the parallel model. Chuine's (2000) unified model, or some modification of it, has been used in several later studies addressing the spring phenology of trees (e.g., Morin et al. 2009; Caffarra et al. 2011b).

As discussed in Hänninen and Kramer (2007), the two approaches are suited for different purposes despite their common idea of unifying the modelling of growth onset in boreal and temperate trees. The choice between the two approaches may even be a matter of taste to some extent. Due to its utmost flexibility, Chuine's (2000) unified model is well suited for studies where the model is fitted to large sets of data without any need to use a priori information in the model formulation. That is often the case when the phenological approach is applied to long-term sets of data.[26] With reference to the discussion presented in Chap. 2, those studies make use of an inductive modelling approach, i.e., the formulation of the model is determined entirely by the information available in the empirical data.

In the present volume, however, the ecophysiological approach is emphasized and therefore a hypothetico-deductive approach is taken, so that a priori information is frequently used. This is demonstrated, for instance, by the frequent use, in various contexts, of the air temperature responses published by Sarvas (1972, 1974) (Figs. 3.7 and 3.13a). Furthermore, even the implications of falsified models are of interest in some cases. In such cases the aim cannot be the fitting of the model to the data. This notion is discussed further in Sect. 8.3.3.

3.4.6.4 The DORMPHOT Model of Growth Onset

Caffarra (2007) and Caffarra et al. (2011a) studied the effects of air temperature and night length on the rest break and bud burst phenology of *Betula pubescens*. In a series of well-interconnected experiments they found several interactions of these environmental factors not addressed in previous models of spring phenology in boreal and temperate trees. Taking Chuine's (2000) unified model (Sect. 3.4.6.3) as their starting point, Caffarra et al. (2011b) introduced the model DORMPHOT, in which the novel interactions are addressed. Most importantly, two novel aspects were introduced into their model.

First, Caffarra et al. (2011a, b) also addressed the effects of environmental conditions on rest induction. Besides long nights, low air temperatures also enhance rest induction in their model. Second, a complicated interaction of chilling, night length, and high air temperatures during dormancy release was introduced into the model. Accordingly, shortening the night length shifts the air temperature response

[26] However, as shown by Linkosalo et al. (2008), Chuine's (2000) unified model, like any other complex model with a high number of parameters, is susceptible to over-parameterisation in these studies.

of the rate of ontogenetic development towards lower air temperatures,[27] and the extent of this shift is affected by previous chilling.

Experimental evidence for the novel and complicated model of Caffarra et al. (2011b) remains scanty. This is especially because rather than being determined on the basis of experimental data, the values of some of the parameters were estimated by fitting the model into field-phenological data, so that the methodological problems of the phenological approach discussed in Sect. 3.4.6.1 were encountered here, too. Furthermore, Caffarra et al.'s (2011a) findings concerning the effects of low temperatures on rest induction are contradictory to Junttila et al.'s (2003) findings for the same species: According to Caffarra et al. (2011a), the rate of rest induction is higher in 10 than in 18 °C, whereas Junttila et al. (2003) found a higher rate in 18 than in 9 °C (see Sect. 3.4.5). Despite these uncertainties, however, the studies of Caffarra et al. (2011a, b) demonstrate how novel phenomena can be introduced into the ecophysiological models of dormancy and growth onset in boreal and temperate trees. Hopefully, their model will be tested in further experimental work, not only with *Betula pubescens* but also with other boreal and temperate tree species.

3.4.6.5 The Alternating Model of Growth Onset

Cannell and Smith (1983) found that the temperature sum required for bud burst in *Picea sitchensis* growing in the UK decreased exponentially with previously accumulated chilling (Fig. 3.22).[28] This relationship has been used subsequently for various purposes in several studies. Cannell et al. (1985) determined the inverse exponential relationship for several clones of *Picea sitchensis* and used the results for studying the possibilities of avoiding frost damage by selection for late bud burst among the clones. In a basically similar approach, the model was also used later for assessing the effects of the climatic warming and increased concentrations of atmospheric CO_2 on the phenology and risk of frost damage in several woody species (Cannell and Smith 1986; Murray et al. 1989, 1994).[29]

Referring to the change in the critical temperature sum causing growth onset, Kramer (1994a) introduced the concept "alternating model" for the inverse exponential relationship (Fig. 3.22). Later on, the model became generally known under this name, so that it was often referred to as one optional model for bud burst in a list of models, together with the sequential model and the parallel model (Chuine et al. 1998). However, as noted in Hänninen and Kramer (2007), the alternating

[27] In the present volume a similar change in the air temperature response was introduced as an extension of Vegis's (1964) theory (Fig. 3.11b in Sect. 3.3.1.2). As no reference to Vegis (1964) is made in Caffarra et al.'s (2011b) study, their model is evidently not explicitly based on Vegis's (1964) theory.

[28] Cannell and Smith (1983) used the concept "thermal time" for the day degree model used in their study.

[29] These studies are discussed in Sect. 8.3.1.3.

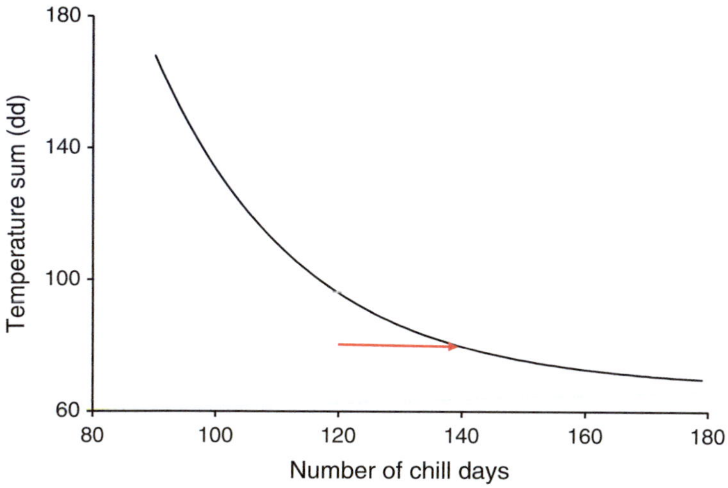

Fig. 3.22 The dependence of the temperature sum required for bud burst on previous accumulation of chilling in 2- to 10-year-old *Picea sitchensis* growing in the UK (Cannell and Smith 1983). The temperature sum is calculated as the accumulation of day degrees (threshold temperature +5 °C) from 1 February and the accumulation of chilling as the number of chill days with the daily minimum temperature below +5 °C after 1 November. This inverse exponential relationship was later dubbed "the alternating model" by Kramer (1994a), and it became known by that name in modelling literature. The *arrow* represents an unrealistic case, where the model predicts occurrence of bud burst during a period of freezing temperatures. See text for details (The figure was plotted on the basis of the equation provided in the original publication. For copyright reasons the original figure including the data points was not redrawn)

model is actually an implication of the parallel model (see below). Both of the two share the same basic ecophysiological assumption, i.e., the gradual effect of chilling on ontogenetic competence (Fig. 3.16b) and, consequently, a similar gradual effect on the rate of ontogenetic development (Fig. 3.17).

The main advantage of the alternating model is its simplicity. The graphic presentation of the exponentially declining temperature sum requirement (Fig. 3.22) is probably a great deal easier to grasp than the more complicated presentation with ontogenetic competence (Fig. 3.16b) and the rate of ontogenetic development (Fig. 3.17). This notion is closely related to the experimental determination of the rate of ontogenetic development. As discussed in Sect. 3.2.1, due to the peculiarities of ontogenetic development as an ecophysiological process, its rate can be quantified only as the reciprocal of the time required for a given point event to occur, i.e., as the reciprocal of the developmental time, Δt. Thus, as the Δt is the measured variable, then its relationship (Fig. 3.19d), or the relationship of the corresponding temperature sum (Fig. 3.22), to the accumulation of chilling is relatively easy to grasp.

Rather than being a dynamic model based on the air temperature responses of rest break and ontogenetic development, the alternating model indicates the prediction of the ecophysiologically analogous parallel model: the time required for growth onset (Fig. 3.19d), and so also the corresponding temperature sum (Fig. 3.22), decrease exponentially with increasing duration of previous chilling. However, this similarity of the two models holds only when the continuous period of chilling is followed by a continuous period of exposure to high temperatures, as in the experiment illustrated in Fig. 3.19. For natural conditions, there is a crucial difference between the two models: regardless of the duration of previous chilling, the alternating model accumulates the temperature sum at the full rate allowed by the prevailing air temperature whenever it rises above the threshold after 1 February, whereas the parallel model addresses the rest status explicitly, i.e., at any given moment the potential rate of ontogenetic development is realised only to the extent allowed by the prevailing state of rest and ontogenetic competence (Eq. 3.13) determined by the accumulated chilling (Fig. 3.16b).

In the alternating model, the number of day degrees accumulated until a given moment does not quantify the state of ontogenetic development because the model does not take into account the timing of the temperature sum accumulation in relation to rest break caused by the accumulation of chilling. This compromises the realism of the alternating model. Consider, for instance, a case where 120 chill days and 80 day degrees are accumulated (Fig. 3.22). Assume further that after that, the daily mean temperature stays below zero for several weeks. The alternating model unrealistically predicts that growth onset would occur during this period of freezing temperatures, because then one point on the exponentially declining curve is attained.[30] However, it is not clear how much this lack of realism would hamper the use of the alternating model. The empirical results of Cannell and Smith (1983) show that the alternating model can be used for predictions, at least in the natural air temperature conditions that prevailed in their study.

However, when an explicit state variable for ontogenetic development is aimed at, then the parallel model and the related procedure of determining the rate of ontogenetic development as the reciprocal of Δt needs to be resorted to, despite the fact that it is intuitively more complicated than the direct approach applied in the alternating model. As stated in Chap. 2 already, the present volume is based on formulating explicit rate and state variables for all the ecophysiological phenomena addressed (Hänninen and Kramer 2007), so that when the ecophysiological phenomenon of the gradual effect of chilling on the rest status is referred to, it is mostly done in terms of the parallel model (Fig. 3.16b).

[30] This illustrative point was made by Tapio Linkosalo.

3.5 Environmental Regulation of Height Growth Cessation[31]

3.5.1 *The Autonomous Theory and the Signal Theory*

The environmental regulation of growth cessation has been actively studied experimentally since the pioneering work of Garner and Allard (1923). In modelling studies, however, growth cessation has received much less attention, so that several phenomena revealed in experimental studies remain to be addressed in modelling work (Tanino et al. 2010; Hänninen and Tanino 2011). Furthermore, compared with the somewhat complicated modelling of growth onset and the related dormancy phenomena (Sects. 3.2, 3.3, and 3.4), the current models of growth cessation are more simple and straightforward (see, however, Delpierre et al. 2009).

In the present volume, these models are discussed within the framework of Koski and Selkäinaho (1982), who subdivided the theories concerning the annual cycle of boreal and temperate trees into two categories. According to the *autonomous theory*, the development takes place along a genetically predetermined path, so that only the rate of development is regulated by air temperature, whereas according to the *signal theory*, a specific short-term signal from the environment is required or else the tree will stay in the same developmental phase.

The developmental phenomena described by various temperature sum models provide a good example of the autonomous theory (Koski and Selkäinaho 1982; see Sect. 3.2.1). No specific signals from the environment are required for the development to take place. The only requirement is that the environmental conditions are in the range promoting development, which in most cases means that the air temperature is above a given threshold. In that way, the rate of development depends on air temperature, but in all other ways, the developmental phenomena take place according to a genetically predetermined sequence. A good example of the signal theory, in turn, is offered by the critical effect of night length on growth cessation discussed in the following section (Koski and Selkäinaho 1982). According to the theory, the tree will not go on to the next developmental phase, i.e., growth cessation and bud set, before it receives the specific short-term environmental signal, i.e., the occurrence of the critical night length, from the environment.

Koski and Selkäinaho's (1982) division explicates an important qualitative difference between the two classes of developmental phenomena and thus improves our understanding of the annual cycle of boreal and temperate trees. Although the dynamic models of the regulation of the annual cycle mainly describe phenomena belonging to the autonomous theory, phenomena belonging to the signal theory can also be readily introduced into these models. This is done by assuming the value of

---

[31] In addition to height growth cessation, the closely linked formation of the terminal bud is also discussed in this section.

Table 3.6 Experimental determination of the critical night length for seedlings of *Pinus banksiana*

Age of seedlings (weeks)	Night length (hours)			
	6	8	10	12
10	–	–	–	+
14	–	+	+	+

(+) = terminal bud formed, (–) = terminal bud not formed
At the age of 10 weeks, the critical night length for the seedlings was between 10 and 12 h, and correspondingly, at the age of 14 weeks it was between 6 and 8 h. See text for details (Vaartaja 1959; published with the permission of the Ecological Society of America)

zero for the rate of development until the specific signal from the environment is received. After that, the rate of development alternates, again dynamically, according to the prevailing environmental factors, typically the air temperature (Fuchigami et al. 1982; Hänninen and Kramer 2007; Delpierre et al. 2009).

3.5.2 The Effects of Night Length

It has been known since the classical work of Garner and Allard (1923) that increasing night length is a crucial environmental factor causing growth cessation in many boreal and temperate trees (Wareing 1956; Dormling et al. 1968; Håbjørg 1972; Heide 1974; Ekberg et al. 1979; Junttila 1980; Junttila and Skaret 1990; Howe et al. 1995). This is the case in trees following the so-called free growth pattern, where new leaf primordia develop in the buds simultaneously with the elongation of the older ones (Junttila and Nilsen 1993).

Each tree genotype has its specific critical night length, NL_{crit}, that is required for growth cessation. The experimental design for revealing the effect of night length is quite simple. After raising the experimental seedlings in common growth-promoting conditions, a subset of the seedlings is transferred into each of a set of growth chambers, and the night length is varied among the chambers. The cessation of elongation growth, or alternatively, bud set, is monitored in the chambers for a predetermined period. Typically, growth continues and no bud set takes place in the growth chambers with short night lengths, whereas the opposite is true in the chambers with long night lengths. The critical night length, NL_{crit}, is determined to lie in the range between the longest non-inductive night length and the shortest inductive night length (Table 3.6). Thus the accuracy of the determination depends on the difference in the night length between two adjacent chambers in the series, which in turn is determined by the number of chambers available.

In the strict sense, the concept of "short-term environmental signal" included in the signal theory of Koski and Selkäinaho (1982) would mean that growth cessation is induced instantaneously as soon as the critical night length is reached for the first time. With real trees, however, a number of nights exceeding the critical value are required (Heide 1974). However, for modelling the effects of night length in natural conditions this does not constitute a problem, for the first inductive night length is followed by successively longer nights.

Among the concepts pertaining to the annual cycle of boreal and temperate trees, the concept of critical night length is perhaps the best established one. Comparisons among different geographical provenances have been an important part of the tradition of research on critical night length. The concept of photoperiodic ecotype, which is essential in these studies, is discussed in Sect. 6.3.3.

3.5.3 The Effects of Air Temperature

3.5.3.1 The Cumulative Effect of Air Temperature and Its Joint Effect with Night Length

Several authors examining the critical night lengths for the growth cessation of various tree species have concluded that several other factors modify the effect of night length (Sect. 3.5.3.2). Koski and Selkäinaho (1982) and Koski and Sievänen (1985), however, emphasized the role of the air temperature, so that rather than referring to it as a modifying factor, they established the concept of joint effect of night length and air temperature. They studied growth cessation with an experimental design where seeds of trees were sown at different times in two greenhouse departments, one with the air temperature at 15 °C and the other at 20 °C. A natural night length pattern was used in both departments. Second-year seedlings were studied with a similar technique, so that seedlings were transferred to the two greenhouse departments from outdoors in early spring and from cold storage in late spring and summer.

With the experimental design used, the correlation prevailing in natural conditions between night length and the accumulation of temperature sum was broken. This facilitated the testing of the two main theories introduced by Koski and Selkäinaho (1982). According to the autonomous theory, the development towards growth cessation progresses without any specific night length signal, so that only the rate of development is regulated by air temperature. Thus, regardless of the sowing time and the air temperature used in the greenhouse, the cessation of growth should take place at a constant accumulation of temperature sum, whereas the night length prevailing at the time of growth cessation varies (Fig. 3.23a).

This prediction was realised in the second-year seedlings of *Pinus sylvestris* (Fig. 3.23b), which follows the fixed growth pattern, where all the primordia develop in the buds in the previous summer before overwintering (Junttila and Nilsen 1993). Thus, contrary to the free growth pattern, no new primordia are developed during the growth, so that in this case the cessation of growth is a point event analogical to the onset of growth. The rate of development is regulated by the air temperature in both cases, so that both point events occur at a constant accumulation of temperature sum, but as the cessation of growth is a later point event in

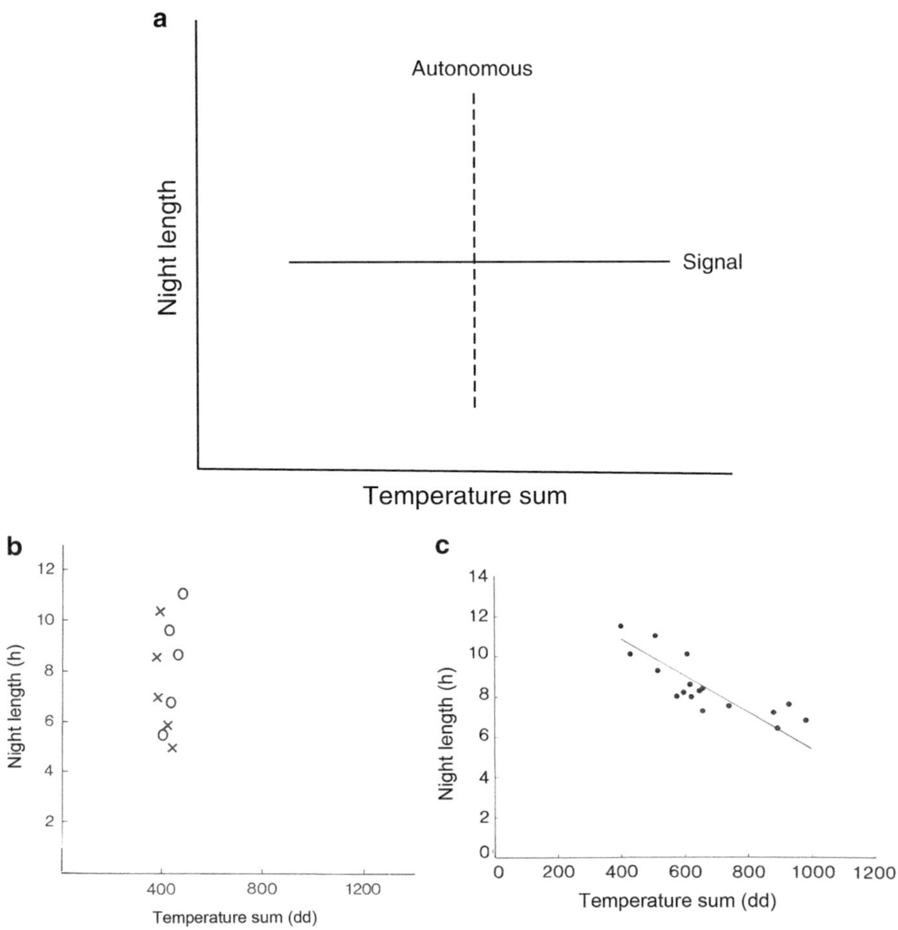

Fig. 3.23 The effects of night length and air temperature on the cessation of height growth in boreal and temperate trees. (**a**) Predictions of the signal theory (*solid line*) and the autonomous theory (*dashed line*). According to the predictions of both theories, only one environmental factor, i.e., prevailing night length (signal theory) or the temperature sum accumulated after growth onset (autonomous theory), regulates growth cessation (Modified from Koski and Selkäinaho 1982). (**b**) The effect of temperature sum observed in second-year seedlings of *Pinus sylvestris* (Koski and Sievänen 1985). (**c**) The joint effect of the two environmental factors observed in second-year seedlings of *Betula pendula* (Koski and Sievänen 1985). Figures redrawn with the permissions of the Finnish Forest Research Institute, currently the Natural Resources Institute Finland, and the Finnish Society of Forest Science (**a**); and the University of Helsinki (**b**, **c**)

the same sequence, it naturally requires a higher temperature sum that the onset of growth does (Raulo and Leikola 1974; Kanninen et al. 1982).

In contrast, the signal theory states that the cessation of growth is regulated by reception of the environmental signal, i.e., passing the critical night length. Thus, regardless of the sowing time and the air temperature used in the greenhouse, the cessation of growth should take place at a constant night length, whereas the accumulated temperature sum varies among the treatments (Fig. 3.23a).

However, this prediction was not realised with any of the species studied (Koski and Selkäinaho 1982; Koski and Sievänen 1985). Rather, with the exception of the second-year seedlings of *Pinus sylvestris*, the night length at growth cessation decreased with increased temperature sum accumulation (Fig. 3.23c). Thus, the critical night length is not a constant but varies according to the air temperatures prevailing in the growing season. These findings are in accordance with earlier results, where a lower value of NL_{crit} was attained when the value was determined at a later phase of the study (Table 3.6). The accumulation of the temperature sum necessarily took place between the early and the late determination of NL_{crit}, so that the earlier result agrees, at least qualitatively, with those of Koski and Selkäinaho (1982) and Koski and Sievänen (1985).

The joint effect model (Fig. 3.23c) is methodologically analogical to the alternating model (Fig. 3.22), i.e., neither of them is a dynamic model using the rate and state variables (Sect. 2.2). Rather, in both models, the critical value of a parameter indicating the occurrence of a given point event decreases as a function of a state variable. As discussed in Sect. 3.4.6.5, the alternating model is mainly replaced in this volume with the ecophysiologically similar parallel model, and the approach of dynamic modelling is adopted with the aid of the concept of ontogenetic competence. For the joint effect model, no such dynamic presentation has been developed, so that the original presentation is used throughout the present volume.

According to the joint effect model, height growth cessation occurs earlier in warm summers than in cool ones. Koski and Sievänen (1985) showed by means of calculations with a restricted set of air temperature data that year-to-year variation in air temperature caused considerable variation in the timing of height growth cessation predicted by the model. This is further demonstrated, with a longer temperature record than the one used by Koski and Sievänen (1985), in Fig. 3.24. Within the 92 years covered by the air temperature data, the timing of height growth cessation predicted by the joint effect model (Fig. 3.23c) varied from the last days of July to late August (Fig. 3.24).

This is in striking contrast with the prediction of the strict night length theory, which states that the cessation of growth takes place on the same calendar day each year. Thus, if the joint effect observed in greenhouse conditions (Fig. 3.23c) regulates the timing of growth cessation in natural conditions, too, then air temperature causes considerable year-to-year variation in the timing of growth cessation. If that is the case, air temperature is more than just a 'modifying' factor, ecologically speaking. Physiologically, night length can be regarded as the main factor because it obviously causes the physiological changes related to the shift from the growth phase to the dormant phase (Junttila 2007).

The notion of the joint effect of night length and temperature sum has been confirmed in several later studies (Partanen and Beuker 1999; Partanen 2004; Viherä-Aarnio et al. 2005). However, the relative importance of air temperature, i.e., the slope of the line in Fig. 3.23c, has varied among the studies. Sometimes a relatively gentle slope, indicating a small effect of the air temperature, has been

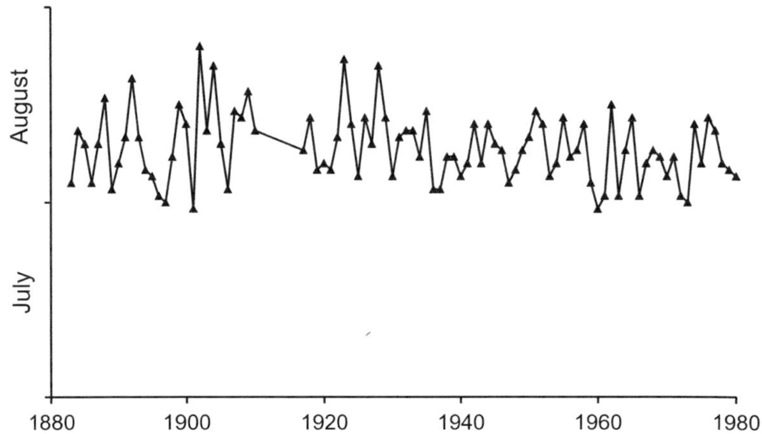

Fig. 3.24 Timing of the growth cessation of *Betula pendula* in 92 years (1883–1980; the years 1911–1916 were excluded because of missing meteorological observations), predicted by the joint effect model (Fig. 3.23c) for Jyväskylä, central Finland

found (Partanen 2004). In those cases, the joint regulation actually approaches the signal regulation by night length (Fig. 3.23a), so that a much smaller year-to-year variation in the timing of growth cessation than the one illustrated in Fig. 3.24 is predicted for those species and provenances.

It should be noted that the temperature sum appearing on the horizontal axis in the joint effect model (Fig. 3.23c) is calculated from the onset of growth. This implies that the environmental factors affecting the timing of growth onset have a delayed impact on the timing of growth cessation: if the timing of growth onset is accelerated or delayed for any reason, then, according to the prediction of the joint effect model, the timing of height growth cessation will change accordingly (Hänninen and Tanino 2011). Such a correlation between spring and autumn phenology was found recently in a warming experiment with two temperate deciduous tree species (Fu et al. 2014) and in analyses of large-scale phenological data sets, both in Europe (Fu et al. 2014) and in eastern United States (Keenan and Richardson 2015).

3.5.3.2 Further Effects of Air Temperature

The effects of air temperature on height growth cessation in boreal and temperate trees are not restricted to the ones modelled by means of the temperature sum model (Fig. 3.23b) or the joint effect model (Fig. 3.23c). As noted in a review by Tanino et al. (2010), it has been known for long that depending on the species and even on the provenance, air temperature has at least two other types of effects on height growth cessation. First, in several species that respond to long nights, high night air temperatures during exposure to long nights accelerate growth cessation. Secondly, there are also species in which height growth cessation is caused solely by exposure to low temperatures, so that night length plays no role (Heide and Prestrud 2005; Heide 2011).

Tanino et al.'s (2010) review shows that the prevailing ecophysiological models do not account for the environmental regulation of height growth cessation for all boreal and temperate tree species and provenances. The unaddressed phenomena can be addressed within the ecophysiological modelling framework adopted in the present volume (Hänninen and Kramer 2007), but it goes without saying that the achievement of models of high realism calls for species- and even provenance-specific experimental studies (Hänninen and Tanino 2011). The published results reviewed by Tanino et al. (2010) provide a good starting point for further model development.

3.6 Modelling the Entire Annual Phenological Cycle

3.6.1 Sarvas's (1972, 1974) Model

Sarvas (1972, 1974) was the first to present a model for the entire phenological cycle of boreal and temperate trees. He divided the annual cycle into three sequential phases, i.e., autumn dormancy (dormancy I), winter dormancy (dormancy II), and the active period (Fig. 3.25a). During each phase, the rate of development is dependent on the air temperature, but the air temperature response varies from phase to phase. Air temperature responses are expressed as accumulation rates of arbitrary developmental units, and for each phase, each genotype has a critical number of units required for completing the phase.

Starting from the beginning of autumn dormancy, the development is first simulated by accumulating chilling units according to the response presented in Fig. 3.13a until a genotype-specific critical value is attained. At this moment, the bud automatically moves on to the phase of winter dormancy. During this phase, the air temperature response of the rate of development is otherwise similar to the air temperature response of the accumulation rate of period units (Fig. 3.7), but the curve saturates at a lower temperature than the PU curve does (response not shown). The arbitrary developmental unit for the phase of winter dormancy is called dormancy unit (DU). When a genotype-specific value of dormancy units has been accumulated, the bud moves automatically to the next phase, i.e., to the active period. During the active period, the air temperature response of the rate of development follows the PU curve (Fig. 3.7). The bud moves on to autumn dormancy, and a new cycle is initiated when the number of accumulated period units attains a genotype-specific critical value.

Sarvas's (1972, 1974) three-phase model for the entire annual cycle (Fig. 3.25a) was the impetus for Koski and Selkäinaho (1982) to introduce the concept of autonomous development (see Sect. 3.5.1). In Sarvas's (1972, 1974) model, no

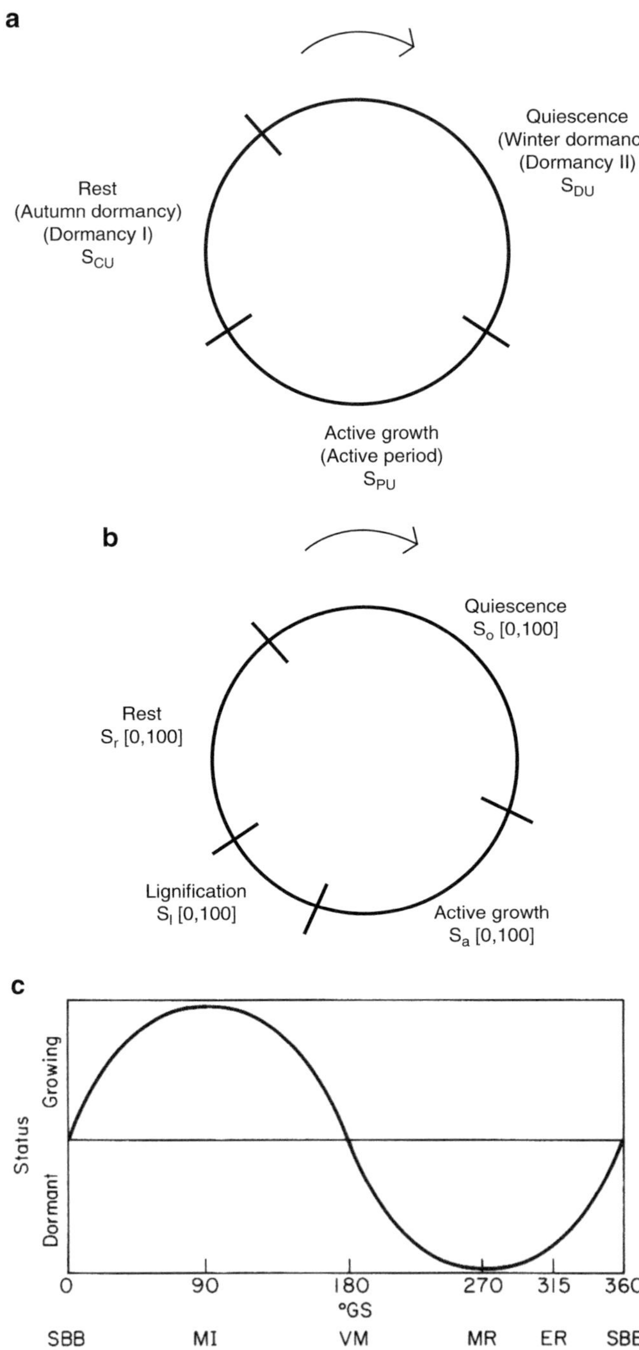

Fig. 3.25 Models for the entire annual phenological cycle of boreal and temperate trees. (**a**) The three-phase model of Sarvas (1972, 1974). The names of the phases in brackets indicate the original concepts used by Sarvas (1972, 1974), and the names without brackets, the corresponding

short-term signals are involved in any phase of the cycle.[32] Rather, the developmental path is otherwise entirely genetically controlled, except that the rate of development is regulated by air temperature according to the three air temperature responses specific to the corresponding three developmental phases. According to this abstract model, the annual phenological (or ontogenetic) cycle of any tree genotype from the boreal or temperate zone can be characterised by the values of the three requirements of arbitrary developmental units. It should be emphasized, however, that according to Sarvas (1972, 1974), this notion is valid only for the strictly defined annual phenological cycle. In other tree attributes, including other seasonal phenomena than the irreversible ontogenetic development, there is more complicated genetic variation.

Together with the models of Richardson et al. (1974), Landsberg (1974), and Cannell and Smith (1983), Sarvas's (1972, 1974) model was a key reference for Hänninen (1990b) in his establishing the generalised modelling framework for growth onset (Fig. 3.15). However, in comparison with Sarvas's (1972, 1974) original model, Hänninen (1990b) made two modifications, which are also followed in the present volume.

In his discussion of various dormancy terms, Sarvas (1974) emphasised that his concept "winter dormancy" (or "dormancy II") is not synonymous with Doorenbos's (1953) concept "imposed dormancy", but in Hänninen's (1990b) interpretation they both are synonymous with the concept "quiescence", which is also adopted in the present volume (Table 3.3 in Sect. 3.3.1.1; Fig. 3.25a). Sarvas (1974) motivated his notion by the fact that microscopic ontogenetic development progresses actively during winter dormancy, even at air temperatures slightly below zero. What is then the "unfavourable environmental condition" that arrests growth

Fig. 3.25 (continued) concepts used in the present volume. The state of development is expressed by accumulated chilling units, S_{CU}, during rest, by accumulated dormancy units, S_{DU}, during quiescence, and by accumulated period units, S_{PU}, during active growth. **(b)** The four-phase model of Kellomäki et al. (1992, 1995), as modified by Hänninen and Kramer (2007). In each of the four phases, the state of development is expressed by a state variable specific to the phase, and all the four state variables are expressed on the percentage scale: S_r state of rest break, S_o state of ontogenetic development, S_a state of active growth, and S_l state of lignification. **c** The Degree Growth Stage (°GS) model (Fuchigami et al. 1982). The state of development during the entire annual cycle is expressed on the °GS scale [0, 360]. The acronyms define the point events as follows: SBB spring bud burst, MI maturity induction point, VM vegetative maturity, MR maximum rest, ER end of rest (Kobayashi and Fuchigami 1983a; redrawn with the permission of Oxford University Press)

[32] Though autumn dormancy belongs to the autonomous theory, it has an effect somewhat similar to the regulation by short-term signals such as night length. After the active period is completed, the development does not proceed further until the 'signal' of air temperatures dropping below +10 °C is received from the environment. This synchronises the development of different tree genotypes that attain autumn dormancy at different times in late summer (Sarvas 1974).

in the case of Doorenbos's (1953) dormancy category "imposed dormancy"? Following this line of reasoning, Sarvas (1974) suggested that the concept "imposed dormancy" be used only when an actively growing tissue stops growing due to unfavourable environmental conditions.

Though there might originally have been a disagreement between the ideas of Doorenbos (1953) and those of Sarvas (1974), the whole-tree modelling approach adopted in the present volume considers both of the concepts "winter dormancy" and "imposed dormancy" to denote the same phenomenon. This phenomenon is referred to as "quiescence" in the present volume: microscopic ontogenetic development takes place within the bud whenever the temperature rises sufficiently, but no bud burst or other visible growth or flowering is seen before the high temperature requirement is met. Thus, as emphasised in Sect. 3.3.1.1, the definition of quiescence "caused by environmental factors unfavourable to growth" refers to the low air temperatures that have prevailed for a prolonged period, for the rising of air temperatures to growth-promoting levels for a short period of time does not cause an onset of visible growth at once. Following these ideas, Hänninen (1990b) gave mathematically exact definitions for the concepts of rest and quiescence (Table 3.4 in Sect. 3.4.1.1).

According to the reasoning of the previous paragraph, it is the air temperature response of the accumulation rate of the dormancy units (DUs) representing dormancy II that should be used in models for quiescence. Hänninen (1990b), however, adopted the air temperature response of the accumulation rate of period units (PUs) (Figs. 3.7 and 3.17), representing the active period in Sarvas's (1972, 1974) model (Fig. 3.25a), for that purpose. There are three reasons for that.

Firstly and most importantly, and similarly to the DU response, the experiments carried out for the PU response addressed microscopic ontogenetic development leading to visible flowering (Table 3.2 in Sect. 3.2.1.3). Designating the period when these phenomena take place as the "active period" contradicts the general practice prevailing in the literature, including the other modelling studies addressed by Hänninen (1990b). According to the general practice, the border between dormancy and the active period is usually defined by the onset of phenomena that are visible to the naked eye. Accordingly, the PU curve (Fig. 3.7) indicates the air temperature response during quiescence rather than during active growth. Secondly, there is only a minor difference between these two responses (see above). Thirdly, Sarvas's (1974) experimental data for the DU curve is quite limited, especially in comparison with Sarvas's (1972) exceptionally strong experimental evidence for the PU curve.

3.6.2 Kellomäki et al.'s (1992, 1995) Model

Sarvas (1972, 1974) deliberately neglected the effects of night length on growth cessation, though these effects were already well established in his time. This was reasonable, because he studied mainly the generative meristems of trees. In order to introduce the effects of night length on growth cessation, Kellomäki et al. (1992, 1995) developed the model further (Fig. 3.25b). Their model simulates the development during the phases of rest and quiescence similarly to Sarvas's (1972, 1974) model, with the slight modifications explained in the previous section. For growth cessation, however, Kellomäki et al. (1992, 1995) adopted the joint effect model (Koski and Selkäinaho 1982; Koski and Sievänen 1985; Fig. 3.23c). Furthermore, they assumed that growth cessation is followed by a phase of lignification requiring exposure to high temperatures, i.e., accumulation of day degrees. When the high temperature requirement of lignification is met, the tree moves on to the phase of rest. Hänninen and Kramer (2007) reformulated the model of Kellomäki et al. (1992, 1995) by introducing the ecophysiological percentage scale for each of the four phases (Fig. 3.25b).

Kellomäki et al.'s (1992, 1995) model of the annual phenological cycle was parameterized with empirical data on several central Finnish tree species (Sarvas 1972, 1974; Koski and Sievänen 1985). Thus, as the model predicts the timing of the four phases of the annual cycle (Fig. 3.25b) for a generalised boreal tree species native to conditions like those in central Finland, it should be regarded as more of a tool for theoretical research and deduction of research hypotheses than a tool for predicting the timing of phenological point events for any given tree species (however, see also Harrington et al. 2010). Examples of this hypothetico-deductive modelling approach, outlined in Sect. 2.1, are given in later chapters.

3.6.3 Fuchigami et al.'s (1982) °GS Model

Taking Sarvas's (1972,1974) model for their starting point, Fuchigami et al. (1982) presented a sophisticated model for the entire annual cycle of temperate deciduous woody plants (Fig. 3.25c). Using the geometrical analogy of the degrees of a circle they established a common scale for the entire annual cycle, which is why their model is called the Degree Growth Stage (°GS) model. Furthermore, in order to visualise the alternation of the phases of growth and dormancy, the model is usually represented by a sine curve (Fig. 3.25c) instead of the circle commonly used in studies with boreal trees (Fig. 3.25a, b; however, see also Fuchigami and Wisniewski 1997).

In the model of Fuchigami et al. (1982), five point events are defined that divide the annual cycle into five segment events (Fig. 3.25c). The first point event is the *spring bud burst*, which is designated numerically as the growth stage $°GS = 0°$. The next point event is the *maturity induction point* at $°GS = 90°$, which denotes the developmental stage where the growing meristems become responsive to the effects of long nights. The first segment event delineated by these two point events is referred to as the *rapid growth phase* ($0° \leq °GS < 90°$). The maturity induction point is followed by the segment event of the *slow-growth and correlative inhibition phase* ($90° \leq °GS < 180°$). During this phase, the growth rate decreases due to the effects of increasing night length, and the bursting of lateral buds is arrested due to correlative inhibition, i.e., due to the hormonal regulation received from the buds or leaves situated at more apical positions. At the point event of *vegetative maturity*

(°GS = 180°), growth stops entirely, and the lateral buds can no more be induced to burst by the experimental removal of the actively growing leaves (Fig. 3.25c).

Vegetative maturity also designates the onset of the first phase of rest, i.e., *the deepening rest phase* ($180° \leq °GS < 270°$). This phase is characterised by an increasing value of the DBB, i.e., an increasing number of days required for bud burst when the rest period is experimentally interrupted by a transfer into high temperature forcing conditions. This phase ends at the point event of *maximum rest* at °GS = 270° (Fig. 3.25c). This point event is followed by the segment event of the *decreasing rest phase* ($270° \leq °GS < 315°$), which is characterised by a decrease of the DBB. When further chilling does not decrease the value of the DBB any longer, the point event *end of rest* (°GS = 315°) is attained. The end of rest is followed by the last segment event of *quiescence*, which ends at bud burst, i.e., with °GS = 360° = 0°.

Within the framework of the conceptual °GS model (Fuchigami et al. 1982; Fig. 3.25c), Kobayashi et al. (1982) and Kobayashi and Fuchigami (1983a, b) presented dynamic simulation models for the effects of air temperature on dormancy development and bud burst in *Cornus sericea*. Using data from a carefully planned set of experiments, they formulated a dynamic simulation model for each phase of dormancy: deepening rest (Kobayashi et al. 1982), decreasing rest (Kobayashi and Fuchigami 1983a), and quiescence (Kobayashi and Fuchigami 1983b). In accordance with the feedback principle introduced in Sect. 2.2, the air temperature response of the rate of development (°GS h^{-1}) was assumed to change during the development, so that besides air temperature, the prevailing value of °GS also appears in the polynomial equations.

Furthermore, Kobayashi et al. (1983) also presented a dynamic frost hardiness model for *Cornus sericea*, likewise within the framework of the conceptual °GS model. Following the principle of integrated models discussed in Sect. 5.3.2, Kobayashi et al.'s (1983) model predicts the rate of hardening or dehardening on the basis of air temperature and the prevailing state of development expressed by the value of the °GS variable.

Unfortunately, the equations presented by Kobayashi et al. (1982, 1983) and Kobayashi and Fuchigami (1983a, b) for *Cornus sericea* are erroneous. This conclusion is unavoidable, for the equations of the four models do not imply meaningful air temperature responses. The model for quiescence (Kobayashi and Fuchigami 1983b), for instance, implies negative values of the rate of development for the entire air temperature range. Contrary to the simulation models developed, however, the empirical results reported in these studies are meaningful and informative. The erroneous modelling is thus especially harmful, as it hampers the use of the empirical results presented; and, as emphasized throughout the present volume, experimental data of that kind are very much needed in modelling the annual cycle of boreal and temperate trees. It goes without saying that the erroneous equations also prevent any comparisons with other models discussed in the present volume. It is to be hoped that the correct equations will still be published despite the long period of time lapsed since the publishing of the original studies.

The problems discussed above should not obscure the merits of the °GS model (Fig. 3.25c), which has formed a framework for experimental studies with the necessity to identify the developmental status of the trees examined. Soon after its introduction, the °GS model became a standard reference in studies of horticultural woody plants, especially among North American scientists (Fuchigami and Nee 1987; Fuchigami and Wisniewski 1997).

3.6.4 Comparison of the Models and Their Further Use as Sub-models

The °GS model (Fig. 3.25c) has a lot in common with the modelling approach adopted in the present volume (Fig. 3.25b). First of all, arbitrary meteorological developmental units are replaced in both approaches with scales explicitly referring to ecophysiological phenomena. Second, the concepts of point event and segment event (Sarvas 1972) are applied in similar ways in both approaches. Third, many of the point events and segment events considered are the same in the two approaches. For instance, the decreasing rest phase is identical to the rest phase of the parallel model (Fig. 3.16b) as formulated by Hänninen (1990b) on the basis of the original models of Landsberg (1974) and Cannell and Smith (1983). Fourth, ecophysiological a priori information is used as a basis of the modelling.

In his general modelling framework, Hänninen (1990b) also addressed the notion of the rest phase consisting of two sub-phases as in the °GS model (Fig. 3.25c). This was readily carried out by assuming that ontogenetic competence first decreases (deepening rest) and then increases (decreasing rest) as a function of the accumulated chilling, in the latter case similarly to the parallel model (Fig. 3.16b). In his literature review, Hänninen (1990b) found no other model besides the °GS model in which the phase of deepening rest was assumed. Such a phase may not occur in evergreen conifers, but a more likely reason for its absence from most of the prevailing models is simply that those models start the simulation of the dynamics of rest at a later phase than the °GS model does (Hänninen and Kramer 2007).

All the three models presented in Fig. 3.25 include the sequential model for the period of dormancy. Accordingly, it is assumed in these models that rest must be completed before ontogenetic development towards growth onset is possible and that at rest completion, full ontogenetic competence is attained abruptly (Fig. 3.16a). This is a simplifying assumption because most of the experimental data show a decreasing value of DBB as a result of increased chilling (Fig. 3.12b; Nienstaedt 1966; Myking and Heide 1995), thus supporting the contradictory concept of gradual rest break described by gradual changes in ontogenetic competence with increased chilling (Figs. 3.16b, c and 3.19d).

Due to the simplifying assumption, a conceptual problem is included in the structure of the otherwise clear and logical °GS model of Fuchigami et al. (1982).

Following the sequential concept, this model also assumes a strict borderline between rest and quiescence at 315 °GS (Fig. 3.25c). Accordingly, no bud burst should be observed before the end of rest, and no changes should be observed in the value of DBB in a re-growth test, except for the slow decrease with the increasing duration of chilling caused by the slow ontogenetic development taking place in the chilling conditions already (Fig. 3.19b). However, as explained in Sect. 3.6.3 above, it is exactly the changes observed in the value of DBB that define the periods of deepening rest and decreasing rest in the °GS model (Fig. 3.25c). This implies that, contrary to the assumption of the sequential model, the seedlings are able in real life to burst buds during the rest period, too. Thus, for instance, if a seedling is moved to high-temperature forcing conditions, say during decreasing rest, at 260 °GS and bud burst is observed in the forcing conditions (as will happen according to the definition of the developmental state 260 °GS), then with reference to the model, the seedling jumps, as it were, directly from 260 °GS to bud burst at 0 °GS. Such jumping is not consistent with the idea of sequential development, which requires that all developmental stages are attained, one by one, during the development.

These problems, caused by the inclusion of the sequential model in the current models for the entire annual phenological cycle (Fig. 3.25), indicate that in order to model the phase of dormancy more realistically, the state of rest should not be included as a sequential part in the annual phenological cycle. Rather, it should be addressed with a separate sub-model, whose output affects the rate of ontogenetic development in the main annual phenological cycle (Hänninen and Kramer 2007).

However, this would complicate the model structure when the model for the annual phenological cycle is used as a sub-model in models of the annual cycle of frost hardiness (Kobayashi et al. 1983; Kellomäki et al. 1992, 1995; Leinonen 1996b). Thus Kellomäki et al.'s (1992, 1995) simplified model (Fig. 3.25b) for the annual phenological cycle is also applied when the annual cycle of frost hardiness is simulated in Chaps. 5 and 8. When the results of these simulations are compared with experimental data, the simulated time of rest completion is best interpreted to correspond to the amount of chilling causing the levelling off of the DBB curve (Figs. 3.12b and 3.19d), even though the adoption of the sequential model as a sub-model of the overall model of Kellomäki et al. (1992, 1995) actually does not predict an exponentially declining DBB curve at all.

3.7 Summary

Phenological point events, such as bud burst and bud set, are visible to the naked eye, so that phenology as such is a well-defined ecophysiological whole-tree attribute of the seasonality of boreal and temperate trees. From the modelling point of view, however, the annual phenological cycle is more problematic than the other whole-tree attributes addressed in Chaps. 4 and 5. This is because the occurrence of the point event predicted by the models is an outcome of invisible developmental phenomena, such as physiological and anatomical changes within the buds, taking place over a prolonged period of time before the occurrence of the point event, and usually there are no observations available on these phenomena. Rather, even though this long-term development is simulated by the time course (s) of one or two state variables, the only test is provided at the end of the simulation by a comparison of the predicted timing of the point event with the observed one. This situation complicates the modelling of the annual phenological cycle in several ways. For instance, evaluating the realism of different competing models is often difficult because models with contrasting ecophysiological assumptions may be equally accurate in tests against field data. For this reason, conceptual clarification of the models and the experimental studies is especially crucial in modelling the annual phenological cycle.

Air temperature is the main environmental factor directly regulating the spring phenology of boreal and temperate trees. The rate of ontogenetic development towards bud burst, growth onset, and flowering increases with rising air temperature, so that the spring phenology of the trees, and of other plants, has traditionally been modelled in terms of temperature sum models. This tradition has mainly been based on data gathered in natural conditions and on statistical model-fitting. For this reason the state variables have been expressed in arbitrary meteorological units, such as day degrees, which do not as such have any direct counterpart in the ecophysiology of the trees.

In this chapter the ecophysiological basis of the temperature sum models was explicated and the original day degree scale of the models was transformed into an ecophysiological percentage scale. The day degree models provide an approximation of the real air temperature response of the rate of development towards the point event modelled. The real response can be readily determined experimentally, but that has been done only rarely. The response determined experimentally by Sarvas (1972) is based on exceptionally comprehensive experimental data, so that it is this response that is mainly referred to in this chapter and elsewhere in the present volume.

Besides the direct regulation of the rate of development towards the springtime phenological events, air temperature also regulates this development indirectly via regulation of the rest status of the buds. After growth cessation and bud set in the autumn, the buds of boreal and temperate trees enter a state of rest, i.e., their bursting and growth are arrested by physiological factors inside the bud. It has been known since the early twentieth century that long-term exposure to chilling temperatures causes rest break, i.e., the removal of the growth-arresting conditions in the buds. Ecophysiological models for the overall regulation of spring phenology by air temperature, i.e., both direct and indirect regulation, have been presented since the 1970s. In this chapter these models are discussed and compared, mainly within a generalised framework, in which each overall model is presented as the synthesis of three sub-models, each addressing its specific ecophysiological phenomenon. The framework facilitates not only the comparison and explication of previously published models but also the development of new ones.

Besides chilling temperatures, night length also has a role in the rest break of the buds. It has been shown experimentally that short nights compensate for lack of chilling in several tree species. In most cases this phenomenon is evidently not ecologically important, for sufficient chilling has usually been accumulated by the spring, when night length decreases to the values causing rest break. However, in very mild climates there might be insufficient chilling, so that in these cases the compensating effect caused by short nights guarantees a timely rest break and subsequent growth onset.

It has also been suggested that in many tree species the effects of night length on rest break are not restricted to compensating for the lack of chilling, so that depending on the species, rest break is regulated by complicated interactions of chilling and night length. In addition to night length, other light-related factors, such as the spectral composition of light, might also have effects on the rest break. However, while these effects of night length and the other light-related factors cannot be ruled out, the experimental evidence for them remains uncertain and in many cases even contradictory.

According to most experimental studies, the chilling requirement of rest completion is met relatively early in autumn. Thus development towards growth onset should commence whenever the air temperature rises to a level promoting ontogenetic development, i.e., approximately above zero, during mild periods in winter. However, model tests carried out with long-term historical sets of phenological and air temperature data have not supported this notion. In these tests, simple models assuming rest completion to take place on a fixed calendar day relatively late in spring have consistently outperformed more complicated models, in which rest break is caused by the accumulation of chilling. Furthermore, some experimental studies, carried out mainly with adult trees, have also arrived at similar conclusions about a relatively late rest completion in spring. The paradox between the traditional concept emphasising the role of chilling and a relatively early rest completion on the one hand and the recent contradictory results on the other hand was discussed in this chapter.

About half a century ago, Vegis (1964) presented a generalised theory of dormancy phenomena in plants. According to Vegis's (1964) theory, changes in the dormancy status of the plant are expressed in the plant's growth-promoting air temperature range, which gets narrower as the dormancy gets deeper and correspondingly wider as the dormancy is gradually released. With evidently only one recent exception, this concept is not included in the present dynamic models, in which the accumulation of chilling and rest break increase the subsequent rate of development towards growth onset at any temperature in the growth-promoting range without widening the range.

Vegis's (1964) theory has been almost completely forgotten in modelling studies addressing the annual phenological cycle of boreal and temperate trees. In this chapter, Vegis's (1964) theory is revisited as a potential explanation for several earlier results, including the paradox in the timing of rest completion discussed in the previous paragraph. The version of Vegis's (1964) theory relevant for boreal and temperate trees was formulated and extended within the overall modelling framework applied in the present volume. It is strongly suggested that Vegis's

(1964) theory should be tested experimentally with several boreal and temperate tree species and provenances in the future.

It has been known since the early twentieth century that night length is a major environmental factor regulating the cessation of height growth and the induction of rest in most boreal and temperate trees. According to the view prevailing in much of the literature, growth cessation occurs when night length attains a critical genotype-specific value in late summer. However, though night length is the primary environmental factor triggering the physiological reactions leading to growth cessation in most boreal and temperate trees, air temperature also plays a crucial role in several of these tree species. In some species, height growth cessation is regulated solely by air temperature, independently of night length. Furthermore, several interactive effects of air temperature have also been found in species in which night length regulates growth cessation. High night temperatures accelerate growth cessation in many such species. Furthermore, in many species growth cessation is accelerated by increased temperature sum accumulation during the growing season. In these cases the critical night length triggering growth cessation decreases with increasing temperature sum, and it is for this reason that the concept of joint effect regulation has been attached to this mode of regulating growth cessation.

Though the number of empirical studies addressing height growth cessation is probably at least as great as that of those addressing growth onset and bud burst, growth cessation has been addressed in modelling studies much less frequently than growth onset. Furthermore, as the few modelling studies addressing growth cessation are quite straightforward, they do not require such conceptual clarification as the modelling studies addressing growth onset do. For these reasons, growth cessation was not addressed at the same length in this chapter as growth onset was. It should be emphasised, however, that this procedural decision does not imply any notion of less further work being needed in modelling the cessation than the onset of growth.

Models for the entire annual phenological cycle in boreal and temperate trees have been presented since the 1970s. These models also address the rest break of the buds, though this physiological phenomenon has no visible manifestation until growth onset or bud burst. The most advanced models for the entire annual phenological cycle address quite a few of the ecophysiological phenomena discussed in this chapter, while others, such as the quantitative induction of dormancy depending on the air temperatures during the induction, remain unaddressed. Furthermore, these models often present generalised rather than species-specific responses. In all, then, the models for the entire annual phenological cycle serve best as a framework for designing further experimental work. Though it goes without saying that a model with absolute realism is beyond the reach of researchers, experimental work with different tree species and even provenances and cultivars is needed to develop the models towards better realism.

List of Symbols

t time

Time-Dependent Environmental Variables

$T(t)$	air temperature (°C)
$T_{mean}(t)$	daily mean air temperature (°C)
$NL(t)$	daily night length (h)

Time-Dependent Rate and State Variables

Ontogenetic Development

$R_{o,pot}(t)$	potential rate of ontogenetic development (% day^{-1})
$R_o(t)$	rate of ontogenetic development (% day^{-1})
$S_o(t)$	state of ontogenetic development (%)
R_{dh}	accumulation rate of degree hour units (dh h^{-1})
$R_{dd}(t)$	accumulation rate of day degree units (dd day^{-1})
$R_{HU}(t)$	accumulation rate of high temperature units (HU day^{-1})
$R_{PU}(t)$	accumulation rate of period units (PU h^{-1})
$R_{FU}(t)$	accumulation rate of forcing units (FU day^{-1})
$R_{FU'}(t)$	accumulation rate of modified forcing units (FU′ day^{-1})
$S_{dd}(t)$	accumulated temperature sum (dd)
$S_{PU}(t)$	accumulated period units (PU)
$S_{DU}(t)$	accumulated dormancy units (DU)
$S_{FU}(t)$	accumulated forcing units (FU)
$S_{FU'}(t)$	accumulated modified forcing units (FU′)

Rest Break

$R_r(t)$	rate of rest break (% day^{-1})
$S_r(t)$	state of rest break (%)
$R_{CU}(t)$	accumulation rate of chilling units (CU day^{-1}, CU h^{-1})
$S_{CU}(t)$	accumulated chilling units

Development During Growing Season

S_a	state of active growth (%)
S_l	state of lignification (%)

Variable Mediating the Effect of Rest Break on the Ontogenetic Development

$C_o(t)$ ontogenetic competence

Model Parameters

Ontogenetic Development

t_0	starting date of the simulations
T_{thr}	air temperature threshold of ontogenetic development (and of accumulation of forcing units)
FU_{crit}	Critical forcing unit sum (FU)
H_{crit}	high temperature requirement of growth onset (HU)
a	steepness of air temperature response of rate of ontogenetic development (and of accumulation rate of forcing units) (°C^{-1})
b	inflexion point of air temperature response of rate of ontogenetic development (and of accumulation rate of forcing units) (°C)
c	Upper asymptote of air temperature response of accumulation rate of forcing units (FU)
NL_{crit}	critical night length of growth cessation (h)

Rest Break

C_{crit}	chilling requirement of rest break (CU)
T_1	minimum air temperature for rest break (°C)
T_2	temperature for maximum rate of rest break (°C)
T_3	maximum air temperature for rest break (°C)

Other Symbols

Ontogenetic Development

dd	day degree unit
dh	degree hour unit
PU	period unit
FU	forcing unit
FU'	modified forcing unit
HU	high temperature unit (general term covering all specific units)
$\Delta t(T)$	developmental time (time required for a given point event to occur in experimental conditions) in temperature T
Δt_{10}	developmental time in the reference temperature 10 °C

Rest Break

CU chilling unit

Variables for a Regrowth Test

BB% bud burst percentage (%)
DBB days required for bud burst (day)

References

Arnold, C. Y. (1959). The determination and significance of the base temperature in a linear heat unit system. *Proceedings of the American Society of Horticultural Science, 74*, 430–445.

Arora, R., Rowland, L. J., & Tanino, K. (2003). Induction and release of bud dormancy in woody perennials: A science comes of age. *HortScience, 38*, 911–921.

Bakkenes, M., Alkemade, J. R. M., Ihle, F., Leemans, R., & Latour, J. B. (2002). Assessing effects of forecasted climate change on the diversity and distribution of European higher plants for 2050. *Global Change Biology, 8*, 390–407.

Basler, D., & Körner, C. (2012). Photoperiod sensitivity of bud burst in 14 temperate forest tree species. *Agricultural and Forest Meteorology, 165*, 73–81.

Blümel, K., & Chmielewski, F.-M. (2012). Shortcomings of classical phenological forcing models and a way to overcome them. *Agricultural and Forest Meteorology, 164*, 10–19.

Boyer, W. D. (1973). Air temperature, heat sums, and pollen shedding phenology of longleaf pine. *Ecology, 54*, 420–426.

Brunner, A., Evans, L. M., Hsu, C.-Y., & Sheng, X. (2014). Vernalization and the chilling requirement to exit bud dormancy: Shared or separate regulation? *Frontiers in Plant Science, 5, Article 732*. doi:10.3389/fpls.2014.00732.

Caffarra, A. (2007). *Quantifying the environmental drivers of tree phenology*. Thesis submitted for the degree of Doctor of Philosophy, School of Natural Sciences, Trinity College, University of Dublin.

Caffarra, A., & Donnelly, A. (2011). The ecological significance of phenology in four different tree species: Effects of light and temperature on bud burst. *International Journal of Biometeorology, 55*, 711–721.

Caffarra, A., Donnelly, A., Chuine, I., & Jones, M. B. (2011a). Modelling the timing of *Betula pubescens* budburst. I. Temperature and photoperiod: A conceptual model. *Climate Research, 46*, 147–157.

Caffarra, A., Donnelly, A., & Chuine, I. (2011b). Modelling the timing of *Betula pubescens* budburst. II. Integrating complex effects of photoperiod into process-based models. *Climate Research, 46*, 159–170.

Campbell, R. K. (1974). Use of phenology for examining provenance transfers in reforestation of Douglas-fir. *Journal of Applied Ecology, 11*, 1069–1080.

Campbell, R. K. (1978). Regulation of bud-burst timing by temperature and photoregime during dormancy. In C. A. Hollis & A. E. Squillace (Eds.), *Proceedings: Fifth North American forest biology workshop* (pp. 19–33). Gainesville: University of Florida, Forestry Department.

Campbell, R. K., & Sugano, A. I. (1975). Phenology of bud burst in Douglas-fir related to provenance, photoperiod, chilling, and flushing temperature. *Botanical Gazette, 136*, 290–298.

Cannell, M. G. R. (1985). Analysis of risks of frost damage to forest trees in Britain. In P. M. A. Tigerstedt, P. Puttonen, & V. Koski (Eds.), *Crop physiology of forest trees* (pp. 153–166). Helsinki: Helsinki University Press.

Cannell, M. G. R. (1989). Chilling, thermal time and the date of flowering of trees. In C. J. Wright (Ed.), *Manipulation of fruiting* (pp. 99–113). London: Butterworths.

Cannell, M. G. R., & Smith, R. I. (1983). Thermal time, chill days and prediction of budburst in *Picea sitchensis*. *Journal of Applied Ecology, 20*, 951–963.

Cannell, M. G. R., & Smith, R. I. (1986). Climatic warming, spring budburst and frost damage on trees. *Journal of Applied Ecology, 23*, 177–191.

Cannell, M. G. R., Murray, M. B., & Sheppard, L. J. (1985). Frost avoidance by selection for late budburst in *Picea sitchensis*. *Journal of Applied Ecology, 22*, 931–941.

Chuine, I. (2000). A unified model for budburst of trees. *Journal of Theoretical Biology, 207*, 337–347.

Chuine, I., Cour, P., & Rousseau, D. D. (1998). Fitting models predicting dates of flowering of temperate-zone trees using simulated annealing. *Plant, Cell and Environment, 21*, 455–466.

Chuine, I., Cour, P., & Rousseau, D. D. (1999). Selecting models to predict the timing of flowering of temperate trees: Implications for tree phenology modelling. *Plant, Cell and Environment, 22*, 1–13.

Chuine, I., Morin, X., & Bugmann, H. (2010). Warming, photoperiods, and tree phenology. *Science, 329*, 277–278.

Chuine, I., de Cortazar-Atauri, I. G., Kramer, K., & Hänninen, H. (2013). Plant development models. In M. D. Schwartz (Ed.), *Phenology: An integrative environmental science* (2nd ed., pp. 275–293). Dordrecht: Springer Science+Business Media B.V.

Chung, M. (1981). Flowering characteristics of Pinus sylvestris L. with special emphasis on the reproductive adaptation to local temperature factor. *Acta Forestalia Fennica, 169*, 1–69.

Clark, J. S., Salk, C., Melillo, J., & Mohan, J. (2014). Tree phenology responses to winter chilling, spring warming, at north and south range limits. *Functional Ecology, 28*, 1344–1355.

Cooke, J. E. K., Eriksson, M. E., & Junttila, O. (2012). The dynamic nature of bud dormancy in trees: Environmental control and molecular mechanisms. *Plant, Cell and Environment, 35*, 1707–1728.

Coville, F. V. (1920). The influence of cold in stimulating the growth of plants. *Journal of Agricultural Research, 20*, 151–160.

Delbart, N., & Picard, G. (2007). Modeling the date of leaf appearance in low-arctic tundra. *Global Change Biology, 13*, 2551–2562.

Delpierre, N., Dufrê, E., Soudani, K., Ulrich, E., Cecchini, S., Boé, J., & François, S. (2009). Modelling interannual and spatial variability of leaf senescence for three deciduous tree species in France. *Agricultural and Forest Meteorology, 149*, 938–948.

Delpierre, N., Vitasse, Y., Chuine, I., Guillemot, J., Bazot, S., Rutishauser, T., & Rathgeber, C. B. K. (2015). Temperate and boreal forest tree phenology: From organ-scale processes to terrestrial ecosystem models. *Annals of Forest Science, 73*, 5–25. doi:10.1007/s13595-015-0477-6.

Dennis, F. G., Jr. (1994). Dormancy – What we know (and don't know). *HortScience, 29*, 1249–1255.

Doorenbos, J. (1953). Review of the literature on dormancy in buds of woody plants. *Mededelingen van de Landbouwhogeschool te Wageningen/Nederland, 53*. 23 p.

Dormling, I. (1989). The role of photoperiod and temperature in the induction and release of dormancy in *Pinus sylvestris* L. seedlings. *Annals of Forest Science, 46*(suppl), 228–232.

Dormling, I., Gustafsson, Å., & von Wettstein, D. (1968). The experimental control of the life cycle in Picea abies (L.) Karst. I. Some basic experiments on the vegetative cycle. *Silvae Genetica, 17*, 44–64.

Ekberg, I., Eriksson, G., & Dormling, I. (1979). Photoperiodic reactions in conifer species. *Holarctic Ecology, 2*, 255–263.

Emmingham, W. H. (1977). Comparison of selected Douglas-fir seed sources for cambial and leader growth patterns in four western Oregon environments. *Canadian Journal of Forest Research, 7*, 154–164.

Erez, A., & Couvillon, G. A. (1987). Characterization of the influence of moderate temperatures on rest completion in peach. *Journal of the American Society for Horticultural Science, 112*, 677–680.

Erez, A., & Lavee, S. (1971). The effect of climatic conditions on dormancy development of peach buds. I. Temperature. *Journal of the American Society for Horticultural Science, 96*, 711–714.

Erez, A., Couvillon, G. A., & Hendershott, C. H. (1979a). Quantitative chilling enhancement and negation in peach buds by high temperatures in a daily cycle. *Journal of the American Society for Horticultural Science, 104*, 536–540.

Erez, A., Couvillon, G. A., & Hendershott, C. H. (1979b). The effect of cycle length on chilling negation by high temperatures in dormant peach leaf buds. *Journal of the American Society for Horticultural Science, 104*, 573–576.

Falusi, M., & Calamassi, R. (1990). Bud dormancy in beech (*Fagus sylvatica* L.). Effect of chilling and photoperiod on dormancy release of beech seedlings. *Tree Physiology, 6*, 429–438.

Farmer, R. E., Jr. (1968). Sweetgum dormancy release: Effects of chilling, photoperiod, and genotype. *Physiologia Plantarum, 21*, 1241–1248.

Fisher, J. I., Richardson, A. D., & Mustard, J. F. (2007). Phenology model from surface meteorology does not capture satellite-based greenup estimations. *Global Change Biology, 13*, 707–721.

Fishman, S., Erez, A., & Couvillon, G. A. (1987a). The temperature dependence of dormancy breaking in plants: Mathematical analysis of a two-step model involving a cooperative transition. *Journal of Theoretical Biology, 124*, 473–483.

Fishman, S., Erez, A., & Couvillon, G. A. (1987b). The temperature dependence of dormancy breaking in plants: Computer simulation of processes studied under controlled temperatures. *Journal of Theoretical Biology, 126*, 309–321.

Fu, Y. H., Campioli, M., Van Oijen, M., Deckmyn, G., & Janssens, I. A. (2012). Bayesian comparison of six different temperature-based budburst models for four temperate tree species. *Ecological Modelling, 230*, 92–100.

Fu, Y. S. H., Campioli, M., Vitasse, Y., De Boeck, H. J., Van den Berge, J., AbdElgawad, H., Asard, H., Piao, S., Deckmyn, G., & Janssens, I. A. (2014). Variation in leaf flushing date influences autumnal senescence and next year's flushing date in two temperate tree species. *PNAS, 111*, 7355–7360.

Fu, Y. H., Piao, S., Vitasse, Y., Zhao, H., De Boeck, H. J., Liu, Q., Yang, H., Weber, U., Hänninen, H., & Janssens, I. A. (2015). Increased heat requirement for leaf flushing in temperate woody species over 1980–2012: Effects of chilling, precipitation and insolation. *Global Change Biology, 21*, 2687–2697.

Fuchigami, L. H., & Nee, C.-C. (1987). Degree growth stage model and rest-breaking mechanisms in temperate woody perennials. *HortScience, 22*, 836–845.

Fuchigami, L. H., & Wisniewski, M. (1997). Quantifying bud dormancy: Physiological approaches. *HortScience, 32*, 618–623.

Fuchigami, L. H., Weiser, C. J., Kobayashi, K., Timmis, R., & Gusta, L. V. (1982). A degree growth stage (°GS) model and cold acclimation in temperate woody plants. In P. H. Li & A. Sakai (Eds.), *Plant cold hardiness and freezing stress. Mechanisms and crop implications* (Vol. 2, pp. 93–116). New York: Academic Press.

Garner, W. W., & Allard, H. A. (1923). Further studies in photoperiodism, the response of the plant to relative length of day and night. *Journal of Agricultural Research, 23*, 871–920.

Gilreath, P. R., & Buchanan, D. W. (1981). Rest prediction model for low-chilling 'Sungold' nectarine. *Journal of the American Society of Horticultural Science, 106*, 426–429.

Grace, J., Allen, S. J., & Wilson, C. (1989). Climate and the meristem temperatures of plant communities near the tree-line. *Oecologia, 79*, 198–204.

Granhus, A., Fløistad, I. S., & Søgaard, G. (2009). Bud burst timing in *Picea abies* seedlings as affected by temperature during dormancy induction and mild spells during chilling. *Tree Physiology, 29*, 497–503.

Håbjørg, A. (1972). Effects of photoperiod and temperature on growth and development of three latitudinal and three altitudinal populations of *Betula pubescens* Ehrh. *Scientific Reports of the Agricultural University of Norway, 51*(2). 27 p.

Häkkinen, R. (1999a). Analysis of bud-development theories based on long-term phenological and air temperature time series: Application to *Betula* sp. leaves. *Finnish Forest Research Institute, Research Papers, 754*. 59 p.

Häkkinen, R. (1999b). Statistical evaluation of bud development theories: Application to bud burst of *Betula pendula* leaves. *Tree Physiology, 19,* 613–618.

Häkkinen, R., Linkosalo, T., & Hari, P. (1995). Methods for combining phenological time series: Application to bud burst in birch (*Betula pendula*) in central Finland for the period 1896–1955. *Tree Physiology, 15*, 721–726.

Häkkinen, R., Linkosalo, T., & Hari, P. (1998). Effects of dormancy and environmental factors on timing of bud burst in *Betula pendula. Tree Physiology, 18*, 707–712.

Hammond, M. W., & Seeley, S. D. (1978). Spring bud development of *Malus* and *Prunus* species in relation to soil temperature. *Journal of the American Society for Horticultural Science, 103*, 655–657.

Hannerz, M. (1999). Evaluation of temperature models for predicting bud burst in Norway spruce. *Canadian Journal of Forest Research, 29*, 1–11.

Hannerz, M., Ekberg, I., & Norell, L. (2003). Variation in chilling requirements for completing bud rest between provenances of Norway spruce. *Silvae Genetica, 52*, 161–168.

Hänninen, H. (1984). A method for identifying the flowering season from aerobiological data. In S. Nilsson & B. Raj (Eds.), *Nordic aerobiology. 5th Nordic symposium on aerobiology, Abisko 1983* (pp. 51–57). Stockholm: Almqvist & Wiksell International.

Hänninen, H. (1987). Effects of temperature on dormancy release in woody plants: Implications of prevailing models. *Silva Fennica, 21*, 279–299.

Hänninen, H. (1990a). Modelling the annual growth rhythm of trees: conceptual, experimental, and applied aspects. In: H. Jozefek (Ed.) Modelling to understand forest functions. *Silva Carelica, 15*, 35–45

Hänninen, H. (1990b). Modelling bud dormancy release in trees from cool and temperate regions. *Acta Forestalia Fennica, 213*, 1–47.

Hänninen, H. (1991). Does climatic warming increase the risk of frost damage in northern trees? *Plant, Cell and Environment, 14*, 449–454.

Hänninen, H. (1995a). Effects of climatic change on trees from cool and temperate regions: An ecophysiological approach to modelling of bud burst phenology. *Canadian Journal of Botany, 73*, 183–199.

Hänninen, H. (1995b). Assessing ecological implications of climatic change: Can we rely on our simulation models? *Climatic Change, 31*, 1–4.

Hänninen, H. (1996). Effects of climatic warming on northern trees: Testing the frost damage hypothesis with meteorological data from provenance transfer experiments. *Scandinavian Journal of Forest Research, 11*, 17–25.

Hänninen, H. (2009). Sarvaksen vuosisyklimalli. In S. Ruotsalainen & J. Häggman (Eds.), *Kutsumuksena metsätiede. Risto Sarvaksen 100-vuotisjuhlakirja* (pp. 53–60). Jyväskylä: Gummerus kirjapaino Oy.

Hänninen, H., & Hari, P. (1996). The implications of geographical variation in climate for differentiation of bud dormancy ecotypes in Scots pine. In P. Hari, J. Ross, M. Mecke (Eds.), Production process of Scots pine; Geographical variation and models. *Acta Forestalia Fennica, 254*, 11–21.

Hänninen, H., & Kramer, K. (2007). A framework for modelling the annual cycle of trees in boreal and temperate regions. *Silva Fennica, 41*, 167–205.

Hänninen, H., & Lundell, R. (2007). Dynamic models in plant ecophysiology. In K. Taulavuori & E. Taulavuori (Eds.), *Physiology of northern plants under changing environment* (pp. 157–175). Trivandrum, Kerala: Research Signpost.

Hänninen, H., & Pelkonen, P. (1988). Effects of temperature on dormancy release in Norway spruce and Scots pine seedlings. *Silva Fennica, 22*, 241–248.

Hänninen, H., & Pelkonen, P. (1989). Dormancy release in *Pinus sylvestris* L. and *Picea abies* (L.) Karst. seedlings: Effects of intermittent warm periods during chilling. *Trees, 3*, 179–184.

Hänninen, H., & Tanino, K. (2011). Tree seasonality in a warming climate. *Trends in Plant Science, 16*, 412–416.

Hänninen, H., Slaney, M., & Linder, S. (2007). Dormancy release of Norway spruce under climatic warming: Testing ecophysiological models of bud burst with a whole-tree chamber experiment. *Tree Physiology, 27*, 291–300.

Hänninen, H., Lundell, R., & Junttila, O. (2013). A fifty-year-old conceptual plant dormancy model provides new insights into dynamic phenology modelling. In: R. Sievänen, E. Nikinmaa, C. Godin, A. Lintunen, & P. Nygren *Proceedings of the 7th international conference on functional-structural plant models, Saariselkä, Finland, 9–14 June 2013*. http://www.metal.fi/fspm2013/proceedings. ISBN 978-951-651-408-9. p. 176.

Harding, P. H., Cochrane, J., & Smith, L. P. (1976). Forecasting the flowering stages of apple varieties in Kent, England, by the use of meteorological data. *Agricultural Meteorology, 17*, 49–54.

Hari, P. (1972). Physiological stage of development in biological models of growth and maturation. *Annales Botanici Fennici, 9*, 107–115.

Hari, P., & Häkkinen, R. (1991). The utilization of old phenological time series of budburst to compare models describing annual cycles of plants. *Tree Physiology, 8*, 281–287.

Hari, P., Leikola, M., & Räsänen, P. (1970). A dynamic model of the daily height increment of plants. *Annales Botanici Fennici, 7*, 375–378.

Harrington, C. A., Gould, P. J., & St. Clair, J. B. (2010). Modeling the effects of winter environment on dormancy release of Douglas-fir. *Forest Ecology and Management, 259*, 798–808.

Heide, O. M. (1974). Growth and dormancy in Norway spruce ecotypes (*Picea abies*) I. Interaction of photoperiod and temperature. *Physiologia Plantarum, 30*, 1–12.

Heide, O. M. (1993a). Daylength and thermal time responses of budburst during dormancy release in some northern deciduous trees. *Physiologia Plantarum, 88*, 531–540.

Heide, O. M. (1993b). Dormancy release in beech buds (*Fagus sylvatica*) requires both chilling and long days. *Physiologia Plantarum, 89*, 187–191.

Heide, O. M. (2003). High autumn temperature delays spring bud burst in boreal trees, counterbalancing the effect of climatic warning. *Tree Physiology, 23*, 931–936.

Heide, O. M. (2011). Temperature rather than photoperiod controls growth cessation and dormancy in *Sorbus* species. *Journal of Experimental Botany, 62*, 5397–5404.

Heide, O. M., & Prestrud, A. K. (2005). Low temperature, but not photoperiod, controls growth cessation and dormancy induction and release in apple and pear. *Tree Physiology, 25*, 109–114.

Heikinheimo, M., & Lappalainen, H. (1997). Dependence of the flower bud burst of some plant taxa in Finland on effective temperature sum: Implications for climate warming. *Annales Botanici Fennici, 34*, 229–243.

Horvath, D. P., Anderson, J. V., Chao, W. S., & Foley, M. E. (2003). Knowing when to grow: Signals regulating bud dormancy. *Trends in Plant Science, 8*, 534–540.

Howe, G. T., Hackett, W. P., Furnier, G. R., & Klevorn, R. E. (1995). Photoperiodic responses of a northern and southern ecotype of black cottonwood. *Physiologia Plantarum, 93*, 695–708.

Höyhtyä, R., & Hänninen, H. (1991). Effect of photon flux density on bud dormancy release in Norway spruce seedlings. *Silva Fennica, 25*, 177–180.

Huikari, O., & Paarlahti, K. (1967). Results of field experiments on the ecology of pine, spruce, and birch. *Communicationes Instituti Forestalis Fenniae, 64*(1), 1–135.

Hunter, A. F., & Lechowicz, M. J. (1992). Predicting the timing of budburst in temperate trees. *Journal of Applied Ecology, 29*, 597–604.

Jones, H. G., Gordon, S. L., & Brennan, R. M. (2015). Chilling requirement of Ribes cultivars. *Frontiers in Plant Science, Article 5*, 767. doi:10.3389/fpls.2014.00767.

Junttila, O. (1970). Effects of stratification, gibberellic acid and germination temperature on the germination of *Betula nana*. *Physiologia Plantarum, 23*, 425–433.

Junttila, O. (1976). Seed germination and viability in five Salix species. *Astarte, 9*, 19–24.

Junttila, O. (1980). Effect of photoperiod and temperature on apical growth cessation in two ecotypes of Salix and Betula. *Physiologia Plantarum, 48*, 347–352.

Junttila, O. (2007). Regulation of annual shoot growth cycle in northern tree species. In K. Taulavuori & E. Taulavuori (Eds.), *Physiology of northern plants under changing environment* (pp. 177–210). Trivandrum, Kerala: Research Signpost.

Junttila, O., & Hänninen, H. (2012). The minimum temperature for budburst in *Betula* depends on the state of dormancy. *Tree Physiology, 32*, 337–345.

Junttila, O., & Nilsen, J. (1993). Growth and development of northern forest trees as affected by temperature and light. In J. N. Alden, S. Ødum, & J. L. Mastrantonio (Eds.), *Forest development in cold climates* (pp. 43–57). New York: Plenum Press.

Junttila, O., & Skaret, G. (1990). Growth and survival of seedlings of various *Picea* species under northern climatic conditions. *Scandinavian Journal of Forest Science, 5*, 69–81.

Junttila, O., Nilsen, J., & Igeland, B. (2003). Effect of temperature on the induction of bud dormancy in ecotypes of *Betula pubescens* and *Betula pendula*. *Scandinavian Journal of Forest Research, 18*, 208–217.

Kalcsits, L. A., Silim, S., & Tanino, K. (2009). Warm temperature accelerates short photoperiod induced growth cessation and dormancy induction in hybrid poplar (*Populus* x spp.). *Trees, 23*, 971–979.

Kalliola, R. (1973). *Suomen kasvimaantiede*. Porvoo: Werner Söderström Osakeyhtiön kirjapaino. 308 p.

Kanninen, M., Hari, P., & Kellomäki, S. (1982). A dynamic model for above-ground growth and dry matter production in a forest community. *Journal of Applied Ecology, 19*, 465–476.

Kasap, O. E., & Alten, B. (2005). Laboratory estimation of degree-day developmental requirements of Phlebotomus papatasi (Diptera: Psychodidae). *Journal of Vector Ecology, 30*, 328–333.

Keenan, T. F., & Richardson, A. D. (2015). The timing of autumn senescence is affected by the timing of spring phenology: Implications for predictive models. *Global Change Biology, 21*, 2634–2641.

Kellomäki, S., Väisänen, H., Hänninen, H., Kolström, T., Lauhanen, R., Mattila, U., & Pajari, B. (1992). A simulation model for the succession of the boreal forest ecosystem. *Silva Fennica, 26*, 1–18.

Kellomäki, S., Hänninen, H., & Kolström, M. (1995). Computations on frost damage to Scots pine under climatic warming in boreal conditions. *Ecological Applications, 5*, 42–52.

Kilpeläinen, A., Peltola, H., Rouvinen, I., & Kellomäki, S. (2006). Dynamics of daily height growth in Scots pine trees at elevated temperature and CO_2. *Trees, 20*, 16–27.

Kobayashi, K. D., & Fuchigami, L. H. (1983a). Modelling temperature effects in breaking rest in red-osier dogwood (*Cornus sericea* L.). *Annals of Botany, 52*, 205–215.

Kobayashi, K. D., & Fuchigami, L. H. (1983b). Modeling bud development during the quiescent phase in red-osier dogwood (*Cornus sericea* L.). *Agricultural Meteorology, 28*, 75–84.

Kobayashi, K. D., Fuchigami, L. H., & English, M. J. (1982). Modeling temperature requirements for rest development in *Cornus sericea*. *Journal of the American Society for Horticultural Science, 107*, 914–918.

Kobayashi, K. D., Fuchigami, L. H., & Weiser, C. J. (1983). Modeling cold hardiness of red-osier dogwood. *Journal of the American Society for Horticultural Science, 108*, 376–381.

Koch, E., Donelly, A., Lipa, W., Menzel, A., & Nekovář, J. (2009). *Final scientific report of COST 725. Establishing a European dataplatform for climatological applications*. Luxembourg: Publications Office of the European Union. 85 p.

Körner, C., & Basler, D. (2010a). Phenology under global warming. *Science, 327*, 1461–1462.

Körner, C., & Basler, D. (2010b). Response. *Science, 329*, 278.

Koski, V. (2009). Siemensadosta vuosisyklimalliin. In S. Ruotsalainen & J. Häggman (Eds.), *Kutsumuksena metsätiede. Risto Sarvaksen 100-vuotisjuhlakirja* (pp. 43–52). Jyväskylä: Gummerus kirjapaino Oy.

Koski, V., & Selkäinaho, J. (1982). Experiments on the joint effect of heat sum and photoperiod on seedlings of Betula pendula. *Communicationes Instituti Forestalis Fenniae, 105*, 1–34.

Koski, V., & Sievänen, R. (1985). Timing of growth cessation in relation to the variations in the growing season. In P. M. A. Tigerstedt, P. Puttonen, & V. Koski (Eds.), *Crop physiology of forest trees* (pp. 167–193). Helsinki: Helsinki University Press.

Kramer, K. (1994a). Selecting a model to predict the onset of growth of *Fagus sylvatica*. *Journal of Applied Ecology, 31*, 172–181.
Kramer, K. (1994b). A modelling analysis of the effects of climatic warming on the probability of spring frost damage to tree species in The Netherlands and Germany. *Plant, Cell and Environment, 17*, 367–377.
Kramer, K., Leinonen, I., Bartelink, H. H., Berbigier, P., Borghetti, M., Bernhofer, C., Cienciala, E., Dolman, A. J., Froer, O., Gracia, C. A., Granier, A., Grünwald, T., Hari, P., Jans, W., Kellomäki, S., Loustau, D., Magnani, F., Markkanen, T., Matteucci, G., Mohren, G. M. J., Moors, E., Nissinen, A., Peltola, H., Sabaté, S., Sanchez, A., Sontag, M., Valentini, R., & Vesala, T. (2002). Evaluation of six process-based forest growth models using eddy-covariance measurements of CO_2 and H_2O fluxes at six forest sites in Europe. *Global Change Biology, 8*, 213–230.
Kubin, E., Kotilainen, E., Poikolainen, J., Hokkanen, T., Nevalainen, S., Pouttu, A., Karhu, J., & Pasanen, J. (2007). *Monitoring instructions of the Finnish National Phenological Network*. Muhos: Finnish Forest Research Institute, Muhos Research Unit. 43 p.
Lana, E. P., & Haber, E. S. (1952). Seasonal variability as indicated by cumulative degree hours with sweet corn. *Proceedings of the American Society for Horticultural Science, 59*, 389–392.
Landsberg, J. J. (1974). Apple fruit bud development and growth; analysis and an empirical model. *Annals of Botany, 38*, 1013–1023.
Landsberg, J. J. (1977). Effects of weather on plant development. In J. J. Landsberg & C. V. Cutting (Eds.), *Environmental effects on crop physiology* (pp. 289–307). London: Academic Press.
Lang, G. A., Early, J. D., Martin, G. C., & Darnell, R. L. (1987). Endo-, para-, and ecodormancy: Physiological terminology and classification for dormancy research. *HortScience, 22*, 371–377.
Lappalainen, H., & Heikinheimo, M. (1992). Relations between climate and plant phenology. In *Part 1. Survey of plant phenological observations in Finland from 1896 to 1965*. Helsinki: Meteorological Publications 20. Finnish Meteorological Institute. 74 p.
Larcher, W. (2003). *Physiological plant ecology. Ecophysiology and stress physiology of functional groups* (4th ed.). Berlin: Springer. 513 p.
Lavender, D. P., Sweet, G. B., Zaerr, J. B., & Hermann, R. K. (1973). Spring shoot growth in Douglas-fir may be initiated by gibberellins exported from the roots. *Science, 182*, 838–839.
Lawrence, E. (Ed.). (2011). *Henderson's dictionary of biology* (15th ed.). Essex: Pearson Education Limited. 758 p.
Leinonen, I. (1996a). Dependence of dormancy release on temperature in different origins of *Pinus sylvestris* and *Betula pendula* seedlings. *Scandinavian Journal of Forest Research, 11*, 122–128.
Leinonen, I. (1996b). A simulation model for the annual frost hardiness and freeze damage of Scots pine. *Annals of Botany, 78*, 687–693.
Leinonen, I., & Hänninen, H. (2002). Adaptation of the timing of bud burst of Norway spruce to temperate and boreal climates. *Silva Fennica, 36*, 695–701.
Lindfors, A., & Kupila-Ahvenniemi, S. (1988). Metabolic and structural responses of pine buds under climate chamber conditions. *Physiologia Plantarum, 72*, 407–413.
Linkosalo, T. (2000). Mutual regularity of spring phenology of some boreal tree species: Predicting with other species and phenological models. *Canadian Journal of Forest Research, 30*, 667–673.
Linkosalo, T., & Lechowicz, M. (2006). Twilight far-red treatment advances leaf bud burst of silver birch (*Betula pendula*). *Tree Physiology, 26*, 1249–1256.
Linkosalo, T., Häkkinen, R., & Hari, P. (1996). Improving the reliability of a combined phenological time series by analyzing observation quality. *Tree Physiology, 16*, 661–664.
Linkosalo, T., Carter, T. R., Häkkinen, R., & Hari, P. (2000). Predicting spring phenology and frost damage risk of *Betula* spp. under climatic warming: A comparison of two models. *Tree Physiology, 20*, 1175–1182.

Linkosalo, T., Häkkinen, R., & Hänninen, H. (2006). Models of the spring phenology of boreal and temperate trees: Is there something missing? *Tree Physiology, 26*, 1165–1172.

Linkosalo, T., Lappalainen, H. K., & Hari, P. (2008). A comparison of phenological models of leaf bud burst and flowering of boreal trees using independent observations. *Tree Physiology, 28*, 1873–1882.

Linkosalo, T., Häkkinen, R., Terhivuo, J., Tuomenvirta, H., & Hari, P. (2009). The time series of flowering and leaf bud burst of boreal trees (1846–2005) support the direct temperature observations of climatic warming. *Agricultural and Forest Meteorology, 149*, 453–461.

Luomajoki, A. (1977). Effects of temperature on spermatophyte male meiosis. *Hereditas, 85*, 33–47.

Luomajoki, A. (1982). Temperature and dates of male meiosis in trees. *Hereditas, 97*, 167–178.

Luomajoki, A. (1984). The tetrad phase of microsporogenesis in trees with reference to the annual cycle. *Hereditas, 101*, 179–197.

Luomajoki, A. (1986a). The latitudinal and yearly variation in the timing of microsporogenesis in *Alnus*, *Betula* and *Corylus*. *Hereditas, 104*, 231–243.

Luomajoki, A. (1986b). Timing of microsporogenesis in trees with reference to climatic adaptation. A review. *Acta Forestalia Fennica, 196*, 1–33.

Luomajoki, A. (1993a). Climatic adaptation of Scots pine (*Pinus sylvestris* L.) in Finland based on male flowering phenology. *Acta Forestalia Fennica, 237*, 1–27.

Luomajoki, A. (1993b). Climatic adaptation of Norway spruce (*Picea abies* (L.) Karsten) in Finland based on male flowering phenology. *Acta Forestalia Fennica, 242*, 1–28.

Luomajoki, A. (1999). Climatic adaptation of Silver Birch (*Betula pendula*) and Downy Birch (*B. pubescens*) in Finland based on male flowering phenology. *Acta Forestalia Fennica, 263*, 1–35.

Madariaga, F. J., & Knott, J. E. (1951). Temperature summations in relation to lettuce growth. *Proceedings of the American Society for Horticultural Science, 58*, 147–152.

Menzel, A., & Fabian, P. (1999). Growing season extended in Europe. *Nature, 397*, 659.

Menzel, A., Sparks, T. H., Estrella, N., Koch, E., Aasa, A., Ahas, R., Alm-Kübler, K., Bissolli, P., Braslavská, O., Briede, A., Chmielewski, F. M., Crepinsek, Z., Curnel, Y., Dahl, Å., Defila, C., Donnelly, A., Filella, Y., Jatczak, K., Måke, F., Mestre, A., Nordli, Ø., Peñuelas, J., Pirinen, P., Remišová, V., Scheifinger, H., Striz, M., Susnik, A., van Lieth, A. J. H., Wielgolaski, F.-E., Zach, S., & Zust, A. (2006). European phenological response to climate change matches the warming pattern. *Global Change Biology, 12*, 1969–1976.

Mikola, J. (1982). Bud-set phenology as an indicator of climatic adaptation of Scots pine in Finland. *Silva Fennica, 16*, 178–184.

Mohr, H., & Schoffer, P. (1995). *Plant physiology*. Berlin: Springer. 629 p.

Morin, X., Lechowicz, M. J., Augspurger, C., O'Keefe, J., Viner, D., & Chuine, I. (2009). Leaf phenology in 22 North American tree species during the 21st century. *Global Change Biology, 15*, 961–975.

Murray, M. B., Cannell, M. G. R., & Smith, R. I. (1989). Date of budburst of fifteen tree species in Britain following climatic warming. *Journal of Applied Ecology, 26*, 693–700.

Murray, M. B., Smith, R. I., Leith, I. D., Fowler, D., Lee, H. S. J., Friend, A. D., & Jarvis, P. G. (1994). Effects of elevated CO_2, nutrition and climatic warming on bud phenology in Sitka spruce (*Picea sitchensis*) and their impact on the risk of frost damage. *Tree Physiology, 14*, 691–706.

Myking, T. (1997). Effects of constant and fluctuating temperature on time to budburst in *Betula pubescens* and its relation to bud respiration. *Trees, 12*, 107–112.

Myking, T. (1998). Interrelations between respiration and dormancy in buds of three hardwood species with different chilling requirements for dormancy release. *Trees, 12*, 224–229.

Myking, T., & Heide, O. M. (1995). Dormancy release and chilling requirement of buds of latitudinal ecotypes of *Betula pendula* and *B. pubescens*. *Tree Physiology, 15*, 697–704.

Myneni, R. B., Keeling, C. D., Tucker, C. J., Asrar, G., & Nemani, R. R. (1997). Increased plant growth in the northern high latitudes from 1981 to 1991. *Nature, 386*, 698–702.

Nienstaedt, H. (1966). Dormancy and dormancy release in white spruce. *Forest Science, 12*, 374–384.
Nienstaedt, H. (1967). Chilling requirements in seven Picea species. *Silvae Genetica, 16*, 65–68.
Nizinski, J. J., & Saugier, B. (1988). A model of leaf budding and development for a mature *Quercus* forest. *Journal of Applied Ecology, 25*, 643–652.
Nygren, M. (1987). Germination characteristics of autumn collected Pinus sylvestris seeds. *Acta Forestalia Fennica, 201*, 1–42.
Overcash, J. P., & Campbell, J. A. (1955). The effects of intermittent warm and cold periods on breaking the rest period of peach leaf buds. *Proceedings of the American Society for Horticultural Science, 66*, 87–92.
Parmesan, C., & Yohe, G. (2003). A globally coherent fingerprint of climate change impacts across natural systems. *Nature, 421*, 37–42.
Partanen, J. (2004). Dependence of photoperiodic response of growth cessation on the stage of development in *Picea abies* and *Betula pendula* seedlings. *Forest Ecology and Management, 188*, 137–148.
Partanen, J., & Beuker, E. (1999). Effects of photoperiod and thermal time on the growth rhythm of *Pinus sylvestris* seedlings. *Scandinavian Journal of Forest Research, 14*, 487–497.
Partanen, J., Koski, V., & Hänninen, H. (1998). Effects of photoperiod and temperature on the timing of bud burst in Norway spruce (*Picea abies*). *Tree Physiology, 18*, 811–816.
Partanen, J., Hänninen, H., & Häkkinen, R. (2005). Bud burst in Norway spruce (*Picea abies*): Preliminary evidence for age-specific rest patterns. *Trees, 19*, 66–72.
Perry, T. O. (1971). Dormancy of trees in winter. *Science, 171*, 29–36.
Picard, G., Quegan, S., Delbart, N., Lomas, M. R., Le Toan, T., & Woodward, F. I. (2005). Bud-burst modelling in Siberia and its impact on quantifying the carbon budget. *Global Change Biology, 11*, 2164–2176.
Pitacco, A. (1990). The "Dynamic Model" of chilling accumulation: Its structure and performance under simulated and real thermal regimes. *Advances in Horticultural Science, 3*, 167–171.
Polgar, C. A., & Primack, R. B. (2011). Leaf-out phenology of temperate woody plants: From trees to ecosystems. *New Phytologist, 191*, 926–941.
Pudas, E., Leppälä, M., Tolvanen, A., Poikolainen, J., Venäläinen, A., & Kubin, E. (2008). Trends in phenology of *Betula pubescens* across the boreal zone in Finland. *International Journal of Biometeorology, 52*, 251–259.
Raulo, J., & Leikola, M. (1974). Tutkimuksia puiden vuotuisen pituuskasvun ajoittumisesta. Summary: Studies on the annual height growth of trees. *Communicationes Instituti Forestalis Fenniae, 81*(2), 1–19.
Réaumur, M de. (1735). *Observations du thermomètre faites à Paris pendant l'année 1735 comparées avec celles qui ont été faites sous la ligne à l' Ile de France, à Alger et en quelques-unes de nos îles de l'Amérique*. Académie royale des sciences – Année 1735.
Repo, T., Hänninen, H., & Kellomäki, S. (1996). The effects of long-term elevation of air temperature and CO_2 on the frost hardiness of Scots pine. *Plant, Cell and Environment, 19*, 209–216.
Repo, T., Kalliokoski, T., Domisch, T., Lehto, T., Mannerkoski, H., Sutinen, S., & Finér, L. (2005). Effects of timing of soil frost thawing on Scots pine. *Tree Physiology, 25*, 1053–1062.
Repo, T., Lehto, T., & Finér, L. (2008). Delayed soil thawing affects root and shoot functioning and growth in Scots pine. *Tree Physiology, 28*, 1583–1591.
Richardson, E. A., Seeley, S. D., & Walker, D. R. (1974). A model for estimating the completion of rest for 'Redhaven' and 'Elberta' peach trees. *HortScience, 9*, 331–332.
Rinne, P. L. H., & van der Schoot, C. (2003). Plasmodesmata at the crossroads between development, dormancy, and defence. *Canadian Journal of Botany, 81*, 1182–1197.
Rohde, A., & Bhalerao, R. P. (2007). Plant dormancy in the perennial context. *Trends in Plant Science, 12*, 217–223.

Romberger, J. A. (1963). *Meristems, growth, and development in woody plants. An analytical review of anatomical, physiological, and morphogenetic aspects*. USDA Technical Bulletin 1292. U.S. Government Printing Office. 214 p.

Rousi, M., & Heinonen, J. (2007). Temperature sum accumulation effects on within-population variation and long-term trends in date of bud burst of European white birch (*Betula pendula*). *Tree Physiology, 27*, 1019–1025.

Ruttink, T., Arend, M., Morreel, K., Storme, V., Rombauts, S., Fromm, J., Bhalerao, R. P., Boerjan, W., & Rohde, A. (2007). A molecular timetable for apical bud formation and dormancy induction in poplar. *The Plant Cell, 19*, 2370–2390.

Salažs, A., & Ievinsh, G. (2004). The impact of the Latvian plant physiologist Auseklis Veğis (1903–1973) in modern natural sciences. *Acta Universitatis Latviensis, Biology, 676*, 7–15.

Salminen, H., & Jalkanen, R. (2015). Modeling of bud break of Scots pine in northern Finland in 1908–2014. *Frontiers in Plant Science, Article 6*, 104. doi:10.3389/fpls.2015.00104.

Samish, R. M. (1954). Dormancy in woody plants. *Annual Review of Plant Physiology, 5*, 183–204.

Sarvas, R. (1967). The annual period of development of forest trees. *Proceedings of the Finnish Academy of Science and Letters, 1965*, 211–231.

Sarvas, R. (1972). Investigations on the annual cycle of development of forest trees. Active period. *Communicationes Instituti Forestalis Fenniae, 76*(3), 1–110.

Sarvas, R. (1974). Investigations on the annual cycle of development of forest trees. II. Autumn dormancy and winter dormancy. *Communicationes Instituti Forestalis Fenniae, 84*(1), 1–101.

Saure, M. C. (1985). Dormancy release in deciduous fruit trees. *Horticultural Reviews, 7*, 239–300.

Schaber, J., & Badeck, F. W. (2003). Physiology-based phenology models for forest tree species in Germany. *International Journal of Biometeorology, 47*, 193–201.

Schwartz, M. D. (2013). Introduction. In M. D. Schwartz (Ed.), *Phenology: An integrative environmental science* (2nd ed., pp. 1–5). Dordrecht: Springer Science+Business Media B.V.

Slaney, M., Wallin, G., Medhurst, J., & Linder, S. (2007). Impact of elevated carbon dioxide concentration and temperature on bud burst and shoot growth of boreal Norway spruce. *Tree Physiology, 27*, 301–312.

Smolander, H. (1984). Measurement of fluctuating irradiance in field studies of photosynthesis. *Acta Forestalia Fennica, 187*, 1–56.

Søgaard, G., Johnsen, Ø., Nilsen, J., & Junttila, O. (2008). Climatic control of bud burst in young seedlings of nine provenances of Norway spruce. *Tree Physiology, 28*, 311–320.

Sorensen, F. C., & Campbell, R. K. (1978). Comparative roles of soil and air temperatures in the timing of spring bud flush in seedling Douglas-fir. *Canadian Journal of Botany, 56*, 2307–2308.

Sutinen, S., Partanen, J., Viherä-Aarnio, A., & Häkkinen, R. (2009). Anatomy and morphology in developing vegetative buds on detached Norway spruce branches in controlled conditions before bud burst. *Tree Physiology, 29*, 1457–1465.

Sutinen, S., Partanen, J., Viherä-Aarnio, A., & Häkkinen, R. (2012). Development and growth of primordial shoots in Norway spruce buds before visible bud burst in relation to time and temperature in the field. *Tree Physiology, 32*, 987–997.

Tanino, K. K., Kalcsits, L., Silim, S., Kendall, E., & Gray, G. R. (2010). Temperature-driven plasticity in growth cessation and dormancy development in deciduous woody plants: A working hypothesis suggesting how molecular and cellular function is affected by temperature during dormancy induction. *Plant Molecular Biology, 73*, 49–65.

Thomas, C. D., Cameron, A., Green, R. E., Bakkenes, M., Beaumont, L. J., Collingham, Y. C., Erasmus, B. F. N., de Siqueira, M. F., Grainger, A., Hannah, L., Hughes, L., Huntley, B., van Jaarsveld, A. S., Midgley, G. F., Miles, L., Ortega-Huerta, M. A., Peterson, A. T., Phillips, O. L., & Williams, S. E. (2004). Extinction risk from climate change. *Nature, 427*, 145–148.

Thuiller, W., Lavorel, S., Araújo, M. B., Sykes, M. T., & Prentice, I. C. (2005). Climate change threats to plant diversity in Europe. *PNAS, 102*, 8245–8250.

Tooke, F., & Battey, N. H. (2010). Flowering newsletter review. Temperate flowering phenology. *Journal of Experimental Botany, 61*, 2853–2862.

Tuhkanen, S. (1980). Climatic parameters and indices in plant geography. *Acta Phytogeographica Suecica, 67*. 105 p.
Ununger, J., Ekberg, I., & Kang, H. (1988). Genetic control and age-related changes of juvenile growth characters in *Picea abies*. *Scandinavian Journal of Forest Research, 3*, 55–66.
Vaartaja, O. (1959). Evidence of photoperiodic ecotypes in trees. *Ecological Monographs, 29*, 91–111.
Van den Driessche, R. (1975). Flushing response of Douglas fir buds to chilling and to different air temperatures after chilling. *British Columbia Forest Service, Research Note, 71*, 1–22.
Vegis, A. (1964). Dormancy in higher plants. *Annual Review of Plant Physiology, 15*, 185–224.
Viherä-Aarnio, A., Häkkinen, R., Partanen, J., Luomajoki, A., & Koski, V. (2005). Effects of seed origin and sowing time on timing of height growth cessation of *Betula pendula* seedlings. *Tree Physiology, 25*, 101–108.
Viherä-Aarnio, A., Sutinen, S., Partanen, J., & Häkkinen, R. (2014). Internal development of vegetative buds of Norway spruce trees in relation to accumulated chilling and forcing temperatures. *Tree Physiology, 34*, 547–556.
Vitasse, Y. (2013). Ontogenic changes rather than difference in temperature cause understory trees to leaf out earlier. *New Phytologist, 198*, 149–155.
Vitasse, Y., & Basler, D. (2013). What role for photoperiod in the bud burst phenology of European beech. *European Journal of Forest Research, 132*, 1–8.
Wang, J. Y. (1960). A critique of the heat unit approach to plant response studies. *Ecology, 41*, 785–790.
Wareing, P. F. (1953). Growth studies in woody species V. Photoperiodism in dormant buds of Fagus sylvatica L. *Physiologia Plantarum, 6*, 692–706.
Wareing, P. F. (1956). Photoperiodism in woody plants. *Annual Review of Plant Physiology, 7*, 191–214.
Weiser, C. J. (1970). Cold resistance and injury in woody plants. *Science, 169*, 1269–1278.
Westergaard, L., & Eriksen, E. N. (1997). Autumn temperature affects the induction of dormancy in first-year seedlings of *Acer platanoides* L. *Scandinavian Journal of Forest Research, 12*, 11–16.
Wielgolaski, F.-E. (1999). Starting dates and basic temperatures in phenological observations of plants. *International Journal of Biometeorology, 42*, 158–168.
Worrall, J., & Mergen, F. (1967). Environmental and genetic control of dormancy in *Picea abies*. *Physiologia Plantarum, 20*, 733–745.
Young, E. (1992). Timing of high temperature influences chilling negation in dormant apple trees. *Journal of the American Society for Horticultural Science, 117*, 271–273.

Chapter 4
The Annual Cycle of Photosynthesis in Evergreen Conifers

Contents

Abstract The hypothetico-deductive modelling framework introduced in Chap. 2 is applied to modelling the annual cycle of photosynthesis in boreal and temperate evergreen conifers. The main emphasis is in the photosynthetic capacity of the needles, i.e., their inherent photosynthesis potential, which is an emergent manifestation of the physiological status of the photosynthetic machinery in the needles. Photosynthetic capacity is usually measured by means of the photosynthetic light curve. Unlike the models of the annual phenological cycle discussed in Chap. 3, models of photosynthetic capacity can be tested continuously on a daily basis. Since 1980, the annual cycle of photosynthetic capacity has been successfully simulated by means of different versions of a fully reversible fluctuating model, in which the changes in photosynthetic capacity follow the changes in air temperature. The

H. Hänninen, *Boreal and Temperate Trees in a Changing Climate*, Biometeorology 3,
DOI 10.1007/978-94-017-7549-6_4

effects of short-term frost are not taken into account in these models, but there is another model available for them. In further work, a synthesis of these models is called for. Moreover, the effects of solar radiation need to be taken into account in further work to attain better model realism, for the seasonality of photosynthetic capacity is largely caused by acclimation to different combinations of air temperature and solar radiation. The evidence for the effects of other environmental factors, such as soil temperature and ground frost, remains controversial, but these factors cannot be ruled out in further model development. Novel simulations using long-term air temperature records as input for the fluctuating model revealed large year-to-year variation in the springtime recovery of photosynthetic capacity.

Keywords Annual photosynthesis cycle • Boreal trees • Carbon dioxide exchange • Computer simulations • Fluctuating model • Model realism • Photosynthetic capacity • P_{max} • Quantum yield • Seasonality • Temperate trees

4.1 The Phenomena and the Measurements Involved

4.1.1 Measuring the Carbon Dioxide Exchange

4.1.1.1 The Leaf, Shoot, and Plant Level

Like all other green plants, boreal and temperate trees are autotrophic organisms. Thus they obtain the energy required for all of their vital processes, i.e., growth, development, reproduction, and defence, by converting, through the process of photosynthesis, the energy of incoming solar radiation into the chemical energy of organic compounds formed by reducing carbon dioxide taken up from the atmosphere. Organic compounds are the basis of life, not only for green plants themselves but also for all heterotrophic organisms, such as animals and saprophytic microbes (Chapin et al. 2002).

In the last few decades, sophisticated equipment has been developed for measuring plant photosynthesis. It is now possible to measure photosynthesis in field conditions, either with small portable equipment (Fig. 4.1a) or with large measuring systems installed in the stands of trees at permanent field stations (Fig. 4.1b, d).

Fig. 4.1 (continued) cuvette at the SMEAR II station (Station for Measuring Ecosystem Atmospheric Relationships; Hari et al. 2008a) in Hyytiälä, southern Finland. On *top* of the cuvette, a light sensor is seen. (**c**) A tower for eddy covariance measurements located at SMEAR II. (**d**) The field laboratory building of SMEAR II. The building houses the gas analysers and computers that control the measurements automatically (The unpublished photographs were provided by Timo Saarinen (**a**) and Juho Aalto (**b**–**d**))

Fig. 4.1 Equipment and facilities used in studies of the CO_2 exchange of trees and other plants. (**a**) Portable photosynthesis-measuring system LI-6400, shown with the measuring cuvette at the front. (**b**) A shoot from a *Pinus sylvestris* tree growing in natural conditions, placed in a measuring

Most of the modern measurement techniques are based on measuring the carbon dioxide exchange between a leaf (or, in the case of conifers, usually a shoot) and air. The shoot to be measured is enclosed in a transparent measuring cuvette, and the amount of CO_2 fixed or released by the shoot is measured (Petäjä et al. 2008). The result indicates the amount of net CO_2 exchange, i.e., gross photosynthesis minus dark respiration.

The amount of CO_2 fixed or released needs to be scaled with respect to the duration of the measurement and the needle area or mass of the measured shoot. Thus, the observed amount of CO_2 exchanged is divided by the duration of the measurement and by either the area or the dry matter of the needles. Today, the former procedure is usually applied, so that the rate of CO_2 exchange is given in units of μmol CO_2 m^{-2} s^{-1}. When the needle mass is measured by dry weight, DW, the amount of CO_2 fixed or released is usually also given in units of mass. The unit of the rate of CO_2 exchange in this case, then, is mg CO_2 g^{-1} DW h^{-1} (Larcher 2003).

To study the environmental responses of photosynthesis and respiration, such environmental factors as light and air temperature are monitored simultaneously with the measurements of CO_2 exchange (Fig. 4.1b).[1] In practice, measuring the environmental factors and the CO_2 exchange of plants involves several methodological problems, such as rising air temperatures in the closed cuvette during the measurement. To solve these problems, both technical skills and an understanding of the physical and physiological phenomena involved are needed. Furthermore, the problems need to be addressed all through the research process, from the planning and constructing of the measurement device to the analysis of the measured data. Permanent field stations for measuring CO_2 exchange, such as the SMEAR stations in Finland (Hari et al. 2008a; Fig. 4.1b–d), are large-scale facilities that are expensive and time-consuming to build. Unlike the portable measuring device (Fig. 4.1a), permanent large-scale field stations are therefore relatively rare facilities in the international scientific community.

4.1.1.2 The Ecosystem Level

In the last couple of decades, CO_2 exchange has increasingly been measured at the ecosystem level, too, by means of the technique of eddy covariance (Suni et al. 2003; Baldocchi 2003). In that technique, the flux of carbon between two major pools is measured.[2] The first pool is the organic carbon contained in all living organisms, mainly trees and other plants, and in dead organic matter in the soil. The

[1] Here the common term 'light' is used to denote the photosynthetic photon flux density, (PPFD), i.e., the photon flux density of incoming solar radiation at the photosynthetically active wavelengths of 400–700 nm, also referred to as PAR (photosynthetically active radiation).

[2] In ecosystem studies, the concept normally used is carbon flux rather than carbon dioxide exchange, which is generally used in plant ecophysiology.

second pool is the CO_2 contained in the atmosphere. For reliable measurements of the carbon flux representing a large area, the concentration of atmospheric CO_2 needs to be measured at great heights, so that high towers are required for carrying out eddy covariance measurements (Hari et al. 2008a; Fig. 4.1c). At any given moment, the rate of net carbon flux, either from the atmosphere to the land ecosystem or vice versa, is measured. In the growing season, the photosynthesis of trees dominates the daytime net carbon flux in boreal and temperate forests, so that it goes from the atmosphere to the land ecosystem, whereas the night-time flux is due to heterotrophic respiration and goes from the land ecosystem to the atmosphere (Kolari et al. 2008).

Eddy covariance provides an efficient tool for studies addressing the cycling of carbon at the ecosystem level, and therefore, due to the general interest in the increased atmospheric concentrations of CO_2, eddy covariance studies are carried out at several locations nowadays as part of the global change research (Baldocchi 2003). In the present volume, results of eddy covariance studies will be reviewed in relation to the annual cycle of photosynthesis in boreal and temperate trees.

4.1.2 The Annual Cycle of Photosynthetic Capacity

4.1.2.1 Photosynthetic Capacity as a Whole-Tree Attribute of Conifer Ecophysiology

Seasonally varying environmental factors affect the photosynthesis of boreal and temperate trees in three basic ways. First, in deciduous trees, the phenology of the leaves, i.e., the timing of bud burst and leaf growth in spring and the timing of leaf senescence in autumn, determine the season during which the incoming solar radiation is intercepted and used for photosynthetic production (Chapin et al. 2002). Thus the annual cycle of photosynthesis in deciduous broad-leaved tree species, and also in *Larix* species, is closely connected to the annual phenological cycle discussed in Chap. 3.

Second, seasonally varying environmental factors have immediate effects on the rate of photosynthesis in trees (Larcher 2003). Most importantly, the photosynthetic photon flux density, PPFD, regulates the rate of CO_2 exchange at any given moment, as indicated by the photosynthetic light curve (Fig. 4.2).

The third effect of seasonally varying environmental factors, especially air temperature, is typical for evergreen conifers growing in the boreal zone and at high altitudes in the temperate zone (Pisek and Winkler 1958; Pelkonen and Hari 1980; Linder and Lohammar 1981; Kolari et al. 2014).[3] This response is a slow change in the physiological photosynthetic properties of the needles and is manifested at the whole-tree level as a change in the rate of photosynthesis. The

[3] For the sake of brevity, these trees will be referred to as "evergreen conifers" or "conifers" in most cases.

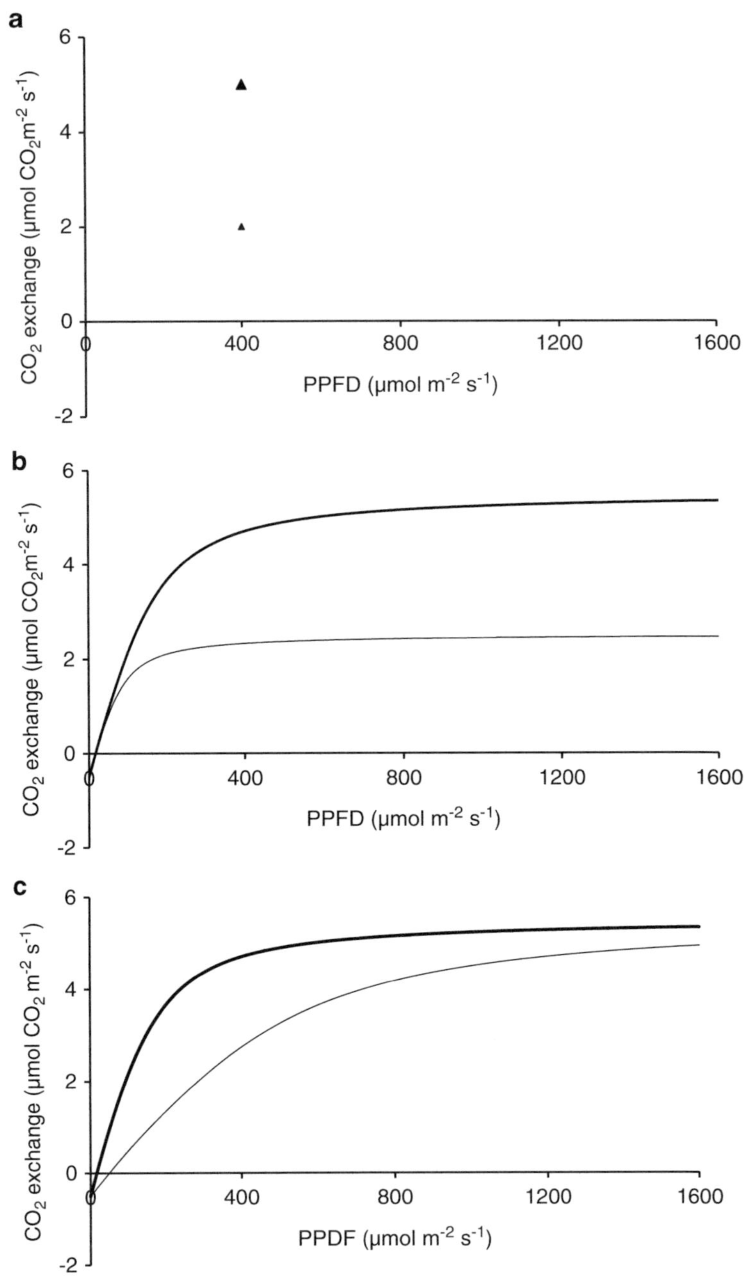

Fig. 4.2 Alternative principles of determining the photosynthetic capacity of evergreen conifers by means of measurements of CO_2 exchange. In each panel, the *thin curves* or *small symbols* represent low photosynthetic capacity and the *thick curves* or *large symbols*, high photosynthetic capacity. The photosynthetic capacity is determined (**a**) by means of measurements under one reference condition (Pisek and Winkler 1958); (**b**) by means of the light-saturated rate of photosynthesis, P_{max}, (Lundell et al. 2008; Ensminger et al. 2008); (**c**) by means of the quantum yield α, i.e., the slope of the photosynthetic light curve at low light levels (Mäkelä et al. 2004; Kolari et al. 2007; see footnote 5 in Sect. 4.2.1.2); and (**d**) by means of both P_{max} and α, so that the rate of photosynthesis increases with increasing photosynthetic capacity proportionally at all light levels (Pelkonen 1980; Pelkonen and Hari 1980)

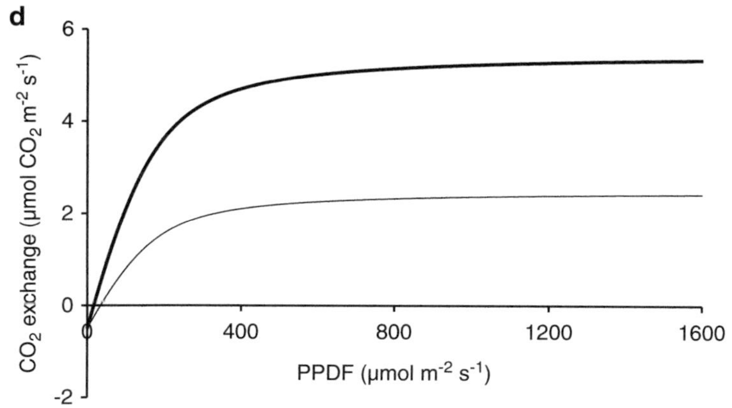

Fig. 4.2 (continued)

response is readily demonstrated with an experiment where seedlings or twigs of an evergreen conifer are transferred from natural conditions into a laboratory at different times in late winter and spring and the rate of CO_2 exchange is subsequently measured after each transfer in standard reference conditions. Despite the same standard measurement conditions, the rate of photosynthesis is typically higher with transfers done in late rather than early spring (Pisek and Winkler 1958; Fig. 4.2a), even though no morphological changes are visible to the naked eye in the needles.

In this volume, the concept *photosynthetic capacity* is used for the seasonally varying inherent photosynthesis potential of the needles. In practice it is determined from the light response of the rate of CO_2 exchange and quantified in terms of the light-saturated rate of photosynthesis P_{max}, (Fig. 4.2b), the quantum yield α, (Fig. 4.2c), or both of these parameters (Fig. 4.2d). In the light of the whole-tree ecophysiological approach adopted in this volume (Sect. 2.4), changes in the photosynthetic capacity observed during the annual cycle are seen as emergent manifestations of the physiological changes taking place in the photosynthetic machinery of the needles.

The effects of air temperature on the annual cycle of photosynthetic capacity were already examined by Pisek and Winkler (1958), who studied *Picea abies* trees growing near Innsbruck in the Alps. The annual pattern of photosynthetic capacity roughly followed the annual pattern of air temperature. Furthermore, when the air temperature dropped below zero, the photosynthetic capacity decreased rapidly. In the respiratory activity, i.e., in the rate of respiration of the twigs measured in standard reference conditions, there was less seasonal variation, but the respiratory activity increased at the time of bud burst, when there was also a transient decrease in the photosynthetic capacity.

The springtime increase in the photosynthetic capacity of conifers is a seasonal event ecologically analogical to bud burst and leaf growth in deciduous species. In both cases the potential for photosynthetic production increases in spring, but in conifers this development is caused by invisible physiological changes in the needles rather than phenological changes, which are so visible in deciduous trees that they characterize the entire landscape. However, the increase in the photosynthetic capacity of conifers usually starts before any signs of bud burst are visible in deciduous species (Kolari et al. 2007; Linkosalo et al. 2008; Lappalainen 2010). Furthermore, unlike the ontogenetic development related to bud burst and leaf growth, the changes in the photosynthetic capacity of conifer needles are reversible (Pelkonen 1980; Bergh et al. 1998; Hänninen and Hari 2002). In this chapter, the main emphasis will be on the annual cycle of the photosynthetic capacity of evergreen conifers, but the immediate effects of seasonally varying environmental factors on the photosynthesis of trees are also considered.

4.1.2.2 The Physiological Basis

The physiological changes in the needles of evergreen conifers, manifesting themselves at the needle or shoot level as changes in photosynthetic capacity, are relatively well understood. These changes can be roughly divided into two main categories, though the classification of any given phenomenon into one or the other category is not always clear. Detailed discussions of these phenomena are outside the scope of the present volume, so that only a brief outline of the mechanisms is provided below. More detailed discussions are provided, e.g., in Demmig-Adams et al. (2008a) and Porcar-Castell et al. (2013).

First, reduced photosynthetic capacity reflects the acclimation of the photosynthetic machinery to high solar radiation combined with the adverse conditions of low air temperature and frozen soil. Such conditions prevent the use of incoming solar radiation for photosynthesis, so that the needles may be damaged by excess radiation. The damage is prevented by reducing the light-harvesting capacity of the chloroplasts (Demmig-Adams et al. 2008b). Naturally, such reduction also reduces the photosynthetic capacity of the needles. Second, photosynthetic capacity may also be reduced by minor needle damage invisible to the naked eye (Linder and Lohammar 1981; Bergh et al. 1998).

Chlorophyll fluorescence measurements are nowadays carried out on a routine basis in research projects addressing the whole-tree ecophysiology of trees or the whole-plant ecophysiology of other plants. This is because these measurements readily yield information on the status of the photosynthetic machinery or, more precisely, the status of Photosystem II (Larcher 2003). Among the several chlorophyll fluorescence parameters available, the most commonly used one is the photochemical efficiency of photosystem II, F_v/F_m, (Lundmark et al. 1998; Repo et al. 2005; Saarinen et al. 2015), but recent developments in measurement equipment have facilitated the introduction of several new chlorophyll fluorescence parameters (Ensminger et al. 2008; Chang et al. 2015). In this chapter, results concerning the parameter F_v/F_m are discussed with reference to the annual cycle of photosynthetic capacity in evergreen conifers.

4.1.3 Implications for Modelling

In this chapter the modelling approach introduced in Chap. 2 is applied to modelling the annual cycle of photosynthetic capacity in evergreen conifers growing in the boreal zone and at high altitudes in the temperate zone (Hänninen and Kramer 2007). Accordingly, the variable state of development, S(t), is specified as the variable *state of photosynthetic capacity*, $S_p(t)$, and the variable rate of development, R(t), correspondingly, as the variable *rate of change of photosynthetic capacity*, $R_p(t)$. $S_p(t)$ indicates the predicted photosynthetic capacity of the needles at a given time instant t. Correspondingly, $R_p(t)$ indicates the predicted rate of change of photosynthetic capacity within the time step used in the calculations, which is usually either an hour or a day.

Contrary to the models of the annual phenological cycle discussed in Chap. 3, the state of development predicted by the models of photosynthetic capacity has a straightforward continuous counterpart in empirical studies. This implies that the predicted course of the state of development can be compared continuously on a day-to-day basis with observations when modelling the annual cycle of photosynthetic capacity. To this end, the state of photosynthetic capacity, $S_p(t)$ predicted by the model is defined to correspond to one of the three definitions of the empirical photosynthetic capacity that are based on the measured photosynthetic light response (Fig. 4.2b–d). Thus the predicted $S_p(t)$ is compared with either the light-saturated rate of photosynthesis, P_{max} (Fig. 4.2b), the quantum yield α, i.e., the slope of the photosynthetic light curve at low light levels (Fig. 4.2c), or both P_{max} and α (Fig. 4.2d). The definition of photosynthetic capacity used by early empirical researchers, i.e., the one defined as the rate of photosynthesis in one standard reference condition (Fig. 4.2a) has not been used in modelling studies, so that it will not be considered further in the present volume either.

In some models, photosynthetic capacity is calculated as a linear or sigmoidal transformation of the modelled state variable. In these cases the state and rate variables are called *state of photosynthetic acclimation*, $S_{pa}(t)$, and *rate of photosynthetic acclimation*, $R_{pa}(t)$, respectively (Mäkelä et al. 2004; Kolari et al. 2007).

4.2 Effects of Air Temperature

4.2.1 Fluctuating Effects of Air Temperature

4.2.1.1 A Fluctuating Model of Photosynthetic Capacity

After the pioneering work of Pisek and Winkler (1958), the annual cycle of the photosynthesis of evergreen conifers has been studied with several tree species. The studies have been carried out in both natural conditions (Pelkonen 1980; Linder and Lohammar 1981; Bauer et al. 1994; Hari and Mäkelä 2003; Kolari et al. 2007, 2014;

Linkosalo et al. 2014) and controlled ones (Bauer et al. 1994; Strand and Lundmark 1995; Repo et al. 2005; Ensminger et al. 2008; Kolari et al. 2014). These and several other empirical studies have increased our understanding of the annual cycle of photosynthesis in conifers. The role of air temperature in regulating the annual cycle of photosynthetic capacity has been repeatedly emphasised in these studies, but the cycle is also affected by other factors, especially incoming solar radiation. The other factors will be discussed in Sect. 4.3.

Pelkonen and Hari (1980) were the first to examine the relationship between air temperature and photosynthetic capacity by means of an ecophysiological dynamic model (Fig. 4.3). Their model is of the following general form:

$$R_p(t) = f\left(T(t), S_p(t)\right) \tag{4.1}$$

where $R_p(t)$ = rate of change of photosynthetic capacity at time instant t, $T(t)$ = air temperature at time instant t, and $S_p(t)$ = state of photosynthetic capacity at time instant t. In their model, Pelkonen and Hari (1980) introduced the feedback principle into the ecophysiological modelling of the annual cycle of boreal and temperate trees (Hänninen and Kramer 2007). According to that principle (Eq. 2.2 in Sect. 2.2), the state of development of a given attribute of the annual cycle, i.e., photosynthetic capacity in this case, affects the rate of development of the same attribute in any given prevailing conditions. This is seen as a change in the air temperature response of $R_p(t)$ as the value of $S_p(t)$ changes (Fig. 4.3a).

In the present volume, the following modified version of Pelkonen and Hari's (1980) model is used:

$$R_p(t) = \left[\frac{100}{S_{p,crit}}\right]\left[\left(\frac{100}{1+100a^{-\left[T(t)-\left(S_p(t)/c\right)\right]}}\right) - \left(\frac{100}{1+100a^{\left[T(t)-\left(S_p(t)/c\right)\right]}}\right)\right] \tag{4.2}$$

where $S_{p,crit}$ = the critical quantity of developmental units required for attaining full photosynthetic capacity (Pelkonen 1981; Hänninen and Hari 2002), a and c are parameters, and the other symbols are as in Eq. 4.1. By definition, the value of $S_p(t)$ is obtained by integrating Eq. 4.2 with respect to time and taking into account the condition that the value of $S_p(t)$ is maximally 100 % (Pelkonen and Hari 1980; Hänninen and Hari 2002):

$$S_p(t) = \min\left(100, \int_{t_0}^{t} R_p(t)dt\right) \tag{4.3}$$

where t_0 = starting moment of the simulation.

The modification of the model made in the present volume is caused by introducing the multiplier $[100/S_{p,crit}]$ into it. In Pelkonen and Hari's (1980) original model, this multiplier is missing. Therefore, the changes in photosynthetic capacity are described in the original model in terms of an arbitrary developmental unit

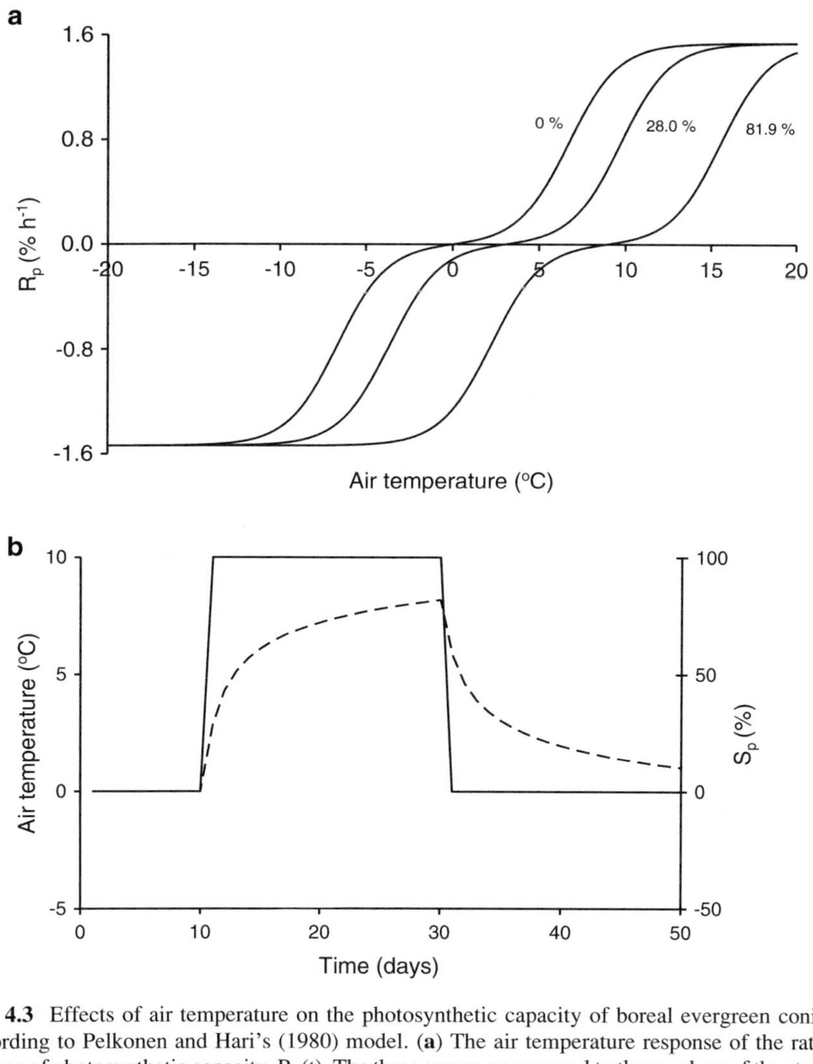

Fig. 4.3 Effects of air temperature on the photosynthetic capacity of boreal evergreen conifers according to Pelkonen and Hari's (1980) model. (**a**) The air temperature response of the rate of change of photosynthetic capacity, $R_p(t)$. The three curves correspond to three values of the state of photosynthetic capacity, $S_p(t)$. (**b**) The pattern of air temperature in a hypothetical experiment (*solid line* and *the vertical axis on the left*) and the course of the state of photosynthetic capacity, $S_p(t)$, predicted by the model for the hypothetical experiment (*dashed line* and the *vertical axis on the right*). The values of both $S_p(t)$ and $R_p(t)$ are given in percentages of the full photosynthetic capacity

corresponding to the rest of the formula in Eq. 4.2, so that the lower and upper asymptotes of the curve of $R_p(t)$ are -100 and $+100$ units h^{-1}, respectively (Pelkonen and Hari 1980), and the value of $S_p(t) = S_{p,crit}$ denotes the attainment of full photosynthetic capacity (Pelkonen 1981). Through the introduction of the multiplier $[100/S_{p,crit}]$, the original arbitrary developmental units are replaced with a percentage scale. Thus the prevailing state of photosynthetic capacity, $S_p(t)$, is expressed as a percentage of the maximum capacity (Fig. 4.3b), and the rate of change of photosynthetic capacity is expressed as a corresponding percentage change per hour (Fig. 4.3a).

The replacement of the arbitrary developmental units with the percentage scale is similar to the corresponding replacement carried out in the models of growth onset and rest break in Chap. 3. As in the cases discussed in Sects. 3.2.1 and 3.3.2, the motivation for the replacement is the facilitation of comparing different tree genotypes with different values of the genotype-specific parameters. In the present volume, the value of $S_{p,crit}$ = 6500 units suggested for southern Finnish *Pinus sylvestris* is used (Pelkonen 1981; Hänninen and Hari 2002). In this way, the upper asymptote of the curve of $R_p(t)$ gets the value of (100 %/6500 units) × (100 units h^{-1}) = 1.54 % h^{-1}, and the lower asymptote gets the same value with a minus sign (Fig. 4.3a). For the other two parameters the values of a = 2 and c = 600 determined for southern Finnish *Pinus sylvestris* were used in the present volume (Pelkonen and Hari 1980).

The air temperature response of $R_p(t)$ is a double sigmoid, having negative values at low air temperatures and positive values at high ones. Due to the feedback involved, the response curve moves to the right, towards higher temperatures, with increasing S_p and correspondingly to the left, towards lower temperatures, with decreasing S_p (Fig. 4.3a).

The ecopysiological phenomena addressed in Pelkonen and Hari's (1980) model are best illustrated by considering the state of photosynthetic capacity, S_p, predicted by the model for a hypothetical experiment with a stepwise change in air temperature (Fig. 4.3b). Assume that the needles have no photosynthetic capacity, i.e., S_p = 0 %, at the beginning of the experiment. With that value of S_p, the rate of change of photosynthetic capacity is zero at the air temperature of 0 °C, i.e., R_p(0 °C, 0 %) = 0 % h^{-1} (Fig. 4.3a, curve on the left). This temperature prevails at the beginning of the hypothetical experiment, so that no recovery of photosynthesis is predicted, i.e., the value of S_p stays at zero during the first days (Fig. 4.3b). When the temperature rises to T = +10 °C on day 11, the initial air temperature response corresponding to S_p = 0 % predicts a positive value of R_p(+10 °C, 0 %) = 1.4 % h^{-1} (Fig. 4.3a, curve on the left), i.e., the recovery of photosynthesis begins as indicated by the increasing value of S_p (Fig. 4.3b). Furthermore, the rate of recovery is near its maximum (Fig. 4.3a).

However, due to the feedback involved in the model, the increase of S_p changes the air temperature response, so that the double sigmoid moves to the right, towards higher temperatures. Thus the value of R_p corresponding to T = +10 °C decreases gradually (Fig. 4.3a), and correspondingly, the increase of S_p levels off (Fig. 4.3b). At the end of the period with T = +10 °C on day 30, the value of the rate of change of photosynthetic capacity goes down to R_p(+10 °C, 81.9 %) = 0.03 % h^{-1} (Fig. 4.3a, curve on the right).

When the air temperature drops back to the initial value of T = 0 °C on day 31, a negative value of R_p is predicted (Fig. 4.3a, curve on the right), and the value of S_p thus starts to decrease (Fig. 4.3b). Furthermore, the rate of decrease of photosynthetic capacity at this time is near its maximum (Fig. 4.3a). However, due to the feedback involved in the model, the decrease of S_p changes the air temperature response, so that the double sigmoid now moves to the left, towards lower temperatures. Thus the absolute value of the negative R_p at the air temperature of $T = 0$ °C becomes smaller (Fig. 4.3a), and the decrease of S_p levels off, similarly to the way its increase levelled off at the end of the period with $T = +10$ °C (Fig. 4.3b).

According to Pelkonen and Hari's (1980) model, the air temperature history determines the photosynthetic capacity prevailing at any given moment in evergreen conifers. In this way the model resembles the temperature sum models used for modelling the springtime phenology, e.g., the timing of bud burst or growth onset, of trees (Sect. 3.2.1). However, as both negative and positive values of the rate of development are involved, Pelkonen and Hari's (1980) model is an example of fluctuating development (Hänninen and Kramer 2007; Sect. 2.2). Rather than progressing on a genetically fixed developmental path, as in the ontogenetic development modelled with the temperature sum models, the state of photosynthetic capacity, $S_p(t)$, fluctuates in a reversible manner according to the fluctuation of the air temperature (Fig. 4.3b). Thus, in Pelkonen and Hari's (1980) model, change in the air temperature is the driving force of the development; in any constant air temperature, the development will stop (Repo 1993; Hänninen and Kramer 2007). However, due to the inertia of the physiological reactions, the changes in $S_p(t)$ are slower than those in the air temperature driving the development (Fig. 4.3b).

Regardless of the parameter values used, Pelkonen and Hari's (1980) model always predicts the recovery of photosynthesis to begin when the air temperature rises above zero. According to Eq. 4.2, photosynthetic capacity continues to increase as long as the air temperature continues to rise. In real trees, however, there is a maximum photosynthetic capacity, and it may be attained at an intermediate air temperature. In other words, for the full recovery of the photosynthetic machinery, summertime temperature maxima are not always needed. This phenomenon is addressed in Pelkonen and Hari's (1980) model in that after full photosynthetic capacity has been attained, the value of the state of photosynthetic capacity, $S_p(t)$, remains unchanged as long as the air temperature does not drop below the point where a decrease of $S_p(t)$ is predicted (Eq. 4.3). According to the parameter values used in the present volume (Pelkonen and Hari 1980; Pelkonen 1981; Hänninen and Hari 2002), full photosynthetic capacity is attained at the temperature of +10.8 °C. In all, then, the model predicts that the photosynthetic machinery of southern Finnish *Pinus sylvestris* trees recovers when the air temperature rises from zero to +10.8 °C in spring.

There is an apparent resemblance between Pelkonen and Hari's (1980) photosynthesis model (Fig. 4.3a) and the post-rest model based on an extension of Vegis's (1964) theory, which addresses the manifestation of the dormancy status of trees in the air temperature response of the rate of ontogenetic development (Fig. 3.11b in Sect. 3.3.1.2). In both models, the air temperature response of the rate of development changes by moving to higher or lower temperatures as the developmental status of the tree changes. However, the ecophysiological interpretation of the changing air temperature responses is different in the two models. This is not only because the post-rest model addresses fixed-sequence development and the photosynthesis model, fluctuating development, but also because in the photosynthesis model, the moving of the air temperature response towards higher temperatures describes the activation of the tree, i.e., increasing photosynthetic capacity, whereas in the post-rest model the corresponding move describes the inactivation, i.e., the deepening of the dormancy status, of the tree.

This comparison also reveals a more general major difference between models of photosynthetic capacity and those of ontogenetic development. The value of $S_p(t)$ predicted by Pelkonen and Hari's (1980) photosynthesis model, for instance, stays relatively unchanged and near its maximum value in summer, i.e., the value of the rate variable, $R_p(t)$, is near zero (Hänninen and Hari 2002). However, the stability of $S_p(t)$ does not indicate inactivity of photosynthesis but a maximal rate of it, facilitated by the unchanged physiological condition of the photosynthetic machinery. In this sense, Pelkonen and Hari's (1980) fluctuating model differs drastically from the fixed-sequence models describing ontogenetic development (Sect. 3.2.1), where a stable value of the state variable, i.e., a near-zero value of the rate variable, indicates inactivity in ontogenetic development.

The difference discussed in the previous paragraph can be understood by examining the state variable in Pelkonen and Hari's (1980) model, i.e., the photosynthetic capacity of plants. The ecophysiological process ultimately addressed by the model is photosynthetic production; therefore, the state variable (photosynthetic capacity) actually indicates the *rate* of photosynthetic production (Fig. 4.2). Consequently, the rate variable in Pelkonen and Hari's (1980) model actually addresses the rate of change of the rate of photosynthetic production, thus being the second derivative of the original process with respect to time. Physically, the second derivative indicates acceleration; therefore, when the second derivative is zero, it indicates that the process, photosynthetic production in this case, progresses at a constant rate.[4]

In winter, sub-zero temperatures prevail, so that Pelkonen and Hari's (1980) model predicts negative values for the state of photosynthetic capacity, $S_p(t)$, (Hänninen and Hari 2002). These negative values may be interpreted to describe the deepness of the inactivity of photosynthesis, i.e., the more negative the value of $S_p(t)$, the longer the time needed for photosynthesis to recover after the air temperature rises above zero (Hänninen and Hari 2002).

Pelkonen and Hari (1980) tested their model (Eq. 4.2) in 1974–1976 with field measurements of *Pinus sylvestris* trees and seedlings growing in natural conditions in southern Finland. They determined the photosynthetic capacity on the basis of the entire photosynthetic light curve (Fig. 4.2d), estimated by means of field measurements in transparent cuvettes, and found that the model generally predicted the observed development of photosynthetic capacity well. In 1976 a clear case of

[4] The short-term fluctuations in the rate of photosynthetic production caused by short-term fluctuations of light are not considered here.

fluctuating development was seen during the last 2 weeks of April, when the photosynthetic capacity decreased because of a cold spell. That decrease was accounted for by the model during the last week of April, but for the previous week starting in mid-April the model predicted higher values of photosynthetic capacity than were observed. According to Pelkonen and Hari (1980), that was due to night frosts occurring at that time. Night frosts are known to impair the photosynthesis of conifers (Sect. 4.2.2), but this phenomenon is not addressed in Pelkonen and Hari's (1980) model, which considers only long-term fluctuating effects of air temperature.

Suni et al. (2003) studied the springtime recovery of photosynthesis in *Pinus sylvestris* and *Picea abies* at five locations representing a large geographical area, extending from southern Sweden to northern Finland and central Siberia. Using results of eddy covariance measurements and measurements of meteorological factors, they tested several hypotheses about the environmental regulation of the recovery of photosynthetic capacity. Pelkonen and Hari's (1980) model provided generally accurate predictions about the recovery, but is was outperformed by a simpler model, where the recovery is regulated by a 5-day running average of the air temperature. The model involving the running average is a simple version of the principle of fluctuating development (Fig. 4.3b) introduced in Pelkonen and Hari's (1980) model, and thus the large-scale eddy variance studies of Suni et al. (2003) provided essential evidence for the principle. In southern Sweden (the southernmost site), however, neither Pelkonen and Hari's (1980) model nor the model of the running average of the air temperature provided accurate predictions for the recovery (Suni et al. 2003).

4.2.1.2 Further Development of the Model

Mäkelä et al. (2004) developed Pelkonen and Hari's (1980) model further. Rather than use the equation of the double sigmoid (Eq. 4.2), they expressed their model with a formulation more commonly used to describe first-order dynamic systems, i.e., systems with one time constant (Repo et al. 1990; Hänninen and Kramer 2007). In Mäkelä et al.'s (2004) model, photosynthetic capacity is calculated in two phases. First, the rate of photosynthetic acclimation, $R_{pa}(t)$, is calculated as follows:

$$R_{pa}(t) = \frac{1}{\tau}\left(T(t) - S_{pa}(t)\right) \tag{4.4}$$

where τ = time constant, T(t) = air temperature at time instant t, and $S_{pa}(t)$ = state of photosynthetic acclimation at time instant t. By definition, the value of $S_{pa}(t)$ is obtained by integrating Eq. 4.4 with respect to time (Eq. 2.4 in Sect. 2.2). The state of acclimation, $S_{pa}(t)$, is expressed in units of temperature. It indicates the temperature to which the photosynthetic machinery is supposed to be acclimated, i.e., when $S_{pa}(t)$ = T(t), then the state of photosynthetic acclimation will stay at this constant value as long as the air temperature stays unchanged. The time constant τ

is defined as the time equal to 63 % of the time required to attain the new constant value of $S_{pa}(t)$ after a stepwise change in the air temperature (Mäkelä et al. 2004).

In Mäkelä et al.'s (2004) model, the quantum yield, α, is taken for a measure of photosynthetic capacity (Fig. 4.2c).[5] In the second phase of using the model, the value of α is calculated as follows:

$$\alpha\left(S_{pa}(t)\right) = \max\left\lfloor c_1\left(S_{pa}(t) - S_{pa,thr}\right), 0\right\rfloor \tag{4.5}$$

where α = modelled value of the quantum yield, $S_{pa,thr}$ = threshold value of $S_{pa}(t)$, and c_1 = coefficient for transforming the value of $S_{pa}(t)$ into the value of α. The value of the parameter $S_{pa,thr}$ indicates the air temperature above which the photosynthetic capacity, i.e., quantum yield α in this case, starts to increase.

It is instructive to compare the models of Pelkonen and Hari (1980) and Mäkelä et al. (2004) with each other, with regard to both their similarities and their differences. For the sake of brevity, these two models are referred to as "the old model" and "the new model", respectively, in the following. The similarities override the differences because both models are special cases of modelling the fully reversible fluctuating development of trees driven by changes in the air temperature. According to both models, the tree attains a stationary state at any constant air temperature (Repo et al. 1990; Repo 1993). In the old model this happens when $T(t) = S_p(t)/c$ (Eq. 4.2), i.e., when the air temperature, $T(t)$, stays at the value where the double sigmoid crosses the horizontal axis (Fig. 4.3a). In the new model this happens, correspondingly, when $T(t) = S_{pa}(t)$ (Eq. 4.4).

The main differences between the two models are the following. First, the air temperature response of the rate variable is sigmoidal in the old model (Eq. 4.2; Fig. 4.3a) but linear in the new model (Eq. 4.4). Second, to describe the inertia of the physiological reactions in the old model, the parameter a is introduced into the model (Eq. 4.2) rather than using the standard formulation applied in systems analysis, i.e., that of introducing the time constant τ in the denominator of the equation of the rate variable, as is done in the new model (Eq. 4.4). These two parameters are not identical, since a decrease in the value of parameter a has the same effect as an increase in the value of τ, i.e., both of these changes slow down the fluctuation of the state variable. Third, the formulation of the new model allows the estimation of the threshold temperature above which the recovery of photosynthesis begins (Eq. 4.5), whereas in the old model the recovery always begins when the temperature rises above zero. Fourth, the new model introduces no upper limit

[5] More precisely, the parameter used by Mäkelä et al. (2004) as a measure of photosynthetic capacity is the quantum yield per unit of internal carbon dioxide concentration. However, this parameter, based on a mechanistic modelling analysis of the empirical measurements, is closely related to the quantum yield based on the photosynthetic light curve (Fig. 4.2c), so that the concept of quantum yield is used for simplicity in the discussion of photosynthetic capacity as examined in Mäkelä et al.'s (2004) model. The same notation is applied in Sect. 4.4, where the year-to-year variation in the springtime recovery of photosynthetic capacity is examined in the light of Kolari et al.'s (2007) model.

to the increasing of the photosynthetic capacity (Eqs. 4.4 and 4.5). This is also the case in the rate equation of the old model (Eq. 4.2) but, as explained above, the old model introduces an additional upper limit by assuming no further increase in photosynthetic capacity after the maximal capacity has been attained (Eq. 4.3).

Mäkelä et al. (2004) combined their model of the annual cycle of photosynthetic capacity (Eqs. 4.4 and 4.5) with another model describing the immediate environmental responses of photosynthesis and the acclimation of photosynthesis to drought (Hari et al. 1986; Mäkelä et al. 1996; Hari and Mäkelä 2003). They fitted the combined overall model to data on meteorological factors and gas exchange of shoots of naturally growing *Pinus sylvestris* measured at the SMEAR I station, situated near the treeline in northern Finland, in 1997. They found that the quantum yield, α, and the light-saturated rate of photosynthesis, P_{max}, were linearly correlated in the measured data. Thus, even though α was used as an indicator of photosynthetic capacity (Fig. 4.2c), its changes were paralleled by corresponding changes in the whole photosynthetic light response (Fig. 4.2d).

According to measurements of photosynthesis in transparent cuvettes, Mäkelä et al.'s (2004) model fitted well to the observed changes in the quantum yield (Fig. 4.4b). This was the case not only during the long recovery period lasting from April to August but also during the period of declining quantum yield in the autumn. Reversals of the recovery, i.e., transient decreases of quantum yield in the spring, were generally also accounted for by the model. In some cases, however, e.g., in the decline occurring in the latter half of May, the model predicted higher quantum yields than were observed (Fig. 4.4b). As suggested by Pelkonen and Hari (1980) concerning their similar results from model testing, these few shortcomings, too, may have been caused by short-term effects of frost (Mäkelä et al. 2004). Those effects will be discussed in Sect. 4.2.2.

The value of $S_{pa,thr} = -4.5$ °C was obtained in the model fitting for the threshold value of the state of photosynthetic acclimation $S_{pa}(t)$ (Eq. 4.5) (Mäkelä et al. 2004). This finding shows that the recovery of photosynthesis in evergreen conifers may begin at subzero temperatures. In this respect, Mäkelä et al.'s (2004) model is a clear improvement, for in Pelkonen and Hari's (1980) model the recovery is expected a priori to start when the air temperature rises above zero.

The lack of an upper threshold parameter, corresponding to the air temperature beyond which the rising of air temperature would cause no further recovery of quantum yield α, was supported in the model fitting, as the measured values of α still continued to increase in August (Fig. 4.4). However, when analysing two sets of multi-year data on *Pinus sylvestris*, one gathered in northern and the other in southern Finland, Kolari et al. (2007) found that the relationship between photosynthetic capacity and the state of photosynthetic acclimation was sigmoidal rather than linear,[6] the latter being expected in Mäkelä et al.'s (2004) model (Eq. 4.5). These findings support the notion that even if there is no definite threshold air temperature where the increase in photosynthetic capacity abruptly ceases, the

[6] Figure 5 in Kolari et al. (2007). For copyright reasons, the figure is not presented here.

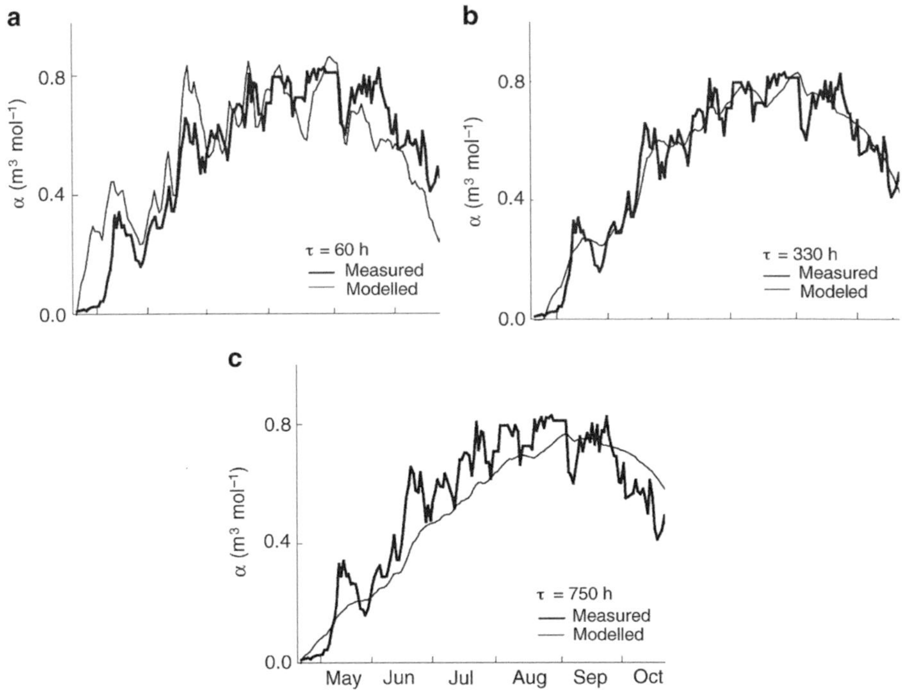

Fig. 4.4 Fitting the model of Mäkelä et al. (2004) to meteorological and photosynthesis data measured in *Pinus sylvestris* trees growing near the treeline at the SMEAR I station (Hari et al. 2008a) in Värriö, northern Finland. The *thick lines* indicate the observed photosynthetic capacity of the shoots, determined by means of measurements in transparent cuvettes. The capacity was determined as the quantum yield α of the photosynthetic *light curve* (Fig. 4.2c), but due to the correlation of α and P_{max}, an increase in α marks a simultaneous rise of the overall level of the photosynthetic light curve (Fig. 4.2d). The *thin line* denotes the modelled photosynthetic capacity, calculated with the fitted values of the parameters in Eqs. 4.4 and 4.5. In addition to (**b**) the fitted optimal value of time constant $\tau = 330$ h, the results are also shown for (**a**) lower than optimal value and for (**c**) higher than optimal value of τ (Mäkelä et al. 2004; redrawn with the permission of Oxford University Press)

increase may level off at the relatively high temperatures typical for the growing season (Kolari et al. 2007).

The model fitting results of Mäkelä et al. (2004) also demonstrated the effect of the time constant τ on the modelled photosynthetic capacity (Fig. 4.4). With the optimal value of $\tau = 330$ h obtained in the model fitting, the pattern of quantum yield α calculated by the model accorded well with the observed pattern (Fig. 4.4b). With lower than optimal values of τ, the quantum yield α predicted by the model fluctuated more than the observed one did, and both the recovery of the quantum yield in the spring and the decline of it in the autumn were predicted to occur earlier than observed (Fig. 4.4a). Correspondingly, with higher than optimal values of τ, the predicted quantum yield α fluctuated less than the observed one did, and the recovery of the quantum yield in the spring and the decline of it in the autumn were predicted to occur later than observed (Fig. 4.4c).

Hari et al. (2009) tested Mäkelä et al.'s (2004) model with temporally independent data, i.e., with data gathered at the same field station but in other years (1998 and 1999) than the 1997 data originally used by Mäkelä et al. (2004) for fitting the model. The values of two parameters specific to the measured shoots were re-estimated with the data for each of the two measurement years, but the other nine parameters were fixed to the values determined when the model was fitted to the 1997 data (Mäkelä et al. 2004). By this use of temporally independent data, the researchers facilitated the examination of the predictive power of the model. The model predicted the seasonal pattern of daily photosynthetic production well for both 1998 (Fig. 4.5a; percentage of explained variance PEV = 93 %) and 1999 (Fig. 4.5b; PEV = 92 %). Furthermore, the daily patterns of the rate of photosynthesis were also predicted accurately, with the PEV values of 94 % for 1998 and 88 % for 1999. The results of Hari et al. (2009) were confirmed by Hari et al. (2008b), who extended the test of Mäkelä et al.'s (2004) model by using temporally independent data for 6 years (2001–2006) gathered both in northern and in southern Finland at the SMEAR I and SMEAR II stations, respectively.

These results from the test with temporally independent data (Hari et al. 2008b, 2009) demonstrated the high predictive power of Mäkelä et al.'s (2004) model. This notion received added emphasis from the fact that such tests of photosynthesis models with temporally independent shoot-level data have been carried out only rarely. In fact, Hari et al.'s (2008b, 2009) tests are evidently the only ones so far that have been carried out with several time scales simultaneously.

Lundell et al. (2008) applied a fluctuating model basically similar to Mäkelä et al.'s (2004) model to examining the annual cycle of photosynthetic capacity in the boreal dwarf shrub *Vaccinium vitis-idea*. According to their model-fitting results, the recovery of photosynthetic capacity in *Vaccinium vitis-idea* begins at the air temperature of −1.8 °C, and full capacity is attained at +10.4 °C. These results show that the reversible fluctuating model can also be applied to other growth forms of boreal woody evergreen plants besides coniferous trees (Lundell et al. 2008).

4.2.1.3 Contradictory Evidence from Wintertime Photosynthesis

Using eddy covariance measurements, Sevanto et al. (2006) detected four periods in winter 2002–2003 where photosynthesis occurred in a *Pinus sylvestris* stand during mild spells in southern Finland. The uptake of carbon dioxide by the shoots of *Pinus sylvestris* was also confirmed by cuvette measurements. Measureable photosynthesis took place in October, February, and during two periods in March. The photosynthesis began when the air temperature rose to +3 or +4 °C, and in February, it was not prevented even during days with preceding night frosts down to −17 °C. Sevanto et al. (2006) also found that photosynthesis started more rapidly in mid-winter than in March.

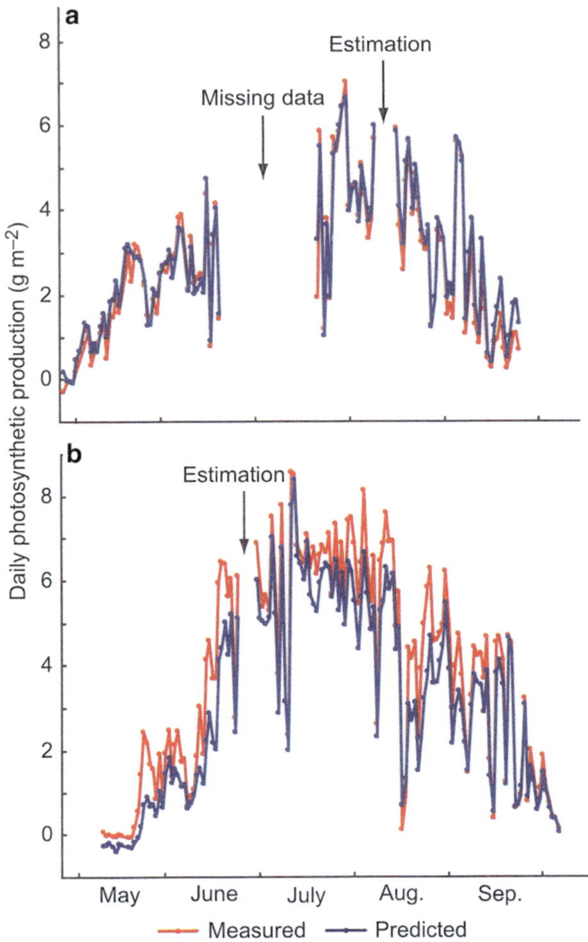

Fig. 4.5 A field test of the photosynthesis model of Mäkelä et al. (2004), using temporally independent data gathered for *Pinus sylvestris* trees growing near the treeline at the SMEAR I station (Hari et al. 2008a) in Värriö, northern Finland, in (**a**) 1998 and (**b**) 1999. With the exception of two shoot-specific parameters, the parameter values were fixed to the values obtained when the model was fitted to meteorological and photosynthesis data gathered at the same station in 1997. The *red curves* indicate the observed daily photosynthetic production based on measurements in transparent cuvettes, and the *blue curves* the corresponding predictions calculated by the model. The gaps in the curves were caused either by missing observations due to technical failures at the measuring station ("Missing data") or by rejection of observations used for estimating the shoot-specific parameters ("Estimation") (Hari et al. 2009; redrawn with the permission of the Boreal Environment Research Publishing Board)

The observations of Sevanto et al. (2006) were confirmed by Hari and Bäck (2008), who studied photosynthesis in *Pinus sylvestris* at the same station as Sevanto et al. (2006) did. Hari and Bäck (2008) found a rapid activation of photosynthesis as a response to the air temperature rising to about +4 °C in early

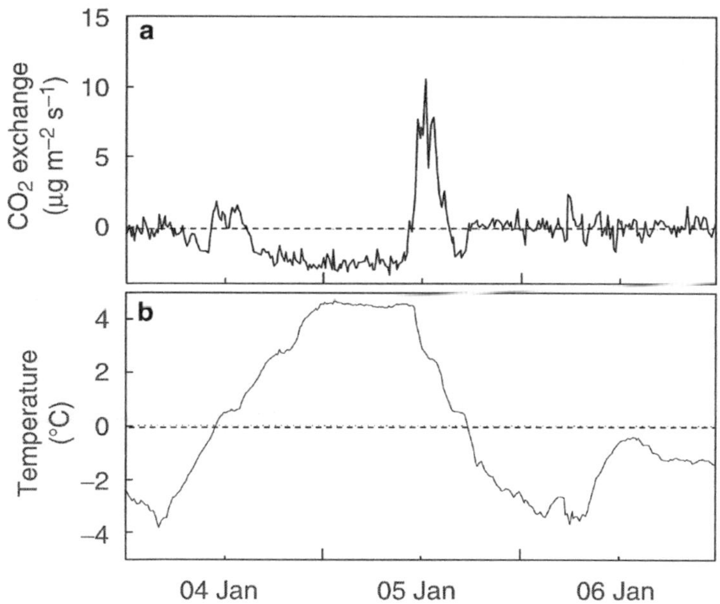

Fig. 4.6 (**a**) The rate of CO_2 exchange in *Pinus sylvestris* shoots and (**b**) the air temperature over three January days at the SMEAR II station (Hari et al. 2008a) in Hyytiälä, southern Finland (Hari and Bäck 2008; redrawn with the permission of Springer Science+Business Media)

January (Fig. 4.6). Furthermore, Hari and Bäck (2008) found interesting differences in the response of photosynthesis to air temperature later in the spring. On 24 April, the air temperature fluctuated a few degrees above zero, with the exception of the last hours of the day (Fig. 4.7b), but no photosynthesis took place on that day (Fig. 4.7a). On 10 May, high levels of photosynthesis were measured (Fig. 4.7c) though there had been frost in the morning and the daytime air temperature fluctuated at approximately the same level as on 24 April (Fig. 4.7d).

Ensminger et al. (2004) studied the seasonality of photosynthesis in *Pinus sylvestris* in central Siberia. On the basis of laboratory measurements with detached twigs they concluded that the trees are able to activate their photosynthesis opportunistically whenever the environmental conditions get favourable in winter. This conclusion is in agreement with the wintertime photosynthesis documented in natural conditions for the same species in southern Finland (Sevanto et al. 2006; Hari and Bäck 2008).

All of these findings contradict the prevailing models of the annual cycle of photosynthetic capacity, which imply that the trees have no remarkable photosynthetic capacity in winter before the air temperature has been rising for a prolonged period (Pelkonen and Hari 1980; Mäkelä et al. 2004; Kolari et al. 2007). Due to the limited amount of incoming solar radiation in winter, this shortcoming does not necessarily cause great errors in the modelled photosynthetic production of the

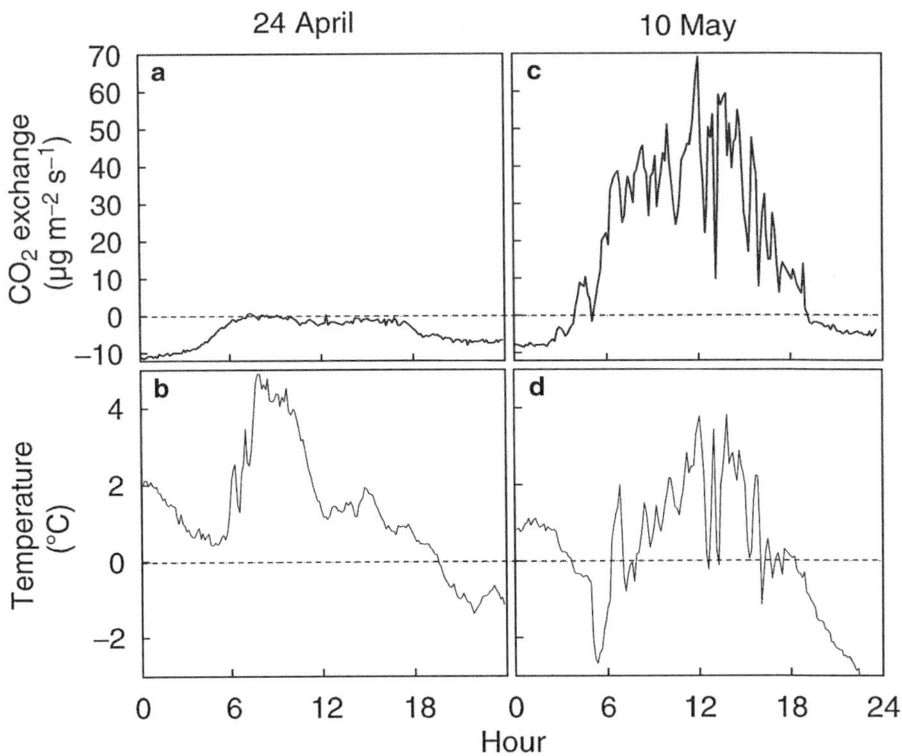

Fig. 4.7 (**a**, **c**) The rate of CO_2 exchange in *Pinus sylvestris* shoots and (**b**, **d**) the air temperature on (**a**, **b**) 24 April and (**c**, **d**) 10 May at the SMEAR II station (Hari et al. 2008a) in Hyytiälä, southern Finland (Hari and Bäck 2008; redrawn with the permission of Springer Science+Business Media)

trees. However, the failure may be a symptom of a lack of realism in the models. This notion is addressed in Sect. 4.3.1, where the effects of light on photosynthetic capacity are discussed.

4.2.2 Effects of Frost

Freezing air temperatures inhibit the dark reactions of photosynthesis and thus cause an abrupt decrease in photosynthetic capacity (Strand and Öquist 1988). Furthermore, a night with frost delays the daily onset of photosynthesis (Polster and Fuchs 1963). However, if the freezing stress is not too severe, the decrease of photosynthetic capacity caused by frost is reversible in basically the same way as the transient decrease of photosynthetic capacity caused by low non-freezing temperatures, which is discussed in Sect. 4.2.1. This was clearly demonstrated by Ensminger et al. (2008), who examined the effects of two intermittent events of

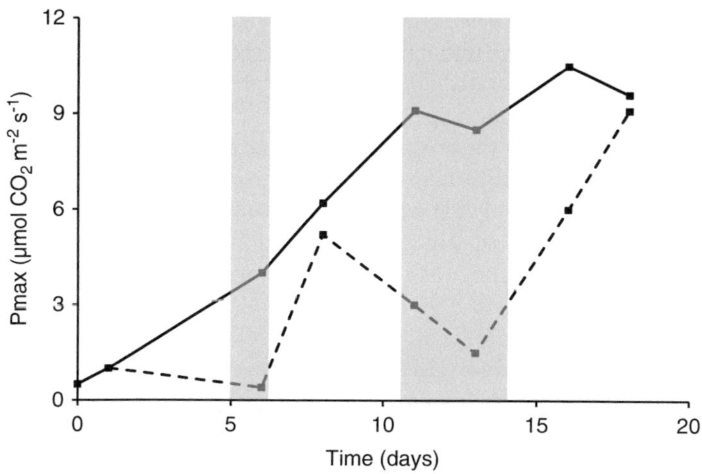

Fig. 4.8 Effects of two intermittent events at air temperature −2 °C on the recovery of photosynthetic capacity in first-year *Pinus sylvestris* seedlings. With the exception of those events, indicated by shading, the air temperature was +15 °C during the 18 days of the experiment. The *continuous line* denotes the control seedlings and the *dashed line* the seedlings exposed to the intermittent events with frost (Ensminger et al. 2008; redrawn with the permission of John Wiley & Sons, Inc.)

2 and 4 days' duration with the air temperature of −2 °C on the recovery of photosynthetic capacity in first-year *Pinus sylvestris* seedlings over 18 days at the air temperature of +15 °C (Fig. 4.8). Both intermittent frost events almost nullified the recovery that had taken place before the exposure to frost, but in both cases a new recovery occurred soon after, so that after a few days the photosynthetic capacity of the frost-treated seedlings was at the same level as that of the control seedlings (Fig. 4.8).

None of the fluctuating models discussed in Sect. 4.2.1 properly addresses the short-term effects of frost. Even so, as explained above, some discrepancies between the predicted and the observed photosynthetic capacity were discussed by the authors and attributed to unaccounted short-term effects of frost (Pelkonen and Hari 1980; Mäkelä et al. 2004). Kolari et al. (2007) examined this notion systematically with their long-term data gathered in 1999–2005 both in Värriö, northern Finland, and in Hyytiälä, southern Finland. They compared the prediction of their fluctuating model, which does not address the short-term effects of frost, either, with the observed photosynthetic capacity and found a biased residual as a function of the minimum air temperature of the preceding night.[7] The results showed that for a day after a night frost, the model predicted a higher photosynthetic capacity than was observed and that the error increased with dropping frost temperatures. However, unless the air temperature dropped below −5 °C, the

[7] Figure 8 in Kolari et al. (2007). For copyright reasons, the figure is not presented here.

adverse effect of a single frost night was no longer distinguishable after 1 day (Kolari et al. 2007). These findings, obtained from long-term data gathered in natural conditions, corroborated the experimental findings (Fig. 4.8) about the short-term effects of frost on the photosynthetic capacity of *Pinus sylvestris*.

The studies discussed in the preceding two paragraphs show that the short-term effects of frost on the photosynthetic capacity of evergreen conifers should also be addressed in the ecophysiological models predicting photosynthetic capacity. The effects of night frosts are actually addressed in the model of Bergh et al. (1998), who modelled the springtime recovery of photosynthesis in *Picea abies* trees growing in northern Sweden. The model predicted the observed recovery quite accurately. However, in Bergh et al.'s (1998) model, only frosts cause decreases in photosynthetic capacity. Otherwise the springtime development of photosynthetic capacity is modelled in terms of an irreversible day degree model, i.e., the reversing effects of other low temperatures than frosts are not accounted for. Therefore Hänninen and Hari (2002) suggested that in further model development, a synthesis of these two models, i.e., the fluctuating long-term effects of air temperature (Pelkonen and Hari 1980) and the short-term effects of night frosts (Bergh et al. 1998), should be considered.

4.3 Effects of Other Environmental Factors

4.3.1 *Effects of Light and Improving the Realism of the Models*

The models based on air temperature-related factors alone have provided quite an accurate description of the springtime recovery of evergreen conifers (Sect. 4.2). However, the decreased photosynthetic capacity is largely a manifestation of the acclimation of the photosynthetic machinery to the adverse conditions of low air temperature and high level of incoming radiation (Sect. 4.1.2.2). This suggests that the prevailing models addressing the effects of air temperature on the annual cycle of photosynthesis probably include an implicit description of the acclimation of evergreen conifers to variations in the light climate as well. The findings of Hari and Bäck (2008), for instance, are in accordance with this notion. They found considerable photosynthesis in southern Finnish *Pinus sylvestris* during mild spells in winter (Fig. 4.6), whereas in approximately similar air temperature conditions in late April, no photosynthesis took place (Fig. 4.7a, b). Even though the effects of incoming solar radiation were not explicitly examined in these studies, these circumstantial findings accord with the notion of the important effect of strong springtime solar radiation on the photosynthetic capacity of evergreen conifers.

Given the relatively good understanding that we have of the physiological mechanisms underlying the annual cycle of photosynthetic capacity (Sect. 4.1.2.2), we will probably see in the future a development of more mechanistic models to explicitly describe the interaction of light and air temperature and the acclimation of conifers to that interaction. Clearly, this would be a considerable improvement of the realism of the models. In such mechanistic model development, the effects of other physiological attributes, such as the nutritional status of the needles, on the susceptibility of conifers to photoinhibition (Strand and Lundmark 1995) may also be addressed.

4.3.2 Effects of Soil Temperature and Ground Frost

Soil temperature and ground frost have a major effect on the functioning of the roots and the water uptake of evergreen conifers growing in the boreal zone (Troeng and Linder 1982). Since the photosynthesis of trees requires water, it seems a priori clear that soil temperature and ground frost also affect the annual cycle of photosynthetic capacity in these trees. However, the empirical evidence for the latter point is contradictory. Several studies carried out in natural conditions have provided evidence against the notion. Thus, when examining the springtime recovery of photosynthetic capacity in *Pinus sylvestris* and *Picea abies* by means of eddy covariance measurements carried out at five locations representing a large geographical area, Suni et al. (2003) tested, among other hypotheses, the hypothesis that soil thaw regulates the timing of the recovery. They found no support for the hypothesis. Indeed, in one case the recovery started 1 month before the soil thaw. Similarly, Kolari et al. (2007) reported, for both northern and southern Finland, that the recovery of photosynthesis in *Pinus sylvestris*, determined on the basis of gas exchange measurements of shoots in cuvettes, started well before the warming up of the soil.

These findings may be partially explained by the availability of liquid water in the otherwise frozen soil. In a study carried out in Finnish Lapland, Sutinen et al. (2009) found that during spring snowmelt, some of the water generated by the strong incoming solar radiation penetrated the partially frozen soil, so that an increase in the water content of the soil and the roots of *Betula pubescens* was observed 2–6 weeks before the air temperature rose above zero.

However, the hypothesis of the effects of soil temperature and ground frost on the recovery of the photosynthetic capacity of evergreen conifers has also found some support in experimental studies. Bergh and Linder (1999) studied the effects of soil warming on the recovery of photosynthesis in 32-year-old *Picea abies* trees in a field experiment in northern Sweden. Snowmelt and soil thaw took place more than a month earlier in the heated plots than in the control ones. The light-saturated rate of photosynthesis, P_{max}, was statistically significantly higher in the heated plots than in the control ones during May, when the soil temperature was 4–5 °C higher in the heated plots than in the control ones. However, there were no differences between the heated plots and the control ones in the photochemical efficiency of photosystem II, F_v/F_m. Bergh and Linder (1999) concluded that in the case of *Picea abies* growing in northern Sweden, the recovery of photosynthesis is mostly regulated by the air temperature and the occurrence of night frost but that soil thawing has some effect on it also.

Repo et al. (2005) studied the effects of soil thaw on several physiological attributes of *Pinus sylvestris* saplings in a rhizotron experiment. Before the initiation of the experimental treatments, the soil was frozen and the air temperature was +4 °C with all experimental saplings. The air temperature was raised to +20/+15 °C (day/night) simultaneously in all the four treatments, but the timing of the soil thaw was varied among them. It was either 14 days before the raising of the air temperature, simultaneous with it, delayed by 14 days, or delayed by 24 days. The delayed soil thaw caused stress to the saplings, as shown by several physiological attributes. Their photosynthetic capacity, as indicated by photochemical efficiency, F_v/F_m, was drastically decreased by the delayed soil thaw. Finally, the delayed soil thaw proved lethal to the seedlings.

Ensminger et al. (2008) examined the effects of cold (+1 °C) and frozen (−2 °C) soil on the recovery of photosynthesis in first-year *Pinus sylvestris* seedlings during 18 days in an experiment where the air temperature was +15 °C in all the treatments. The recovery of photosynthetic capacity, as indicated by the light-saturated rate of photosynthesis, P_{max}, was slowed down by both the cold and the frozen soil.

In conclusion, the effects of soil temperature and ground frost on the recovery of photosynthesis remain unclear. However, on the basis of experimental results (Bergh and Linder 1999; Repo et al. 2005; Ensminger et al. 2008), the effects of the soil temperature and the timing of soil thaw cannot be ruled out, though measurements in natural conditions (Suni et al. 2003; Kolari et al. 2007) do not support the significance of such effects. Therefore, the effect of the timing of soil thaw, and possibly also the effect of the soil temperature, may need to be also addressed in the dynamic modelling of the annual cycle of photosynthetic capacity in evergreen conifers.

4.4 Year-to-Year Variation in the Recovery of Photosynthesis

There is wide year-to-year variation in the springtime development of photosynthetic capacity. Pelkonen (1980) studied the recovery of southern Finnish *Pinus sylvestris* shoots in natural conditions during the three consecutive years of 1974–1976. He defined the state of photosynthetic capacity, S_p, on the basis of the overall level of the photosynthetic light curve, i.e., by both P_{max} and α (Fig. 4.2d). In 1974 and 1975, the value of S_p increased almost monotonically from about 20 April onwards, and full capacity was attained on about 20 May and 10 May, respectively. The almost monotonic increase of S_p in these 2 years was caused by a relatively steady rise of air temperature in April and early May.

In 1976, the air temperatures dropped in late April, causing a marked reversal of the recovery. Thus the value of S_p was considerably lower at the beginning of May

1976 than it had been in the other two years, when the approximate value of $S_p = 50$ % had been attained by that time. Similarly, Hari et al. (2009) found a marked difference between two consecutive years in the starting time of the recovery of photosynthesis in *Pinus sylvestris* trees growing near the treeline in Värriö, northern Finland. In 1998, photosynthetic production started in early May (Fig. 4.5a), whereas in 1999 it started several weeks later (Fig. 4.5b).

The full extent of the year-to-year variation in the springtime development of photosynthetic capacity driven by air temperature is best illustrated by simulations with long-term meteorological data. Such a simulation was carried out for the present volume with Kolari et al.'s (2007) model, using the same long-term meteorological data from Jyväskylä, central Finland, as were used in Chaps. 1 and 3. The simulations were carried out using the daily mean air temperature as input.

In their analysis of empirical photosynthesis data measured in transparent cuvettes, Kolari et al. (2007) determined the photosynthetic capacity as the quantum yield α of the photosynthetic light curve (Fig. 4.2c; however, see footnote 5 in Sect. 4.2.1.2). In order to remove the variation among the measured shoots from the analysis, the quantum yield of each was scaled by dividing the observed value of α by the average quantum yield α_{10} calculated from measurements of the same shoot whenever the air temperature was in the range of +9 – +11 °C. This relative photosynthetic capacity, α/α_{10}, was also used in modelling by Kolari et al. (2007), and the same procedure was applied here. The state of photosynthetic capacity, $S_{pa}(t)$, was first calculated with Eq. 4.4, using the value of 8.3 days for the time constant τ. Subsequently, the relative photosynthetic capacity, α/α_{10}, was modelled to depend sigmoidally on S_{pa} according to the parameter values estimated by Kolari et al. (2007) for southern Finnish *Pinus sylvestris*:

$$\alpha/\alpha_{10} = \left[\frac{1.36}{1 + \exp\left(-0.27\left(S_{pa}(t) - 5.47\right)\right)}\right] \tag{4.6}$$

Air temperatures may drop considerably in different phases of the annual temperature cycle. This phenomenon underlines the importance of the reversal of the recovery of photosynthetic capacity, as indicated by the simulations carried out for 1886 and 1891 (Fig. 4.9a). February was relatively mild in 1891, so that a slight recovery was predicted for that winter month already. In March, however, the temperatures dropped, so that the photosynthetic capacity was near zero in late March, as was also the case at that time of the year 5 years earlier, in 1886 (Fig. 4.9a). March and most of April were exceptionally warm in 1886 but close to average in 1891. Thus a considerable difference developed between these two years in the photosynthetic capacity predicted for that period (Fig. 4.9a).

However, in 1891 the air temperatures rose steadily in April, whereas in 1886 they dropped drastically towards the end of the month. Accordingly, a rapid reversal of the recovery occurred in 1886, so that the photosynthetic capacity at the end of April was again at the same level in both years (Fig. 4.9a). In May, the air

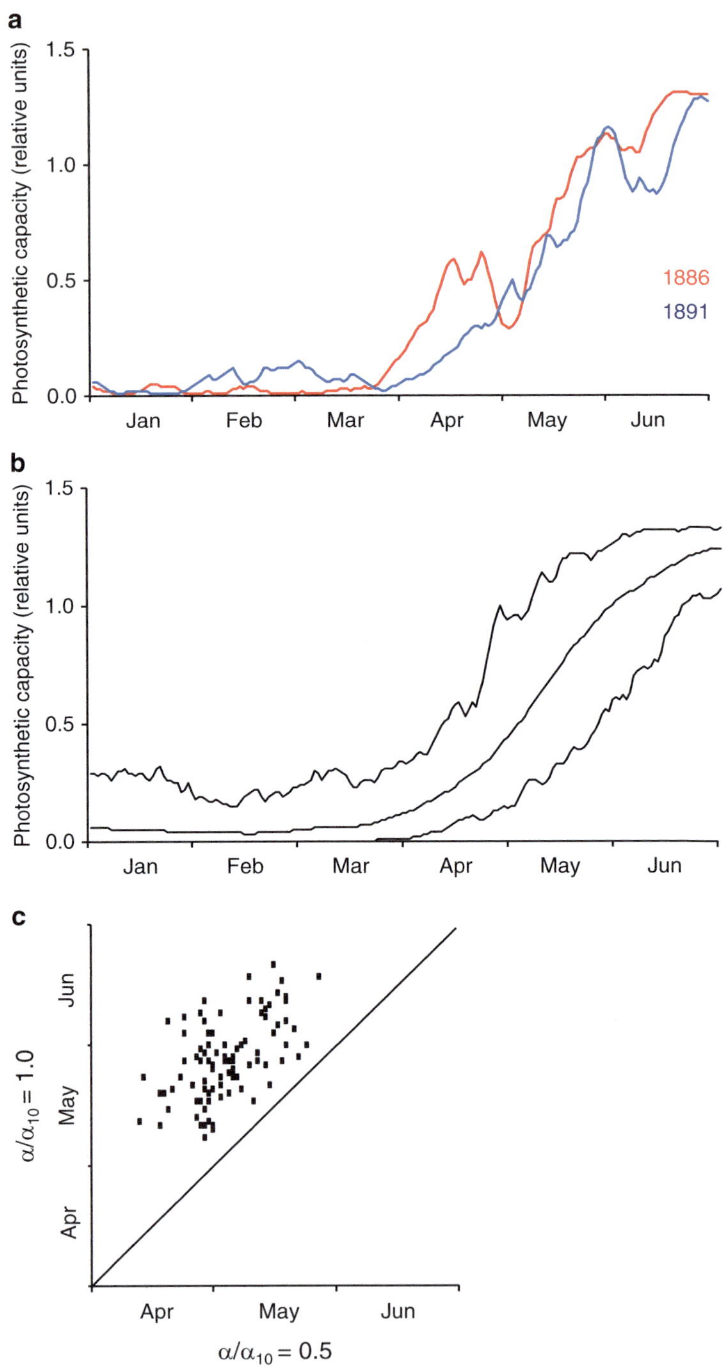

Fig. 4.9 Simulated year-to-year variation in the springtime air-temperature-driven recovery of photosynthetic capacity in *Pinus sylvestris* trees in Jyväskylä, central Finland, over 92 years (1883–1980; the years 1911–1916 were excluded because of missing meteorological

temperatures rose quite steadily in both years, so that no drastic differences developed between the years in the photosynthetic capacity during that month. At the end of May there were practically no differences, but then the 1891 air temperatures dropped. Thus the photosynthetic capacity was again higher in 1886 than in 1891 for most of June, until a common level was attained again at the end of June (Fig. 4.9a).

On the average, the model predicts some, even if low, photosynthetic capacity even for the winter months (Fig. 4.9b), for the empirical observations that are behind it have shown that air temperatures between −10 and −5 °C are sufficient to sustain a low photosynthetic capacity (Kolari et al. 2007). These values are, however, so small as to be negligible. Kolari et al. (2007) concluded that the recovery starts at air temperatures above −4 °C, which is near the threshold value of $S_{pa,thr} = -4.5$ °C suggested by Mäkelä et al. (2004) on the basis of analyses of data gathered during one year (Sect. 4.2.1.2). During the mildest winters, however, the relative photosynthetic capacity α/α_{10} is in the range of 0.1–0.2 (Fig. 4.9b). Thus photosynthesis may start in the warmest winters if there is enough light. On the average, the recovery of photosynthetic capacity starts at the beginning of April and is almost completed by the end of May, as shown by the attainment of the value of $\alpha/\alpha_{10} = 1.0$ for the relative photosynthetic capacity (Fig. 4.9b). Thus, even if the attainment of the highest photosynthetic capacities in early summer were not required in the interpretation of the results of the modelling, the predicted recovery would still take two months on average.

The time of attaining a given value of the relative photosynthetic capacity, α/α_{10}, in the spring varies by about one and a half months among different years (Fig. 4.9b). The year-to-year variation of air temperature, and therefore also that of the predicted temperature-driven recovery of photosynthesis, is further attested by the finding that the attainment of a given level of α/α_{10} does not provide any information about how the recovery will continue. For instance, after the level of $\alpha/\alpha_{10} = 0.5$, the date of which varies by about one and a half months, has been attained (Fig. 4.9c, horizontal axis), it will take anything from a few days to more than a month, varying from year to year, to attain the value $\alpha/\alpha_{10} = 1.0$ (Fig. 4.9c, difference between the 1:1 line and the data point value on the vertical axis). From the point of view of the photosynthetic production of trees, then, the climate is not only variable but also unpredictable.

Fig. 4.9 (continued) observations). The simulations were carried out with the sigmoidal version of the model of Kolari et al. (2007) (see text for details). (**a**) The simulated development of relative photosynthetic capacity, α/α_{10}, in 1886 (*red curve*) and in 1891 (*blue curve*). (**b**) Year-to-year minimum (*lower curve*), mean (*middle curve*) and maximum (*upper curve*) values of the daily α/α_{10} over the 92 years. (**c**) The relationship between the early and the late progress of the recovery of photosynthetic capacity over the 92 years. In the plot, each *square* represents one year, so that the day when the value of $\alpha/\alpha_{10} = 0.5$ was first attained is represented by the *horizontal axis* and the later day when the corresponding $\alpha/\alpha_{10} = 1.0$ was first attained is represented by the *vertical axis*. To facilitate the comparison, the straight 1:1 line describes the theoretical case in which both values of α/α_{10} were attained on the same date

When interpreting the results indicated in Fig. 4.9, one should remember that the simulations addressed only the fluctuating effects of air temperature; the potential effects of other factors, such as frost (Sect. 4.2.2), light (Sect. 4.3.1), soil temperature, and the timing of soil thaw (Sect. 4.3.2) were not addressed in the simulations. Furthermore, as discussed in Sect. 4.2.1.3, coniferous trees may start to photosynthesize even during brief (e.g., one-day) mild periods in winter, even though the current models of photosynthetic capacity predict that these periods are too short for the initiation of photosynthesis. Thus, the aspects of wintertime photosynthesis not accounted for by the model now applied may further increase the year-to-year variation in the initiation of photosynthesis from the variation predicted in Fig. 4.9.

4.5 Summary

Photosynthetic capacity is a relatively well-defined ecophysiological whole-tree attribute of the seasonality of boreal evergreen conifers. In broad terms it indicates the inherent photosynthesis potential of the needles, thus being an emergent manifestation of the physiological status of the photosynthetic machinery in the needles. Photosynthetic capacity is usually measured by means of the photosynthetic light curve. It is usually defined, then, as equal either to the light-saturated rate of photosynthesis or to the quantum yield, i.e., the slope of the photosynthetic light curve at low light levels. The physiological mechanisms of change in the photosynthetic capacity are relatively well understood, but a detailed discussion of them is beyond the scope of this volume.

Over the last decades, the annual cycle of photosynthetic capacity has been modelled with two types of dynamic ecophysiological models. The older type of model addresses the long-term fluctuating effects of air temperature (Pelkonen and Hari 1980). This study deserves special attention, also because it introduced the principle of fluctuating development into the modelling of the annual cycle of boreal and temperate trees. The model has been reformulated later on, but the original idea of fluctuating effects of air temperature has not been dropped.

In the more recent type of model, these fluctuating effects are approximated with a simple irreversible day degree model, but in addition, the reversing effects of short-term frosts on the recovery of photosynthesis are also explicitly addressed (Bergh et al. 1998). These simple models have described the annual cycle of photosynthetic capacity, especially the springtime recovery of photosynthesis, relatively accurately. As discussed in Sects. 7.1.2 and 8.5.2, both of these model types have also been used as sub-models in large ecosystem models in studies of the carbon balance and growth of tree stands.

Despite their relatively high accuracy, neither of the current model types of the photosynthetic capacity in evergreen boreal conifers evidently has particularly high realism. That is because neither of them explicitly addresses the effects of a combination of high incoming solar radiation with low air temperatures, which is known to be an essential factor in the physiological mechanisms behind the changes in photosynthetic capacity observed at the whole-tree level. Given the relatively good understanding that we have of the physiological mechanisms underlying the annual cycle of photosynthetic capacity, we will probably see in the future a development of more mechanistic models to explicitly describe the interactions of light and air temperature and the acclimation of conifers to these interactions.

Among the other environmental factors, soil temperature and ground frost are the most important ones potentially affecting the annual cycle of photosynthetic capacity. The evidence for their effects is somewhat contradictory, but evidently these factors need to be taken into account in further modelling studies, at least with some species and environments.

List of Symbols

t time

Time-Dependent Environmental Variables

$T(t)$ air temperature (°C)

Time-Dependent Rate and State Variables

$R_p(t)$ Rate of change of photosynthetic capacity ($\% \, h^{-1}$, $\% \, day^{-1}$)
$S_p(t)$ State of photosynthetic capacity (%)
$R_{pa}(t)$ Rate of photosynthetic acclimation ($°C \, h^{-1}$, $°C \, day^{-1}$)
$S_{pa}(t)$ State of photosynthetic acclimation (°C)

Model Parameters

t_0 starting moment of the simulations
a parameter affecting the rate of change of photosynthetic capacity
c parameter defining the photosynthetic capacity in any constant temperature
$S_{p,crit}$ critical threshold for full recovery of photosynthetic capacity
τ time constant (h, day)
c_1 slope of the dependence of quantum yield on state of photosynthetic acclimation
$S_{pa,thr}$ threshold temperature for initiation of recovery of photosynthesis (°C)

Other Symbols

α	quantum yield per unit of internal carbon dioxide concentration ($m^3\ mol^{-1}$)
α_{10}	average α in temperatures +9 – +11 °C ($m^3\ mol^{-1}$)
α/α_{10}	relative photosynthetic capacity
P_{max}	light saturated rate of photosynthesis ($\mu mol\ m^{-2}\ s^{-1}$)

References

Baldocchi, D. D. (2003). Assessing the eddy covariance technique for evaluating carbon dioxide exchange rates of ecosystems: Past, present and future. *Global Change Biology, 9*, 479–492.

Bauer, H., Nagele, M., Comploj, M., Galler, V., Mair, M., & Unterpertinger, E. (1994). Photosynthesis in cold acclimated leaves of plants with various degrees of freezing tolerance. *Physiologia Plantarum, 91*, 403–412.

Bergh, J., & Linder, S. (1999). Effects of soil warming during spring on photosynthetic recovery in boreal Norway spruce stands. *Global Change Biology, 5*, 245–253.

Bergh, J., McMurtrie, R. E., & Linder, S. (1998). Climatic factors controlling the productivity of Norway spruce: A model-based analysis. *Forest Ecology and Management, 110*, 127–139.

Chang, C. Y., Unda, F., Zubilewich, A., Mansfield, S. D., & Ensminger, I. (2015). Sensitivity of cold acclimation to elevated autumn temperature in field-grown *Pinus strobus* seedlings. *Frontiers in Plant Science, Article 6*, 165. doi:10.3389/fpls.2015.00165.

Chapin, F. S., III, Matson, P. A., & Mooney, H. A. (2002). *Principles of terrestrial ecosystem ecology*. New York: Springer. 436 p.

Demmig-Adams, B., Adams, W. W., III, & Mattoo, A. K. (Eds.). (2008a). *Photoprotection, photoinhibition, gene regulation, and environment* (Advances in photosynthesis and respiration, Vol. 21). Dordrecht: Springer. 380 p.

Demmig-Adams, B., Ebbert, V., Zarter, C. R., & Adams, W. W., III. (2008b). Characteristics and species-dependent employment of flexible versus sustained thermal dissipation and photoinhibition. In B. Demmig-Adams, W. W. Adams III, & A. K. Mattoo (Eds.), *Photoprotection, photoinhibition, gene regulation, and environment* (Advances in photosynthesis and respiration, Vol. 21, pp. 39–48). Dordrecht: Springer.

Ensminger, I., Sveshnikov, D., Campbell, D. A., Funk, C., Jansson, S., Lloyd, J., Shibistova, O., & Öquist, G. (2004). Intermittent low temperatures constrain spring recovery of photosynthesis in boreal Scots pine forests. *Global Change Biology, 10*, 995–1008.

Ensminger, I., Schmidt, L., & Lloyd, J. (2008). Soil temperature and intermittent frost modulate the rate of recovery of photosynthesis in Scots pine under simulated spring conditions. *New Phytologist, 177*, 428–442.

Hänninen, H., & Hari, P. (2002). Recovery of photosynthesis of boreal conifers during spring: A comparison of two models. *Forest Ecology and Management, 169*, 53–64.

Hänninen, H., & Kramer, K. (2007). A framework for modelling the annual cycle of trees in boreal and temperate regions. *Silva Fennica, 41*, 167–205.

Hari, P., & Bäck, J. (2008). General. In P. Hari & L. Kulmala (Eds.), *Boreal forest and climate change* (pp. 172–176). Dordrecht: Springer Science + Business Media B.V.

Hari, P., & Mäkelä, A. (2003). Annual pattern of photosynthesis in Scots pine in the boreal zone. *Tree Physiology, 23*, 145–155.

Hari, P., Mäkelä, A., Korpilahti, E., & Holmberg, M. (1986). Optimal control of gas exchange. *Tree Physiology, 2*, 169–175.

Hari, P., Nikinmaa, E., Vesala, T., Pohja, T., Siivola, E., Lahti, T., Aalto, P., Hiltunen, V., Ilvesniemi, H., Keronen, P., Kolari, P., Grönholm, T., Palva, L., Pumpanen, J., Petäjä, T., Rannik, Ü., & Kulmala, M. (2008a). SMEAR network. In P. Hari & L. Kulmala (Eds.), *Boreal forest and climate change* (pp. 46–56). Dordrecht: Springer Science + Business Media B.V.

Hari, P., Kolari, P., Bäck, J., Mäkelä, A., & Nikinmaa, E. (2008b). Photosynthesis. In P. Hari & L. Kulmala (Eds.), *Boreal forest and climate change* (pp. 231–242). Dordrecht: Springer Science + Business Media B.V.

Hari, P., Hänninen, H., Berninger, F., Kolari, P., Nikinmaa, E., & Mäkelä, A. (2009). Predicting boreal conifer photosynthesis in field conditions. *Boreal Environment Research, 14*(suppl. A), 19–28.

Kolari, P., Lappalainen, H. K., Hänninen, H., & Hari, P. (2007). Relationship between temperature and the seasonal course of photosynthesis in Scots pine at northern timberline and in southern boreal zone. *Tellus, 59B*, 542–552.

Kolari, P., Pulkkinen, M., Kulmala, L., Pumpanen, J., Nikinmaa, E., Mäkelä, A., Vesala, T., & Hari, P. (2008). Forest ecosystem CO_2 exchange. In P. Hari & L. Kulmala (Eds.), *Boreal forest and climate change* (pp. 351–382). Dordrecht: Springer Science + Business Media B.V.

Kolari, P., Chan, T., Porcar-Castell, A., Bäck, J., Nikinmaa, E., & Juurola, E. (2014). Field and controlled environment measurements show strong seasonal acclimation in photosynthesis and respiration potential in boreal Scots pine. *Frontiers in Plant Science,* 5, *Article 717*. doi:10.3389/fpls.2014.00717.

Lappalainen, H. (2010). Role of temperature in the biological activity of a boreal forest. Helsinki: *Finnish Meteorological Institute Contributions 84*. 52 p.

Larcher, W. (2003). *Physiological plant ecology. Ecophysiology and stress physiology of functional groups* (4th ed.). Berlin: Springer. 513 p.

Linder, S., & Lohammar, T. (1981). Amount and quality of information on CO_2-exchange required for estimating annual carbon balance of coniferous trees. In: Linder, S. (Ed.), Understanding and predicting tree growth. *Studia Forestalia Suecica, 160*, 73–87.

Linkosalo, T., Lappalainen, H. K., & Hari, P. (2008). A comparison of phenological models of leaf bud burst and flowering of boreal trees using independent observations. *Tree Physiology, 28*, 1873–1882.

Linkosalo, T., Heikkinen, J., Pulkkinen, P., & Mäkipää, R. (2014). Fluorescence measurements show stronger cold inhibition of photosynthetic light reactions in Scots pine compared to Norway spruce as well as during spring compared to autumn. *Frontiers in Plant Science,* 5, *Article 264*. doi:10.3389/fpls.2014.00264.

Lundell, R., Saarinen, T., Åström, H., & Hänninen, H. (2008). The boreal dwarf shrub *Vaccinium vitis-idaea* retains its capacity for photosynthesis through the winter. *Botany, 86*, 491–500.

Lundmark, T., Bergh, J., Strand, M., & Koppel, A. (1998). Seasonal variation of maximum photochemical efficiency in boreal Norway spruce stands. *Trees, 13*, 63–67.

Mäkelä, A., Berninger, F., & Hari, P. (1996). Optimal control of gas exchange during drought: Theoretical analysis. *Annals of Botany, 77*, 461–467.

Mäkelä, A., Hari, P., Berninger, F., Hänninen, H., & Nikinmaa, E. (2004). Acclimation of photosynthetic capacity in Scots pine to the annual cycle of temperature. *Tree Physiology, 24*, 369–376.

Pelkonen, P. (1980). The uptake of carbon dioxide in Scots pine during spring. *Flora, 169*, 386–397.

Pelkonen, P. (1981). *Recovery and cessation of CO_2 uptake in Scots pine at the beginning and at the end of the annual photosynthetic period.* Helsinki: *University of Helsinki, Department of Silviculture. Research Notes* 30. 95 p.

Pelkonen, P., & Hari, P. (1980). The dependence of the springtime recovery of CO_2 uptake in Scots pine on temperature and internal factors. *Flora, 169*, 398–404.

Petäjä, T., Siivola, E., Pohja, T., Palva, L., & Hari, P. (2008). On field measurements. In P. Hari & L. Kulmala (Eds.), *Boreal forest and climate change* (pp. 20–46). Dordrecht: Springer Science + Business Media B.V.

Pisek, A., & Winkler, E. (1958). Assimilationsvermögen und Respiration der Fichte (Picea excelsa Link) in verschiedener Höhenlage und der Zirbe (Pinus cembra L.) an der alpinen Waldgrenze. *Planta, 51*, 518–543.

Polster, H., & Fuchs, S. (1963). Winterassimilation und –atmung der Kiefer (*Pinus sylvestris* L.) im mitteldeutschen Binnenlandklima. *Archiv für Forstwesen, 12*, 1011–1023.

Porcar-Castell, A., Olascoaga Gracia, B., Juurola, E., & Hari, P. (2013). Acclimation of photosynthesis to the environment: An optical respective. In P. Hari, K. Heliövaara, & L. Kulmala (Eds.), *Physical and physiological forest ecology* (pp. 125–137). Dordrecht: Springer Science +Business Media.

Repo, T. (1993). *Impedance spectroscopy and temperature acclimation of forest trees*. Joensuu: *University of Joensuu, Faculty of Forestry. Research Notes* 9. 53 p.

Repo, T., Mäkelä, A., & Hänninen, H. (1990). Modelling frost resistance of trees. In: Jozefek, H. (Ed.) *Modelling to understand forest functions. Silva Carelica 15*, 61–74.

Repo, T., Kalliokoski, T., Domisch, T., Lehto, T., Mannerkoski, H., Sutinen, S., & Finér, L. (2005). Effects of timing of soil frost thawing on Scots pine. *Tree Physiology, 25*, 1053–1062.

Saarinen, T., Rasmus, S., Lundell, R., Kauppinen, O.-K., & Hänninen, H. (2015). Photosynthetic and phenological responses of dwarf shrubs to the depth and properties of snow. *Oikos, 125*, 364–373.

Sevanto, S., Suni, T., Pumpanen, J., Grönholm, T., Kolari, P., Nikinmaa, E., Hari, P., & Vesala, T. (2006). Wintertime photosynthesis and water uptake in a boreal forest. *Tree Physiology, 26*, 749–757.

Strand, M., & Lundmark, T. (1995). Recovery of photosynthesis in 1-year-old needles of unfertilized and fertilized Norway spruce (*Picea abies* (L.) Karst.) during spring. *Tree Physiology, 15*, 151–158.

Strand, M., & Öquist, G. (1988). Effects of frost hardening, dehardening and freezing stress on *in vivo* chlorophyll fluorescence of seedlings of Scots pine (*Pinus sylvestris* L.). *Plant, Cell and Environment, 11*, 231–238.

Suni, T., Berninger, F., Vesala, T., Markkanen, T., Hari, P., Mäkelä, A., Ilvesniemi, H., Hänninen, H., Nikinmaa, E., Huttula, T., Laurila, T., Aurela, M., Grelle, A., Lindroth, A., Arneth, A., Shibistova, O., & Lloyd, J. (2003). Air temperature triggers the recovery of evergreen boreal forest photosynthesis in spring. *Global Change Biology, 9*, 1410–1426.

Sutinen, R., Vajda, A., Hänninen, P., & Sutinen, M.-L. (2009). Significance of snowpack for root-zone water and temperature cycles in subarctic Lapland. *Arctic, Antarctic, and Alpine Research, 41*, 373–380.

Troeng, E., & Linder, S. (1982). Gas exchange in a 20-year-old stand of Scots pine. I. Net photosynthesis of current and one-year-old shoots within and between seasons. *Physiologia Plantarum, 54*, 7–14.

Vegis, A. (1964). Dormancy in higher plants. *Annual Review of Plant Physiology, 15*, 185–224.

Chapter 5
The Annual Cycle of Frost Hardiness

Contents

Abstract The hypothetico-deductive modelling framework introduced in Chap. 2 is applied to modelling the annual cycle of frost hardiness in boreal and temperate trees. The concept of frost hardiness is straightforward as such, denoting the threshold temperature that distinguishes between the damaging low-temperature range and the undamaging one. In practice, however, the concept of frost hardiness needs to be specified in regard to the tree tissue addressed and the methods used in assessing it. Similarly to the models of the annual cycle of photosynthetic capacity discussed in Chap. 4, the models of the annual cycle of frost hardiness can be tested continuously on a daily basis. The main environmental factors regulating the annual cycle of frost hardiness are air temperature and night length. These environmental factors regulate the annual cycle of frost hardiness directly and also indirectly by

H. Hänninen, *Boreal and Temperate Trees in a Changing Climate*, Biometeorology 3,
DOI 10.1007/978-94-017-7549-6_5

affecting the annual phenological cycle, which in turn changes the direct response. This complexity calls for development of integrated models that address both the fixed-sequence aspects of the phenological cycle and the fluctuating aspects of frost hardiness as such. Integrated models have been developed since the 1970s, and two of them are discussed in detail here. Both models are quite accurate, but there is still room for considerable improvement in their realism. Novel simulations using long-term air temperature records as input for an integrated model revealed large year-to-year variation in the annual cycle of frost hardiness.

Keywords Annual frost hardiness cycle • Boreal trees • Computer simulations • Dehardening • Frost damage • Hardening • Integrated model • Model realism • Seasonality • Stationary frost hardiness • Temperate trees

5.1 The Phenomena and the Measurements Involved

5.1.1 Frost Hardiness as a Whole-Tree Attribute of Tree Ecophysiology

Even in common language, the meaning of the frost hardiness of trees is readily understood. The concept denotes the threshold temperature that delineates the damaging low temperature range from the undamaging one. Accordingly, damage occurs when the air temperature drops below the prevailing frost hardiness of the tree (Weiser 1970; Sakai and Larcher 1987). Boreal and temperate trees exhibit an annual cycle of frost hardiness, i.e., their frost hardiness changes seasonally (Fuchigami et al. 1982; Repo 1992; Fig. 5.1). During the active growth phase, the trees are susceptible to frost. Depending on the species and provenance or cultivar, their frost hardiness is only a few degrees Celsius at that time of the year. In late summer and autumn, the process of frost hardening, i.e., an increase in the hardiness of the trees, takes place. Frost hardening is a relatively slow process, so that it needs to start several weeks before the air temperatures drop to the low values typical for winter (Weiser 1970; Sakai and Larcher 1987; Fig. 5.1).

Though the frost hardiness of trees may fluctuate somewhat in winter, the trees generally retain their frost hardiness so that on any given day there is a safety margin between their frost hardiness and the daily minimum air temperature. Thus, as the daily minimum temperatures may be dozens of degrees below zero, the mid-winter frost hardiness of boreal tree species is often in the range of −50 to −100 °C. At the time of their maximal frost hardiness, the tissues of some conifers tolerate the temperature of liquid nitrogen, i.e., −196 °C (Sakai and Weiser 1973;

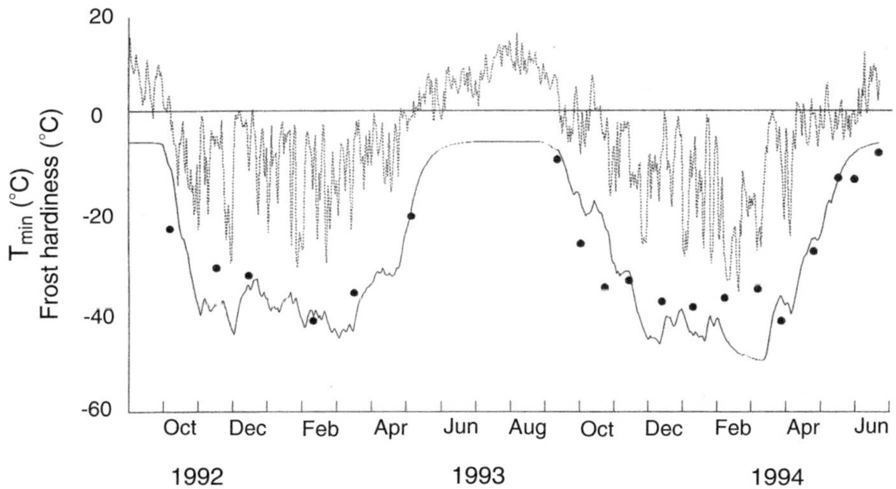

Fig. 5.1 Seasonal development of the daily minimum temperature, T_{min} (*upper curve*) and needle frost hardiness of *Pinus sylvestris* in Mekrijärvi, eastern Finland. The *black circles* indicate the observed needle frost hardiness as measured by Repo et al. (1996) by means of the electrolyte leakage method and the *lower curve* the prediction of a frost hardiness model (Sect. 5.3.2.2) (Modified from Hänninen et al. 1996; published with the permission of the Finnish Forest Research Institute, currently the Natural Resources Institute Finland, and the Finnish Society of Forest Science)

Sutinen et al. 1992). In spring, the process of dehardening, i.e., decrease in the hardiness of the trees, takes place (Fuchigami et al. 1982; Fig. 5.1).[1]

Southern tree species, provenances, or cultivars often suffer from frost damage if transferred too far to the North. The heavy frosts of the northern mid-winter easily exceed the hardening capacity of southern trees. But other times of the year may also be detrimental to transferred trees. Thus, even relatively slight summer frosts may damage or destroy trees adapted to warmer conditions. The critical time is often either autumn or spring. This is because the autumn hardening and the spring dehardening must be synchronized with the falling and rising of the air temperature in these seasons. Thus, both delayed hardening in autumn and premature dehardening in spring may be detrimental to the tree (Bannister and Neuner 2001; Aitken and Hannerz 2001).

The wide year-to-year variation in seasonal air temperature patterns emphasized in the previous chapters sets additional requirements to the climatic adaptation of boreal and temperate tree species. Trees transferred over long distances often

[1] The concepts "frost hardening" and "dehardening" are often replaced with "cold acclimation" and "deacclimation", respectively. In the present volume, the former concepts are used because the concept "acclimation" is reserved for use in a general sense referring to any physiological changes whereby the tree adjusts its physiology to meet the requirements of the prevailing environmental conditions.

survive and grow for several years but finally get damaged or killed as a result of a year with climatic conditions harsher than average (Rousi 1985). The effects of the year-to-year variation in air temperature on the annual cycle of frost hardiness are discussed in Sect. 5.4.

Though the physiological basis of the frost hardiness of trees is relatively well understood, it is outside the scope of the present volume, so that the reader is referred to the several reviews available on that topic (e.g., Sakai and Larcher 1987; Sutinen et al. 2001; Sutinen 2007; Holliday 2009). However, even when the concept of frost hardiness is used exclusively at the whole-tree level, as it is in the present volume, several aspects of it need to be specified whenever the phenomenon is addressed in ecophysiological research.

First, the tree tissue to be addressed has to be specified. In boreal and temperate trees, the roots are generally less hardy than the shoots (Zhang et al. 2003; Räisänen et al. 2009), but there is also considerable variation among the various tissues of the shoot. For instance, the buds are usually less hardy than the cambium (Sakai and Larcher 1987). This is readily understood in evolutionary terms, for the loss of a few vegetative buds is not necessarily detrimental to the tree, but the destruction of the cambium is. Second, the degree of damage used in the determination of frost hardiness has to be specified. Especially when the trees are in the mid-winter state of frost hardiness, there can be a difference of dozens of degrees between the temperature causing the first slight symptoms of damage and the temperature killing the tree (Repo et al. 1994). Third, several aspects of the frost hardiness test, such as the rates of cooling and thawing and the duration of the minimum stress temperature in the artificial freezing test, need to be specified. It is crucial to consider all these specifications when interpreting the results of frost hardiness measurements, especially when comparing those from different studies.

5.1.2 Methods for Assessing Frost Hardiness

The methods used for assessing the frost hardiness of boreal and temperate trees have been reviewed, among others, by Burr et al. (2001), here, only a brief overview is provided. Almost all of the methods available are based on artificial freezing of the samples to be examined.

One family of methods is based on the phenomenon of supercooling, i.e., the lowering of the temperature of a solution below freezing point without formation of ice crystals, and the heat liberated when the water in the solution finally freezes (Wisniewski 1995; Quamme 1995). In the Differential Thermal Analysis, DTA, the tissue samples to be examined are exposed to a freezing treatment with a gradually dropping temperature. The temperature of the tested tissue is monitored during the programmed freezing with thermocouples. Additionally, the temperature of inert dry reference material, sharing the same thermal properties as the tree tissue tested, is similarly monitored. During controlled freezing, the temperature of the tree tissue and that of the reference material drop in a similar manner until the moment when the extracellular water, i.e., the water located outside the cells in the tree tissue, freezes. At this moment the high-temperature exotherm, HTE, appears, i.e., there is a rapid transient rise in the temperature of the tested tree tissue while no such rise is seen in the temperature of the reference material.

With further freezing, the temperature of the tree tissue and that of the reference material drop similarly again until another rapid transient rise is seen in the temperature of the tree tissue but not in the temperature of the reference material. This low-temperature exotherm, LTE, is caused by the freezing of the intracellular water in the tree tissues. Intracellular freezing is lethal to the cells of trees or any other plants, and therefore the temperature corresponding to the LTE is taken as the measure of the frost hardiness of the tested tree tissue.

Unlike the DTA, most of the other methods applying artificial freezing are based on subdividing the material to be tested, e.g., seedlings, twigs, or single needles, into treatment groups and exposing the different groups to different stress temperatures. Treatments with more than one stress temperature are needed for obtaining the stress temperature response to be used in assessments of frost hardiness. In practice a priori information, such as the results obtained in previous tests, may be used for defining the stress temperatures to be applied. In controlled freezing, several freezer chambers are used (Fig. 5.2a). A sample of the material to be tested is placed in each of the freezers, where an air temperature program consisting of three phases is run (Fig. 5.3). First, in the cooling phase the temperature is lowered at a constant rate. After attaining the predefined stress temperature specific for the chamber, the air temperature is kept constant for a predefined period of time. Then the thawing phase is started. The air temperature is raised at a constant rate, which is usually the same as the cooling rate in the first phase.

The whole sequence is run with each of the programmable freezers, with the stress temperature, i.e., the minimum temperature reached in the treatment, varying among the freezers. Optimally, the other specifications of the temperature program, i.e., the rate of cooling, the duration of the exposure to the stress temperature, and the rate of raising the temperature in the thawing phase, are the same in all of the programmable freezers. Accordingly, the temperature program lasts longer with low stress temperatures than with high ones (Fig. 5.3).

After the artificial freezing, the degree of damage caused by the different stress temperatures is assessed and a damage index is calculated for each stress temperature. There are several methods available for that purpose. In the most elementary way, whole seedlings, shoots of seedlings detached from the root system, or twigs are used in the test (Fig. 5.2a) and the damage index is determined on the basis of the mortality of the seedlings or twigs observed by monitoring them, typically in a greenhouse, after the controlled freezing (Fig. 5.2b). The proportion of dead seedlings is plotted against the stress temperature, and a sigmoidal response function is fitted to the data. By means of the response curve, an LT_{50}-value, i.e., the stress temperature that is lethal for 50 % of the tested material, is determined and taken as an estimate of the average frost hardiness in the population examined (Fig. 5.4a).

Besides the mortality of the twigs or seedlings, non-lethal damages can also be assessed by means of various forms of visual damage scoring (VDS) after the freezing treatments. For instance, the degree of browning in the needles or in the stem tissue (Räisänen et al. 2009) can be assessed. In that case, the determination of frost hardiness is based on the stress temperature response of a VDS index calculated from these assessments.

Currently, the most frequently used method for determining the degree of damage after controlled freezing is evidently the electrolyte leakage test (Sutinen et al. 1992; Räisänen et al. 2009; Hänninen et al. 2013). It is an electrophysiological method based on the phenomenon where electrolytes escape from the intracellular to the extracellular space whenever a plant cell gets damaged. The electrolytes escape due to injuries in the plasma membrane, which is the actual site of freezing injury (Sutinen et al. 2001). These electrolytes are extracted into water, and the conductivity of the solution is subsequently measured. The higher the conductivity value, the

a

Fig. 5.2 (**a**) A freezing chamber used in measuring the frost hardiness of tree seedlings or organs. The plastic bags contain shoots of *Pinus sylvestris* seedlings. To prevent root damage caused by the freezing, the shoots have been detached from the root system. (**b**) Visible damage caused to *Pinus sylvestris* seedlings by artificial freezing. To prevent wilting, the cut ends of the detached shoots are kept in water (The unpublished photographs were provided by Risto Rikala)

b

Fig. 5.2 (continued)

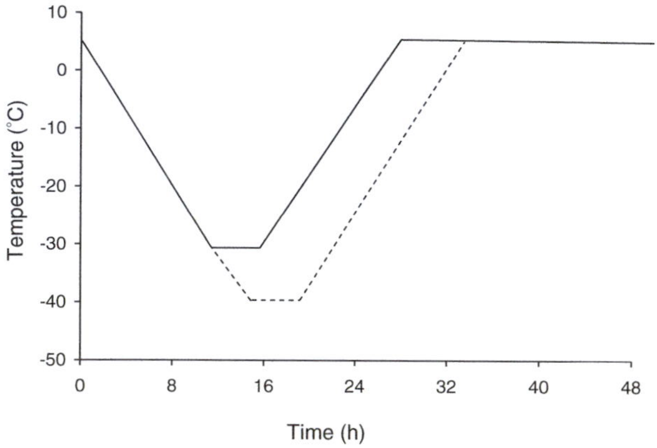

Fig. 5.3 The principle of the artificial freezing test used in the experimental determination of the frost hardiness of trees. For the sake of clarity, the air temperature programme is shown only for two treatments with their corresponding stress temperatures, i.e., −30 °C and −40 °C. In this example, the cooling and thawing rates are −2.9 °C h^{-1} and the duration of the stress temperature is 4 h

heavier the damage. In order to get comparable results, the maximal conductivity of the solution, i.e., the value obtained after the tissue has been completely destroyed, needs to be known. To this end, when the conductivity after the stress temperature exposure has been measured, the examined tissues are destroyed, e.g., by

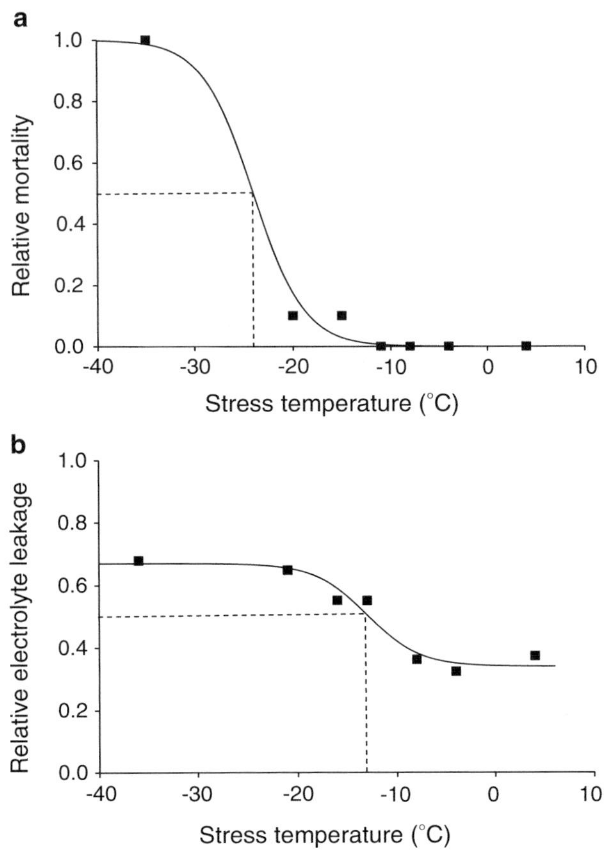

Fig. 5.4 Determination of the frost hardiness of first-year *Betula pendula* seedlings by means of an artificial freezing test: the stress temperature response of (**a**) relative mortality and (**b**) relative electrolyte leakage of the seedlings. The measurements were carried out in Suonenjoki, central Finland, on 13 October 1997. The frost hardiness estimates based on the mid-points of the fitted sigmoidal curves are −24 °C and −13 °C, respectively, as determined on the basis of mortality and the electrolyte leakage method. The frost hardiness value of −24 °C is referred to as the LT_{50}-value of the seedling population examined, as that temperature is lethal for 50 % of the seedlings (Modified from Luoranen 2000; published with the permission of the Finnish Forest Research Institute, currently the Natural Resources Institute Finland, and the Finnish Society of Forest Science)

autoclaving the samples, and then the conductivity is measured again. After that, the relative electrolyte leakage, REL, is calculated by dividing the conductivity value obtained in the first measurement with the value obtained in the second one. Lastly, the frost hardiness of the tissues is determined on the basis of the stress temperature response of the REL (Fig. 5.4b) in a way basically analogous to the mortality test (Fig. 5.4a).

Generally, the electrolyte leakage test detects relatively small sub-lethal damages. Accordingly, it generally indicates a lower degree of hardiness for the tested material than the method based on mortality (Fig. 5.4). With any method, however, other points of the sigmoidal response curve than the mid-point can also be used for estimating frost hardiness. In practical applications used in seedling nurseries, for instance, the air temperature causing slight damage to the most susceptible individual seedlings is usually of more interest than the temperature causing the death of a half of the seedling population. In those cases, frost hardiness can be determined, e.g., so as to correspond to the stress temperature causing an increase of 10 % in the REL value obtained in the electrolyte leakage test.

In the last few decades, methods based on measuring the electrical impedance of the tree tissue have been developed and used in studies of the frost hardiness of trees. A crucial contribution to these studies has been made in Finland by Tapani Repo and his co-workers (Repo and Pelkonen 1986; Repo et al. 1994, 2000; Zhang et al. 2003). More recently, they have used impedance methods for studying the root morphology of trees as well (Chao et al. 2010, 2011). Since the details of the method are outside the scope of the present volume and can be readily found in the publications referred to, only a brief outline is offered here.

Like the measurement of electrolyte leakage, the impedance method is an electrophysiological method. Rather than extracting the electrolytes into a solution, however, the impedance method is used on tree tissue that is intact. Two electrodes are pushed through the tissue to be examined, typically the stem of a tree seedling, and alternating electric current is led to the tissue. In early applications of the method, only a single frequency of the current, typically $f = 1$ kHz, was used. In this way, an electrical impedance value is obtained, measured in ohms (Ω), i.e., the same unit as that of electrical resistance. As the purpose of the measurement is to identify the properties of the material forming the tree tissue, the measured value of electrical impedance needs to be calibrated by multiplying it by the cross-sectional area of the measured stem (m^2) and dividing it by the distance between the electrodes (m). In this way, the value of specific impedance, z, with the unit of Ωm, is obtained (Repo and Pelkonen 1986; Rikala and Repo 1987; Repo 1991).[2]

As a result of damage, the value of the specific impedance decreases as electrolytes escape to the extracellular space. Accordingly, the specific impedance is measured both before and after the controlled freezing, and a specific impedance difference, Δz, is calculated as the difference between the results obtained from the two measurements. Finally, similarly to the mortality and electrolyte leakage tests (Fig. 5.4), a damage index, in this case Δz, is plotted against the stress temperature used, and frost hardiness is determined by means of fitting the sigmoidal curve to the data. Usually, frost hardiness has been determined as equal to the stress temperature corresponding to the value $\Delta z = -10\ \Omega$m of the specific impedance difference (Repo and Pelkonen 1986; Rikala and Repo 1987; Repo 1991).

More recently, the method based on Δz has been replaced by the method of impedance spectroscopy (Repo et al. 1994, 2000; Zhang et al. 2003). That is a

---

[2] The older method is described here because part of the results discussed later in this chapter were obtained by means of it.

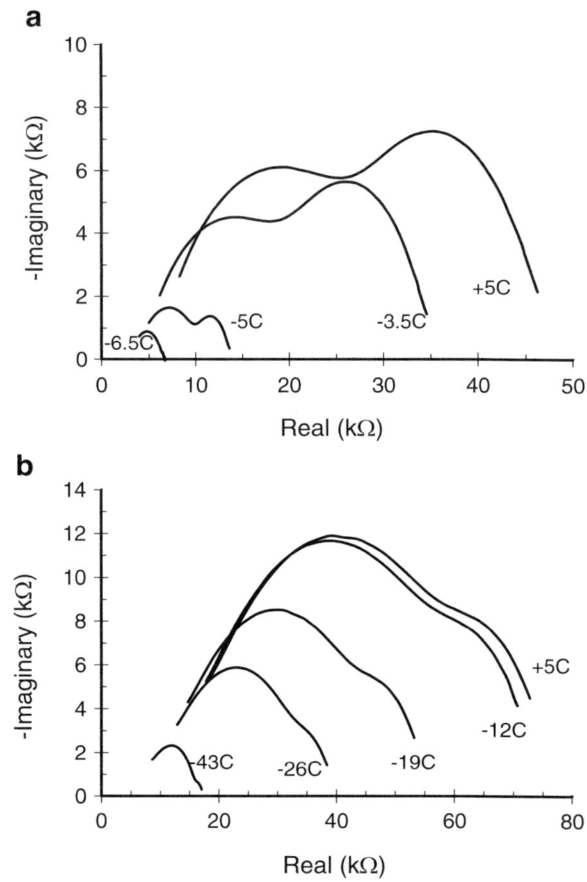

Fig. 5.5 Impedance spectra of (**a**) unhardened and (**b**) hardened stems of second-year *Pinus sylvestris* seedlings. In each spectrum, the frequency of the alternating current used in the measurement varies from f = 80 Hz at the right end to f = 1 MHz at the left end of the spectrum. Before the impedance measurements, the seedlings were exposed to the temperatures indicated at the right end of each spectrum (Repo et al. 1994; Burr et al. 2001; redrawn with the permission of Oxford University Press)

method in which alternating current, with several frequencies rather than just a single one, is led to the tree tissue or any other material examined. A wide range of frequencies, typically from 1 daHZ to 1 MHz, is used in the measurement. With all frequencies used, the measurement produces values for two variables, i.e., the real part and the imaginary part of the impedance. The impedance spectrum is obtained by plotting the imaginary part against the real part across the frequencies used (Fig. 5.5).

Frost damage changes the impedance spectrum of tree tissue. For example, when unhardened second-year *Pinus sylvestris* seedlings were exposed to the stress temperatures of −5 °C or −6.5 °C, the average impedance spectrum of their stems changed considerably from the one measured after exposure to the control temperature of +5 °C (Fig. 5.5a). With hardened *Pinus sylvestris* seedlings, a roughly similar change was observed after exposing the seedlings to the stress temperature of −43 °C (Fig. 5.5b).

The measured impedance spectra are analysed and interpreted by means of equivalent electrical circuit models (Repo et al. 1994, 2000). Accordingly, the

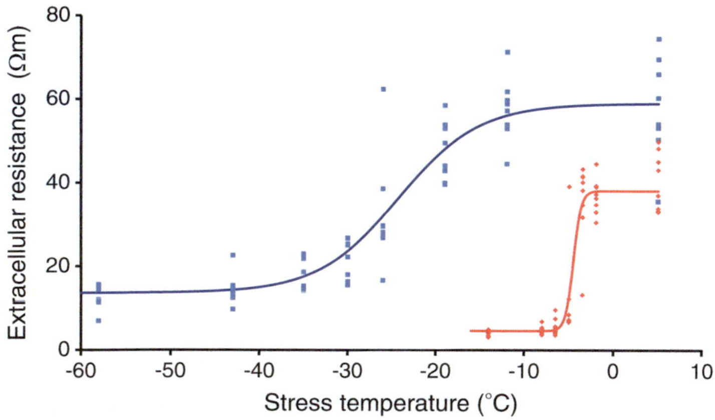

Fig. 5.6 Determination of the frost hardiness of stems of second-year *Pinus sylvestris* seedlings by means of artificial freezing and impedance spectroscopy. The frost hardiness is determined by plotting the value of one parameter obtained from the impedance spectroscopy, i.e., the extracellular resistance, against the stress temperature and fitting a sigmoidal response *curve* to the data. Two examples, one for hardened seedlings (*blue*) and one for unhardened ones (*red*), are shown. The frost hardiness estimated on the basis of the mid-points of the two sigmoidal curves is approximately −25 °C for the hardened seedlings and −5 °C for the unhardened ones (Repo et al. 1994; Burr et al. 2001; published with the permission of Oxford University Press)

values of a number of parameters are estimated by fitting a circuit model to the data forming the impedance spectrum. The values of some of the estimated parameters can be interpreted to describe physiological attributes of the tree tissues examined. One of the parameters, i.e., the extracellular resistance, has been found to reliably reflect the damage done by exposure to frost; so, like the other indices of frost damage (Fig. 5.4), the response of the extracellular resistance to the stress temperature is currently used in determining the frost hardiness of trees (Fig. 5.6).

In conclusion, impedance spectroscopy provides a powerful tool for obtaining versatile information on the physiological status of the tree by means of quite simple measurements taken at the whole-tree level. In this sense, impedance spectroscopy resembles the measurements of chlorophyll fluorescence carried out in studies concerned with photosynthesis (Sect. 4.1.2.2). For some reason, however, impedance spectroscopy has been used for assessing the frost hardiness of trees only by relatively few researchers so far.

Regardless of the damage index used in assessing the frost hardiness of trees, there is a considerable difference in the response to the stress temperature between unhardened and hardened trees. With unhardened trees, the range from the temperature causing the first signs of damage to the temperature causing total damage is narrow, and there is relatively little variation among the samples exposed to a given stress temperature, but with hardened trees, the damaging temperature range is wider, and so is the variation (Fig. 5.6). This implies that the experimental determination of frost hardiness is less precise during maximal hardiness in winter than during other parts of the annual cycle.

5.1.3 Implications for Modelling

In this chapter the modelling approach introduced in Chap. 2 is applied to modelling the annual cycle of frost hardiness (Hänninen and Kramer 2007). Accordingly, the variable state of development, S(t), is specified as the variable *state of frost hardiness*, $S_h(t)$, and the variable rate of development, R(t), correspondingly as the variable *rate of change in frost hardiness*, $R_h(t)$. $S_h(t)$ indicates the predicted frost hardiness at a given time instant t with the dimension of temperature (°C). Correspondingly, $R_h(t)$ indicates the predicted rate of hardening or dehardening within the time step used in the calculations, with the dimension of temperature per time (in most cases °C day^{-1}).

One peculiarity of frost hardiness as a quantitative ecophysiological attribute is that its amount increases as the value of its negative indicator, state of frost hardiness, $S_h(t)$, decreases. Therefore, whenever $S_h(t)$ or any other frost hardiness-related variable to be introduced later is referred to, the minus sign is acknowledged in the text, so that a decrease in $S_h(t)$ indicates increasing frost hardiness and vice versa. A reminder of this procedure is added to the text in some cases where a danger of confusion may prevail.

Similarly to the modelling of the photosynthetic capacity discussed in Chap. 4, the state of development predicted by the models of frost hardiness has a straightforward counterpart in empirical studies. In each modelling study one needs only to specify the variable state of frost hardiness, $S_h(t)$, to correspond to the specifications of the empirically measured frost hardiness applied when gathering the data for model development and testing (tree tissue addressed and method used for assessing the frost hardiness; see Sects. 5.1.1 and 5.1.2 above). After that the predicted and measured frost hardiness can be readily compared with each other continuously on a day-to-day basis.

Though the most important environmental factors regulating the frost hardiness of boreal and temperate trees are air temperature and night length (Aronsson 1975; Christersson 1978; Bigras et al. 2001; Zhang et al. 2003), it is also constrained by the annual phenological cycle discussed in Chap. 3, so that the trees as a rule have a limited hardening potential during the phase of active growth (Weiser 1970; Fuchigami et al. 1982; Larcher 2003; Junttila 2007). Thus, in any comprehensive examination of the effects of environmental factors on the annual cycle of frost hardiness, the annual phenological cycle needs to be considered as well. In other words, both the direct effects and the indirect effects, i.e., the effects caused indirectly via the effects on the annual phenological cycle, of environmental factors on the frost hardiness need to be considered.

In this way the picture gets relatively complicated, as the key driving factors of the frost hardiness dynamics, i.e., air temperature and night length, are also the key driving factors of the annual phenological cycle, as discussed in Chap. 3. Before addressing this complexity in Sect. 5.3, the direct effects of air temperature and night length on the frost hardiness of trees are first discussed in Sect. 5.2 for the sake of illustration, without explicitly considering the annual phenological cycle.

5.2 Direct Environmental Regulation of Frost Hardiness

5.2.1 *Fluctuating Effects of Air Temperature*

5.2.1.1 A Fluctuating Model of Frost Hardiness

Repo and Pelkonen (1986) did an experimental study on the effects of air temperature on the dehardening of first-year *Pinus sylvestris* seedlings. The study was carried out in a systems analysis framework, and on the basis of the results, Repo et al. (1990) presented a dynamic model for predicting the frost hardiness of the seedlings on the basis of the daily minimum air temperature. Repo et al. (1990) also addressed the rehardening of the seedlings, i.e., a new increase in hardiness as a result of dropping air temperatures after the hardiness has decreased earlier as a result of rising air temperatures (Repo 1991).

Repo et al.'s (1990) model is outdated now and has been replaced with the more comprehensive approaches to be discussed in Sect. 5.3. Nevertheless, following the historical approach adopted throughout the present volume, Repo et al.'s (1990) model is discussed in detail below. Besides acknowledging the early pioneering work, the historical approach is needed here because the elements forming the old and simple model are included in the modern, more complicated ones. Thus, a discussion of the old model facilitates the grasping of the modern models.

Ecophysiologically, Repo et al.'s (1990) frost hardiness model is similar to Pelkonen and Hari's (1980) model of photosynthetic capacity, discussed in Sect. 4.2.1.1. Both models simulate fluctuating development (Hänninen and Kramer 2007). However, Repo et al.'s (1990) frost hardiness model was not explicitly based on Pelkonen and Hari's (1980) photosynthesis model. Rather, it was formulated in terms of the standard concepts of systems analysis. It is based on the following three assumptions, all of which are supported by experimental evidence (Repo et al. 1990).

Assumption (*1*). After a stepwise rise in air temperature, the hardiness of trees decreases exponentially, i.e., the dehardening is rapid at first but then levels off, so that a constant hardiness is attained. Repo et al. (1990) referred to this constant as the stable value of frost hardiness. Later on, the concept of stationary frost hardiness has been used (Kellomäki et al. 1992, 1995; Leinonen 1996a; King and Ball 1998), and this concept is also adopted in the present volume.

Assumption (*2*). Stationary frost hardiness depends on air temperature (Fig. 5.7).

Assumption (*3*). When the air temperature drops during dehardening, rehardening takes place. Despite some limitations discussed later in this chapter, the changes in frost hardiness were assumed to be fully reversible.

On the basis of these three assumptions, Repo et al. (1990) introduced the following model for the effects of air temperature on the frost hardiness of boreal and temperate trees (Fig. 5.8):

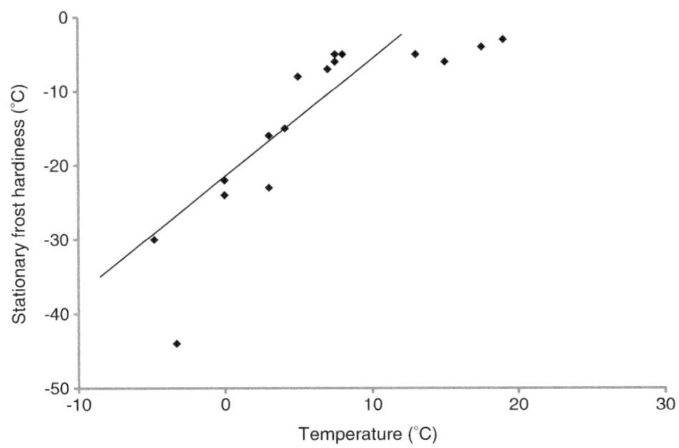

Fig. 5.7 Dependence of the stationary frost hardiness of stems of first-year *Pinus sylvestris* seedlings on the air temperature during dehardening. Stationary frost hardiness indicates the constant level of hardiness attained when the seedlings are subjected to the air temperature indicated by the *horizontal axis* for a prolonged period of time. The *diamonds* stand for measured frost hardiness as determined by means of the specific impedance difference method (see Sect. 5.1.2) and the *line* for a linear response fitted to the data (Repo et al. 1990; redrawn with the permission of the University of Eastern Finland)

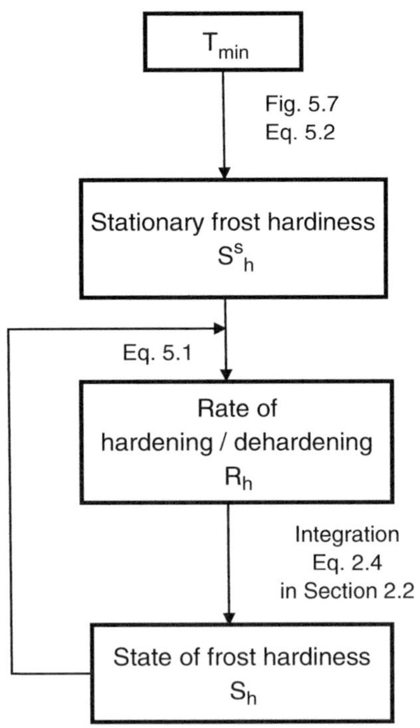

Fig. 5.8 Relations included in Repo et al.'s (1990) fluctuating model of frost hardiness. T_{min} = daily minimum air temperature. The *arrow* on the *left* represents the feedback included in the model (Sect. 2.2)

$$R_h(t) = \left(\frac{1}{\tau}\right)\left(S_h^s(t) - S_h(t)\right) \tag{5.1}$$

where $R_h(t)$ = rate of change of frost hardiness, i.e., rate of hardening or dehardening at time instant t, $S_h(t)$ = state of frost hardiness at time instant t, $S^s{}_h(t)$ = stationary frost hardiness at time instant t, and τ = time constant. As a first approximation, the stationary frost hardiness $S^s{}_h(t)$ was assumed to depend linearly on the daily minimum air temperature (Fig. 5.7):

$$S_h^s(t) = a \cdot T_{\min}(t) + b \tag{5.2}$$

where $T_{min}(t)$ = daily minimum air temperature during day t, a = slope and b = intercept of the linear dependence indicated in Fig. 5.7, respectively.

The model defined by Eqs. 5.1 and 5.2 can be verbally described in this way: at any given moment, the frost hardiness is changing towards the stationary frost hardiness, which depends linearly on the daily minimum air temperature. The rate of hardening or dehardening is directly proportional to the difference between the stationary and the prevailing frost hardiness and inversely proportional to the time constant. According to the prediction of the model, changes in frost hardiness follow changes in air temperature with a delay.[3]

Repo et al.'s (1990) model involves several simplifying assumptions (Repo 1993). First, the assumption of the full reversibility of changes in frost hardiness is not realistic. Most importantly, the trees have only a limited potential for hardening during active growth, even when exposed to low air temperatures that generally cause hardening during other phases of the annual cycle (Weiser 1970; Fuchigami et al. 1982). Second, the effects of night length on hardening are not addressed in the model. Third, even though Repo et al. (1990) discussed the nonlinear dependency of frost hardiness on changes in the air temperature, they did not address it in their model, where stationary frost hardiness was assumed to depend linearly on the daily minimum air temperature (Fig. 5.7).

Despite its strong simplifications, Repo et al.'s (1990) model predicted the annual cycle of frost hardiness in naturally growing *Pinus sylvestris* trees with moderate accuracy. This finding is once again a good demonstration of the importance of the concepts of model realism and model accuracy, introduced in Sect. 2.4 and discussed several times in various chapters of the present volume. In the case of Repo et al.'s (1990) model, the moderate accuracy is not an indication of moderate realism. However, as suggested by the experimental results supporting the model's assumptions, the model identifies some part of the environmental regulation of frost hardiness in the examined seedlings in a realistic way. Thus its deficiency in realism comes mainly from its insufficient coverage, i.e., from what is lacking, not so much from the inclusion of any erroneous or non-causal parts (see Sect. 2.4). That is why

[3] See Fig. 4.3b in Sect. 4.2.1.1 for a similar prediction of Pelkonen and Hari's (1980) model of photosynthetic capacity.

Repo et al.'s (1990) model provides a good first approximation and starting point for the further model development discussed in Sect. 5.3.

5.2.1.2 Comparisons with Photosynthetic Capacity and Other Physiological Attributes

Mäkelä et al.'s (2004) model of the annual cycle of photosynthetic capacity of boreal conifers is a close relative of Repo et al.'s (1990) frost hardiness model (Eqs. 5.1 and 5.2). In the former, however, the dynamic part of the modelling is carried out with reference to air temperature, so that the value of the theoretical state variable, i.e., state of photosynthetic acclimation, $S_{pa}(t)$, is given in terms of air temperature (Eq. 4.4 in Sect. 4.2.1.2). Accordingly, the stationary state is attained whenever the condition $S_{pa}(t) = T(t)$ is met. In this approach the value of the ecophysiological variable to be compared with empirical measurements, e.g., the quantum yield of photosynthesis, is calculated as a transformation of the predicted value of $S_{pa}(t)$ (Eq. 4.5 in Sect. 4.2.1.2).

In Repo et al.'s (1990) frost hardiness model, the calculation sequence is the other way round. Accordingly, the transformation into the final ecophysiological attribute, frost hardiness in this case, is already completed when the stationary frost hardiness is calculated (Fig. 5.8; Eq. 5.2), and the value of the rate variable is subsequently calculated in terms of frost hardiness. Despite this difference, the basic idea is the same as in Mäkelä et al.'s (2004) model. According to both models, each constant air temperature has a corresponding constant physiological state of the tree. In the structure of Mäkelä et al.'s (2004) model, the role of air temperature is emphasized, as the model has been explicitly constructed on the idea of trees' acclimation to changes in air temperature. As will be seen in Sect. 5.3, Repo et al.'s (1990) formulation is more general because it allows the simultaneous modelling of the effects of several environmental factors.

In boreal conifers, the annual cycles of frost hardiness and of photosynthetic capacity have much in common (Repo et al. 2006; Hänninen and Kramer 2007). Accordingly, both ecophysiological attributes follow, by and large, the concept of fluctuating development, and for both attributes, air temperature is a crucial environmental factor regulating the development. Thus, the trees deharden and their photosynthetic capacity increases in spring as a result of rising air temperatures; and correspondingly, the trees harden and their photosynthetic capacity decreases in autumn as a result of dropping air temperatures. This simple model provides quite an accurate prediction for the seasonal changes in photosynthetic capacity in several cases (Sect. 4.2.1), but for the corresponding changes in frost hardiness it gives a first approximation only (see Sect. 5.3).

Repo et al. (2006) examined the relationship between photosynthetic capacity and frost hardiness in *Pinus sylvestris* saplings growing in whole-tree chambers under either current or elevated air temperature conditions. Their results show that the relationship between these two ecophysiological attributes is different in spring than in autumn. This is indicated by a hysteresis loop in the plot between these two

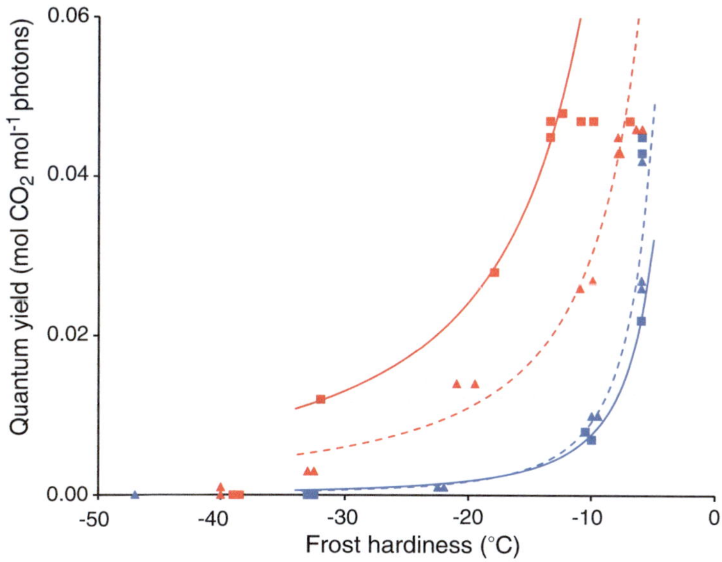

Fig. 5.9 Dependence of the photosynthetic capacity, determined as the quantum yield of the photosynthetic light response (Fig. 4.2c in Sect. 4.1.2.1), on the needle frost hardiness of *Pinus sylvestris* saplings grown in whole-tree chambers in current (*squares* and *continuous curves*) and elevated (*triangles* and *dashed curves*) air temperature conditions. Measurements were carried out during the dehardening phase in spring (*red*) and the hardening phase in autumn (*blue*). Needle frost hardiness was determined by means of the electrolyte leakage method (Repo et al. 2006; redrawn with the permission of John Wiley & Sons, Inc.)

variables, i.e., with a given level of needle frost hardiness, a higher photosynthetic capacity was observed during dehardening in spring than during hardening in autumn (Fig. 5.9).

This illustrative result elaborates the principle of whole-tree ecophysiology discussed in Sect. 2.4. The ecophysiological quantities measured at the whole-tree level are emergent manifestations of the basic physiological properties existing and phenomena occurring at the anatomical, cellular, and molecular levels. Thus a correlation between any two whole-tree attributes suggests that they may partly share a common physiological basis. This goes for photosynthetic capacity and frost hardiness as well, but as shown by the hysteresis loop, the basis of these two whole-tree attributes is not entirely the same. Thus, a given value of frost hardiness corresponds to a different overall physiological condition in spring than in autumn, although from the point of view of frost tolerance, the condition in the two seasons is similar if not identical.

The correlation of frost hardiness with physiological attributes, such as electrolyte leakage, is used in several methods to assess the frost hardiness and damage (Sect. 5.1.2). When these methods are applied, mathematical models are frequently fitted to the data (Figs. 5.4, 5.5, and 5.6). It should be noted that in this methodological context, the concept of "model" is used in a meaning different from that in the dynamic modelling discussed in various chapters of the present volume, where the model is used to predict and explain the seasonality of various physiological tree attributes on the basis of the seasonality of environmental factors. In this context, then, the aim is to understand the environmental responses of the trees sufficiently to predict the seasonal phenomena of the trees without any additional physiological measurements (Fuchigami et al. 1982; Hänninen and Kramer 2007).

Taulavuori et al. (1997) took a different approach to the modelling of the frost hardiness of *Pinus sylvestris* seedlings. In addition to the environmental factors air temperature and night length, they also included in their model one physiological attribute (the pH of the cell effusate) as an explanatory factor. In a sense, then, their model is a mixture of the models used for assessing the frost hardiness on the basis of its correlation with other physiological attributes and those explaining the frost hardiness dynamics by means of environmental factors. It goes without saying that while Taulavuori et al.'s (1997) model is interesting as such, it does not facilitate any prediction of the seasonality of frost hardiness without additional physiological measurements.

5.2.2 Combined Effects of Air Temperature and Night Length

It has been known for a long time that in addition to air temperature, night length is another major environmental factor affecting the seasonality of frost hardiness in boreal and temperate trees (Aronsson 1975; Christersson 1978; Zhang et al. 2003). As a review of the vast body of literature discussing the combined effects of air temperature and night length on the seasonality of frost hardiness is beyond the scope of the present volume, the reader is referred to the several reviews available (e.g. Sakai and Larcher 1987; Bigras et al. 2001; Greer et al. 2001; Zhang 2001). Here, only two alternative theories are briefly addressed that are essential for the ecophysiological models to be discussed later in this chapter.

In the classical theory, night length and air temperature are assumed to affect the hardening of boreal and temperate trees sequentially in late summer and autumn (Weiser 1970; Fuchigami 1982; Kellomäki et al. 1992, 1995); Bigras et al. (2001). This theory is therefore referred to as *the sequential theory* in the present volume. It identifies two or three sequential phases of hardening. During the first phase, a long night length is the essential factor causing hardening in the trees. This effect is closely related to the effects of long nights on the cessation of growth (Sect. 3.5). During the second phase, a low air temperature is the essential driving factor causing hardening. Sometimes a third phase of hardening, caused by extremely low freezing air temperatures, is also identified (Weiser 1970).

According to the alternative theory, air temperature and night length affect frost hardiness additively, i.e., the prevailing hardiness is obtained by summing the hardening effect of the prevailing air temperature and that of the prevailing night length (Chen and Li 1978; Greer 1983; Leinonen et al. 1995, 1996; Leinonen 1996a). In the present volume, this theory is referred to as *the additive theory*. In comparison with the vast body of literature related to the sequential theory, there are relatively few studies addressing the additive theory.

Both the sequential and the additive theory mainly address the hardening phase of boreal and temperate trees. According to the prevailing view, dehardening in spring is mainly driven by rising air temperatures (Aronsson 1975; Leinonen 1996a).

Besides air temperature and night length, several other environmental factors are known to affect the frost hardiness of boreal and temperate trees. These include the fertility of the soil (Aronsson 1980; Luoranen et al. 2008), drought (Chen and Li 1978; Colombo et al. 2001), air pollutants (Taulavuori et al. 2005a), and UV radiation (Taulavuori et al. 2005b). However, the effects of these other environmental factors have evidently not been addressed so far in terms of the dynamic modelling approach adopted in the present volume, so that despite their potential importance, the other environmental factors will not be discussed any further in the present volume.

5.3 Overall Environmental Regulation of Frost Hardiness

5.3.1 Changing Environmental Responses During the Annual Cycle

The environmental responses of frost hardiness in boreal and temperate trees change during the annual cycle. To start with, the principle of fluctuating development as such presupposes a change in the air temperature response of hardening and dehardening (Eqs. 5.1 and 5.2). Accordingly, a given air temperature may cause either hardening or dehardening, depending on whether the prevailing value of the state of frost hardiness, $S_h(t)$, is above or below the stationary frost hardiness determined by the prevailing air temperature (Fig. 5.7).

However, this feedback principle (Fig. 5.8), introduced in Sect. 2.2, does not take into account the indirect regulation caused by the prevailing state of the annual phenological cycle. During the active growth phase, little hardening takes place even in conditions causing hardening during later phases of the cycle (Weiser 1970; Fuchigami et al. 1982; Larcher 2003; Junttila 2007). Furthermore, the environmental responses of hardening and dehardening change not only between the active growth phase and the dormant phase but also within the dormant phase. During the rest phase the trees typically retain their hardiness in many conditions that would cause dehardening during quiescence. This was shown decades ago for *Prunus persica* (Edgerton 1954; Proebsting 1963) and for *Viburnum plicatum tomentosum* and *Acer negundo* (Irving and Lanphear 1967).

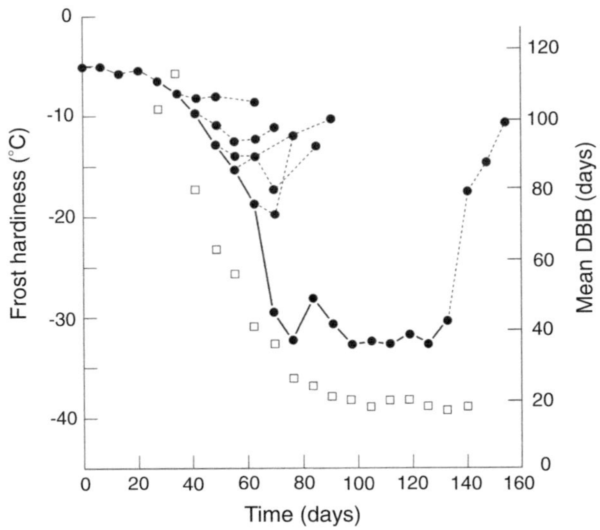

Fig. 5.10 Effects of air temperature and night length treatments on rest break and needle frost hardiness in first-year *Pseudotsuga menziesii* seedlings. The experiment was started with actively growing seedlings by exposing them to hardening conditions with long-night (14 h) treatments. The air temperature (day/night) in the hardening conditions was 20/15 °C during the first 28 days and 10/3 °C after that. The white squares indicate mean days to bud burst (mean DBB, *right-hand vertical axis*) at the time the seedlings were transferred from the hardening conditions to a regrowth test in growth-promoting forcing conditions. No bud burst was observed after the transfers done during the first 28 days. The *black circles* indicate needle frost hardiness (*left-hand vertical axis*) as determined by means of the electrolyte leakage method. The *dashed line* indicates the initial hardening during the first 28 days at 20/15 °C and the continuous line the subsequent hardening at 10/3 °C. Subgroups of the seedlings were transferred from the hardening conditions into dehardening conditions with 2-h night length and 21/18 °C air temperature at six time instants. The black circles on dotted lines indicate the development of frost hardiness in the dehardening conditions (Leinonen et al. 1995; redrawn with the permission of Oxford University Press)

Leinonen et al. (1995) studied the effects of high air temperatures on the frost hardiness of first-year seedlings of *Pseudotsuga menziesii* during the dormant phase (Fig. 5.10). They first induced growth cessation and initial hardening by exposing the seedlings to long-night and high-temperature conditions for 28 days. Subsequent exposure to a chilling treatment of 10/3 °C (day/night) in long-night conditions caused both rest break and rapid hardening, so that the maximum level of hardiness was attained approximately 20 days before rest completion, the latter being indicated by the levelling off of the DBB curve (Fig. 5.10). As could be expected on the basis of the discussion presented in Sect. 5.2.1.1, rapid exponential dehardening occurred when the fully hardened quiescent seedlings were transferred to short-night and high-temperature 21/18 °C (day/night) conditions at the end of the experiment (Fig. 5.10).

However, a different pattern of hardiness development was observed when a similar transfer was carried out during days 30–60, when the seedlings were hardening and the process of rest break was going on (Fig. 5.10). At this time, the hardening first continued despite the transfer to the high-temperature conditions, but the rate was reduced in comparison with that observed in the chilling conditions. However, with the exception of the first transfer, when only a few degrees of hardening had taken place, the development was subsequently reversed, so that the seedlings started to deharden at the high-temperature conditions. In this way, approximately the same level of hardiness was finally attained as when the transfer was carried out at the end of the experiment with fully hardened quiescent seedlings (Fig. 5.10).

The air temperature response of frost hardiness also changes during the phase of quiescence. This was shown by Leinonen et al. (1997), who studied the dehardening of *Pinus sylvestris* saplings in whole-tree chambers under elevated temperature conditions. Two different temperature treatments, with approximately the same mean temperature, were applied. In one of them, a constant air temperature of +6 °C was used, and in the other, the air temperature fluctuated between +1 and +11 °C. Both treatments caused dehardening in the needless of the saplings. However, superimposed on the overall dehardening, rehardening of the needless took place each time the temperature was lowered in the fluctuating treatment. This happened despite the average air temperature being approximately the same in the two treatments. This observation led Leinonen et al. (1997) to conclude that high temperatures cause dehardening in two different ways. In addition to the direct short-term fluctuating effects discussed in Sect. 5.2.1, sufficiently high air temperatures cause an indirect effect, which is related to the ontogenetic development towards growth onset.[4]

5.3.2 *Integrated Models of Frost Hardiness*

5.3.2.1 Historical Overview

The studies discussed in the previous section call for development of frost hardiness models in which both the direct effects of environmental factors and the indirect effects mediated by the annual phenological cycle are addressed. The direct effects often belong to the category of fluctuating development and the indirect ones to the category of fixed-sequence development. Following Hänninen and Kramer (2007), frost hardiness models addressing both categories simultaneously are referred to as integrated models in the present volume (Fig. 5.11a).

The first integrated model was evidently introduced by Winter (1973) for the dehardening of temperate *Malus domestica* trees. Kobayashi et al. (1983)

[4] It is assumed here that the ontogenetic development of the buds towards growth onset is also reflected in the dehardening of the previous year's needles. While the reasoning of Leinonen et al. (1997) is otherwise plausible, there is evidently no direct experimental evidence for this reflection.

introduced an integrated model for frost hardiness in temperate *Cornus sericea*. Their frost hardiness model is based on the Degree Growth State (°GS) annual cycle model discussed in Sect. 3.6.3. Accordingly, the rates of hardening and dehardening are calculated in their model by means of polynomial equations where the prevailing growth stage, °GS, is an explaining factor in addition to air temperature.[5] Cannell et al. (1985) presented a model for predicting the autumn hardening of *Picea sitchensis* in Scotland. Even though their model does not address the annual phenological cycle as such, it includes an idea somewhat similar to the concept of the integrated model. Besides addressing the short-term fluctuating effects of air

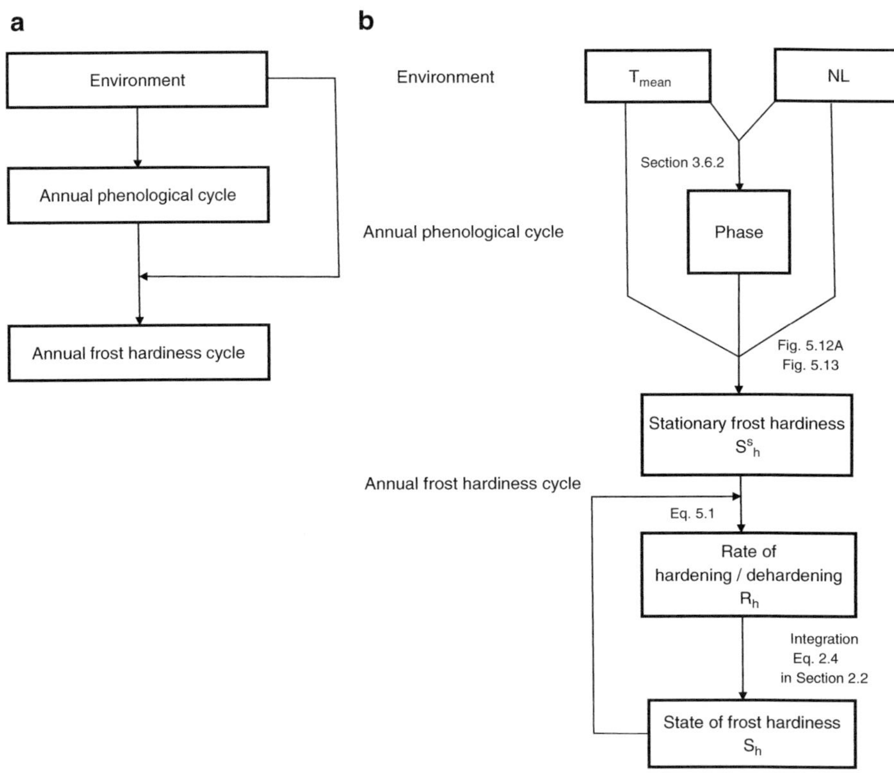

Fig. 5.11 Outlines of integrated models of tree frost hardiness. (**a**) General principle of the models, (**b**) relations included in Kellomäki et al.'s (1992, 1995) model, and (**c**) relations included in Leinonen's (1996a) model. T_{mean} = daily mean temperature, NL daily night length, and T_{min} daily minimum temperature. Note that the final steps of the calculations in both integrated models are identical with those in Repo et al.'s (1990) fluctuating model (Fig. 5.8). With all models, the extent of frost damage is calculated by comparing the T_{min} with the predicted daily value of the state of frost hardiness (relation not shown in the figure)

[5] Unfortunately, the use of this integrated model of frost hardiness is prevented by the erroneous equations discussed in Sect. 3.6.3.

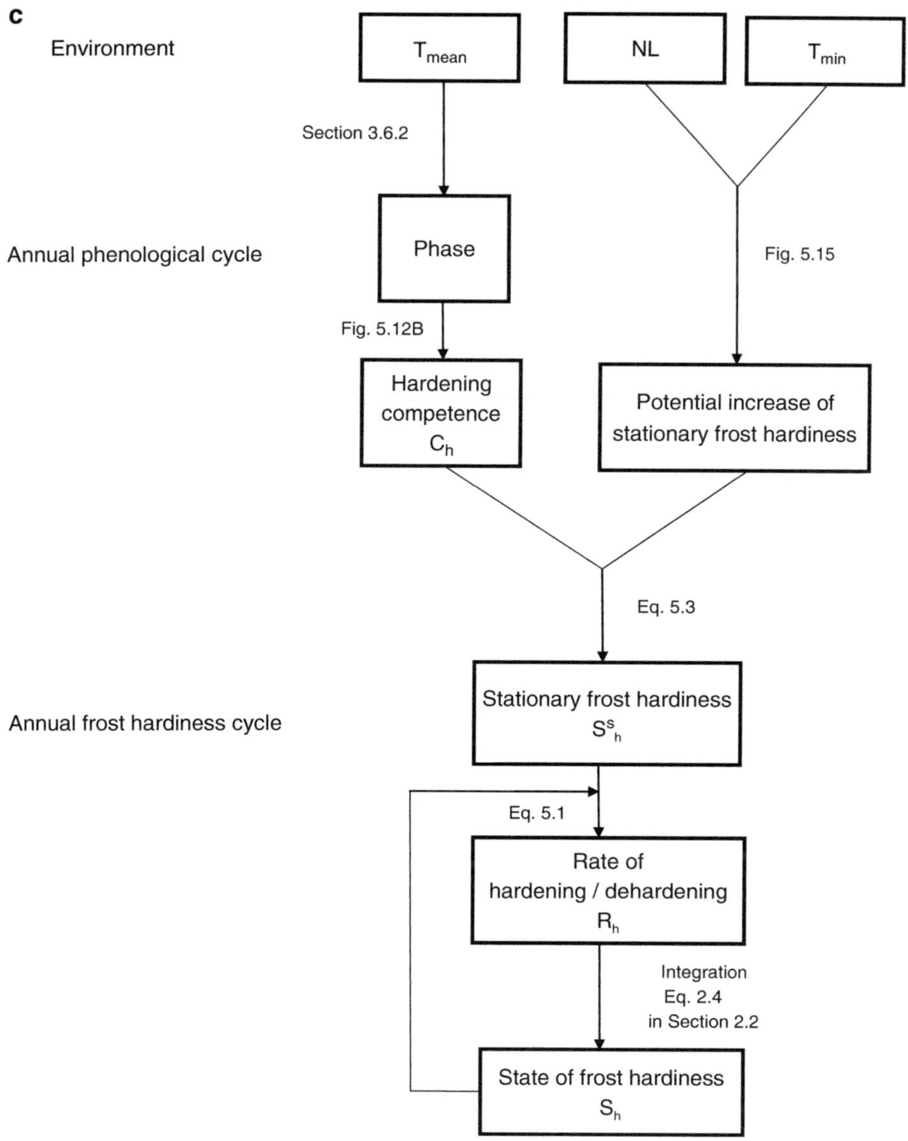

Fig. 5.11 (continued)

temperature, it posits an irreversible hardening caused by the passing of a critical night length or exposure to frost.

In terms of systems analysis, Repo et al.'s (1990) model is a first-order model, i.e., the inertia of the changes in frost hardiness is modelled by means of one time constant (Eq. 5.1). Leinonen et al. (1995) introduced a second-order model, i.e., a model with two time constants, for the frost hardiness of boreal and temperate trees.

Rather than the exponential changes implied by the first-order model, the second-order model implies a sigmoidal change in frost hardiness after a stepwise change in the air temperature (Hänninen and Kramer 2007). Thus there is an initial phase of slow change, followed by an intermediate phase of rapid change, and a final phase of slow change. Furthermore, under certain circumstances the second-order model predicts a delayed response to the stepwise change in air temperature, so that the direction of change in frost hardiness does not change immediately at the time of the change in air temperature (Leinonen et al. 1995; Hänninen and Kramer 2007).

The second-order model developed by Leinonen et al. (1995) provided a good account of the delayed dehardening response of *Pseudotsuga menziesii* seedlings to the high air temperatures during rest (Fig. 5.10, results of the model test not shown). However, as the dehardening of the quiescent seedlings followed the prediction of the first-order model, i.e., it was immediate and exponential at the end of the experiment (Fig. 5.10), Leinonen et al. (1995) introduced an integrating component into their model. Accordingly, the effect of the second time constant vanishes during rest break, so that the second-order model reduces to the first-order one at the beginning of the phase of quiescence. This sophisticated model provided a good account of all the experimental results in Leinonen et al.'s (1995) study. However, evidently due to the high requirement of detailed data for the parameterization, Leinonen et al.'s (1995) model has evidently not been used in later studies.

Below, two integrated models developed for boreal trees are discussed in detail (Figs. 5.11b, c; Kellomäki et al. 1992, 1995; Leinonen 1996a). Both of these models have also been used for assessing the effects of the projected climate change; those studies are discussed in Chap. 8. As Leinonen's (1996a) model is the more recent one and is a further development of the older model of Kellomäki et al. (1992, 1995), the main emphasis in this chapter is on Leinonen's (1996a) model.

Despite many essential differences between these two models, they have several commonalities as well. First of all, both are based on integrating Kellomäki et al.'s (1992, 1995) model of the entire annual phenological cycle (Fig. 3.25b in Sect. 3.6.2) with the concept of stationary frost hardiness introduced by Repo et al. (1990). Thus, as in Repo et al.'s (1990) fluctuating model (Fig. 5.8), the rate of change of frost hardiness is also calculated in the two integrative models by comparing the stationary frost hardiness with the prevailing state of frost hardiness by means of Eq. 5.1 (Fig. 5.11b, c). There are marked differences, however, among the three models in the calculation of the stationary frost hardiness used in the Eq. 5.1.

5.3.2.2 Kellomäki et al.'s (1992, 1995) Integrated Model

In Kellomäki et al.'s (1992, 1995) model of the annual phenological cycle, discussed in Sect. 3.6.2, four sequential phases are assumed: active growth, lignification, rest, and quiescence. In regard to frost hardiness, these phases determine the calculation of the value of the stationary frost hardiness, $S^{s}_{h}(t)$ (Figs. 5.11b and

5.12a). During active growth, $S^s_h(t)$ is assumed to stay at a constant maximum level, $S^s_{h,min}(t)$, indicating a minimum level of hardiness. For the next two phases, the sequential theory of the effects of night length and air temperature on frost hardening is applied (Sect. 5.2.2). Accordingly, $S^s_h(t)$ is modelled to depend on night length during the lignification phase and on air temperature during the rest phase (Fig. 5.12a).

For both relationships, a piecewise-linear response is determined.[6] During lignification, $S^s_h(t)$ starts to decrease from its maximal value $S^s_{h,min}$ when the night length exceeds the lower threshold value of 8 h, and the maximal effect of night length on $S^s_h(t)$ is attained when the night length attains the upper threshold value of 12 h (Fig. 5.13a). Correspondingly, during the rest phase the value of $S^s_h(t)$ decreases from its maximal value $S^s_{h,min}$ whenever the daily mean temperature drops below the upper threshold value of +10 °C, and the maximal effect of air temperature on $S^s_h(t)$ is attained whenever the daily mean air temperature attains the lower threshold value of −10 °C (Fig. 5.13b). The air temperature response assumed for the rest phase is also assumed for the phase of quiescence; so, during the dormant phases of rest and quiescence, frost hardiness is modelled otherwise similarly to Repo et al.'s (1990) fluctuating model, except that instead of the linear response to the daily minimum temperature used by Repo et al. (1990) (Fig. 5.7), a piece-wise linear response to the daily mean temperature is used in Kellomäki et al.'s (1992, 1995) model (Fig. 5.13b).

As indicated in Sect. 3.6.2, Kellomäki et al.'s (1992, 1995) model of the annual phenological cycle was parameterized with data on several central Finnish tree species. Thus, because the same phenological model is also the basis for Kellomäki et al.'s (1992, 1995) model for frost hardiness, the latter model predicts the frost hardiness of a generalised boreal tree species native to conditions such as those in central Finland. Furthermore, the values of some parameters are mainly based on educated guesses rather than solid empirical evidence for any species (Kellomäki et al. 1992, 1995). Therefore the model should be regarded more as a tool for theoretical research and deduction of research hypotheses (Sect. 2.2) than a tool for predicting the frost hardiness of any given species.

Despite this background, the model predicted the needle frost hardiness of *Pinus sylvestris* saplings growing in central Finland surprisingly accurately in an independent test (Fig. 5.14a). It was less accurate in predicting the maximal winter hardiness than in predicting hardiness in other parts of the annual cycle. It is hard, though, to evaluate the importance of this discrepancy, for as discussed in Sect. 5.1.2, the measurement of frost hardiness is less exact in winter than in other parts of the annual cycle, so that the validity of the data points used in the test for winter hardiness remains unclear.

---

[6] The threshold values for both responses appear as parameters in the model of Kellomäki et al. (1992, 1995), but what is referred to here is the numerical values determined for the parameters in their study (Fig. 5.13).

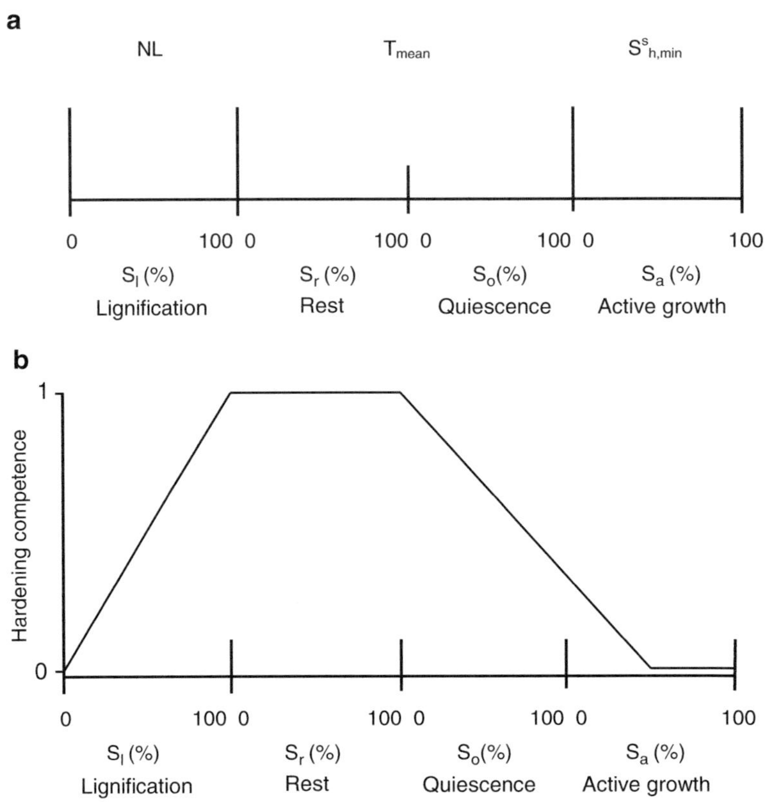

Fig. 5.12 A schematic presentation of the effects of the annual phenological cycle on the annual cycle of frost hardiness in two integrated models of tree frost hardiness. Both models involve the following properties: first, the annual ontogenetic cycle is divided into four phases represented by their respective state variables, all of which are expressed on a percentage scale (Kellomäki et al. 1992, 1995; Hänninen and Kramer 2007; see Sect. 3.6.2): S_l (state of lignification), S_r (state of rest), S_o (state of ontogenetic development, i.e., the state variable for the phase of quiescence), and S_a (state of active growth). Second, the rate of hardening or dehardening is determined by the value of stationary frost hardiness $S^s{}_h$ and the time constant τ together (Eq. 5.1). Third, the effects of the annual phenological cycle on tree frost hardiness are mediated by calculating the value of stationary frost hardiness, $S^s{}_h$, as dependent on the prevailing phase of the annual phenological cycle. (**a**) In Kellomäki et al.'s (1992, 1995) model, a sequential effect of night length, NL, and daily mean air temperature, T_{mean}, on frost hardiness is assumed. During the lignification phase, $S^s{}_h$ depends on NL (Fig. 5.13a), and during the phases of rest and quiescence, on T_{mean} (Fig. 5.13b). During active growth, $S^s{}_h$ stays at its maximum level, $S^s{}_{h,min}$, indicating the minimum stationary frost hardiness. (**b**) In Leinonen's (1996a) model, an additive effect of NL (Fig. 5.15a) and daily minimum air temperature, T_{min}, (Fig. 5.15b) on the potential increase of stationary frost hardiness is assumed, and the effect of the annual phenological cycle is modelled in terms of the concept of hardening competence, C_h. Changing as indicated in the figure, C_h determines the degree to which the additive effect of NL and T_{min} on the potential increase of stationary frost hardiness is realised at each point of the annual phenological cycle (Eq. 5.3). For an overall outline of the models, see Fig. 5.11

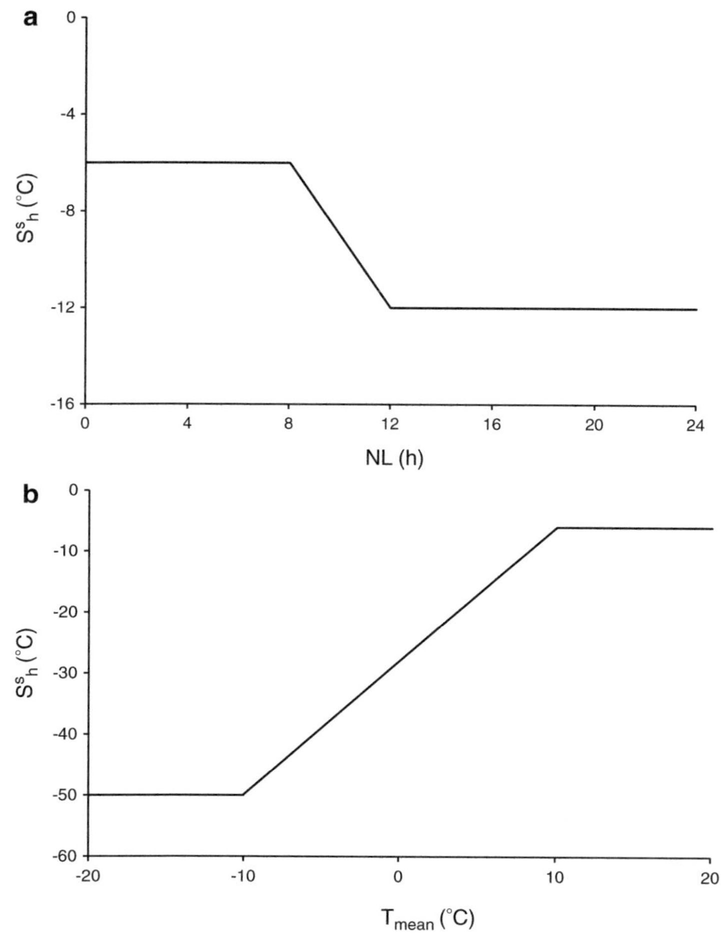

Fig. 5.13 Dependence of the stationary frost hardiness, S^s_h, on (**a**) night length, NL, during lignification and (**b**) on the daily mean temperature, T_{mean}, during rest and quiescence. Note that due to the negative scale included in the frost hardiness variables, the stationary frost hardiness increases as the negative value of the variable S^s_h decreases (Kellomäki et al. 1992; redrawn with the permission of the Finnish Forest Research Institute, currently the Natural Resources Institute Finland, and the Finnish Society of Forest Science). For further explanation, see text and Figs. 5.11b and 5.12a

Kellomäki et al. (1992, 1995) introduced their frost hardiness model as a sub-model for their gap-type forest stand model. In simulations with the stand model, the frost damage occurring in any individual tree is calculated by comparing the daily value of frost hardiness predicted for the tree with the daily minimum air temperature. In this comparison, the change in the air temperature response of damage is addressed. Accordingly, the temperature range from non-damaging to

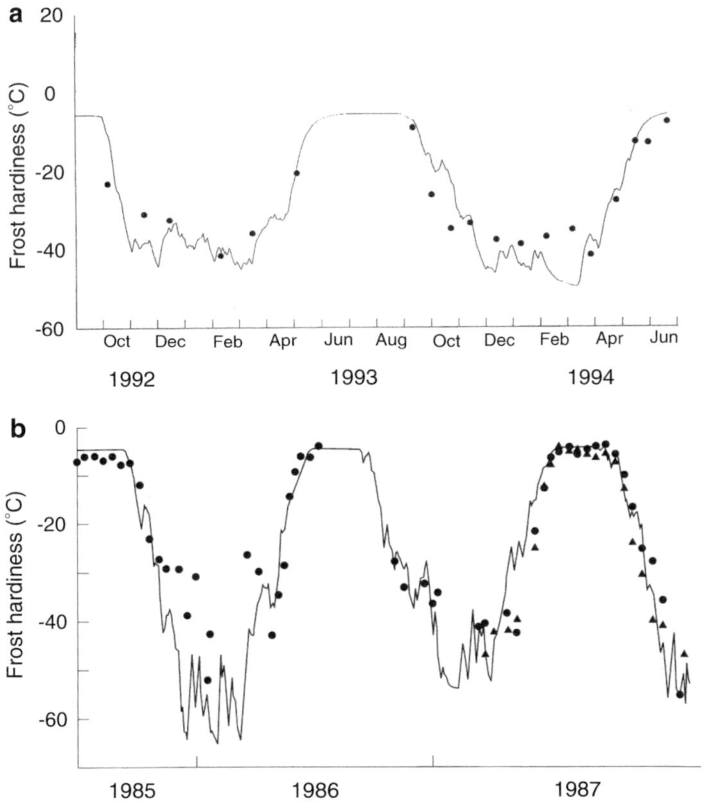

Fig. 5.14 Independent tests of two integrated frost hardiness models against measured empirical data on *Pinus sylvestris* saplings growing in natural conditions in eastern Finland. (**a**) The frost hardiness predicted by Kellomäki et al.'s (1992, 1995) model (*line*) and the observed needle frost hardiness as measured by Repo et al. (1996) by means of the electrolyte leakage method at Mekrijärvi in 1992–1994 (*black circles*) (Modified from Hänninen et al. 1996, and published with the permission of the Finnish Forest Research Institute, currently the Natural Resources Institute Finland, and the Finnish Society of Forest Science). (**b**) The frost hardiness predicted by Leinonen's (1996a) model (*line*) and the observed frost hardiness as measured by Repo (1992) by means of the specific impedance difference method (*black circles*) or visual damage scoring (*black triangles*) at Suonenjoki in 1985–1987 (Leinonen 1996a; redrawn with the permission of Oxford University Press)

lethal temperature is narrow with unhardened trees and gets wider as the hardening proceeds (Fig. 5.6). The effects of frost damage on the survival and growth of the trees are modelled by means of a specific growth multiplier (Kellomäki et al. 1992, 1995). This approach facilitates the up-scaling of modelled frost damages from individual trees to stands.

5.3.2.3 Leinonen's (1996a) Integrated Model

Retaining the basic principle of integrating the fixed-sequence and the fluctuating aspects of the annual cycle, Leinonen (1996a) developed Kellomäki et al.'s (1992, 1995) model further in several respects. To start with, he formulated and parameterized the model explicitly for the needle frost hardiness of *Pinus sylvestris*. The sub-models of the annual phenological cycle he parameterized with Raulo and Leikola's (1974), Repo's (1992), Kramer's (1994), and Leinonen's (1996b) empirical results and, correspondingly, the sub-models describing the direct environmental regulation of frost hardiness with Aronsson's (1975), Christersson's (1978), Repo and Pelkonen's (1986), Repo et al.'s (1990), Repo's (1992), and Leinonen et al.'s (1996), (1997) empirical results. Furthermore, he introduced a sub-model for the needle loss caused by frost damage and parameterized it with original data presented in Leinonen (1996a).

In addition to the species-specific parameterization, Leinonen's (1996a) model introduced four new ecophysiological aspects that were not included in the older model by Kellomäki et al. (1992, 1995). First, in the simulation of the effects of air temperature on the stationary frost hardiness, the daily minimum air temperature is used (Fig. 5.15b) instead of the daily mean temperature used in Kellomäki et al.'s (1992, 1995) model (Fig. 5.13b). Second, Leinonen's (1996a) model replaced the sequential effect of night length and air temperature on the stationary frost hardiness adopted in Kellomäki et al.'s (1992, 1995) model with an additive effect of these factors. Third, rather than changing the calculations of the stationary frost hardiness abruptly at the borders of the four phases (Fig. 5.12a), Leinonen's (1996a) model mediated the effects and constraints of the annual phenological cycle on the frost hardiness cycle by means of a specific [0, 1]-multiplier, i.e., the hardening competence, C_h (Figs. 5.11c, 5.12b). Thus, the stationary frost hardiness appearing in Eq. 5.1 is calculated in Leinonen's (1996a) model as follows[7]:

$$S_h^s(t) = S_{h,\min}^s + C_h(t) \cdot [\Delta H_T(t) + \Delta H_{NL}(t)] \tag{5.3}$$

where $S^s_h(t)$ = stationary frost hardiness, $S^s_{h,min}$ = minimal stationary frost hardiness (maximum value of the variable) prevailing during the latter half of the active growth phase, $C_h(t)$ = hardening competence, $\Delta H_T(t)$ = potential increase of the stationary frost hardiness caused by air temperature (Fig. 5.15b), and $\Delta H_{NL}(t)$ = corresponding potential increase caused by night length (Fig. 5.15a); with the exception of the parameter $S^s_{h,min}$, the values of all variables are given for time

[7] Equation 5.3 provides an illustrative example of the general principle of indirect environmental regulation of the annual cycle introduced by Eq. 2.3 in Sect. 2.2: the state of development of one attribute of the annual cycle (phenological cycle) affects the rate of development of another attribute (frost hardiness). In this way the daily mean air temperature, T_{mean}, affects the seasonality of frost hardiness via its effects on the annual phenological cycle (Fig. 5.11c). Hardening competence mediates this effect (Fig. 5.12b), so that it is analogical to the concept of ontogenetic competence introduced in Sect. 3.4.1.1, where the regulation of spring phenology is discussed.

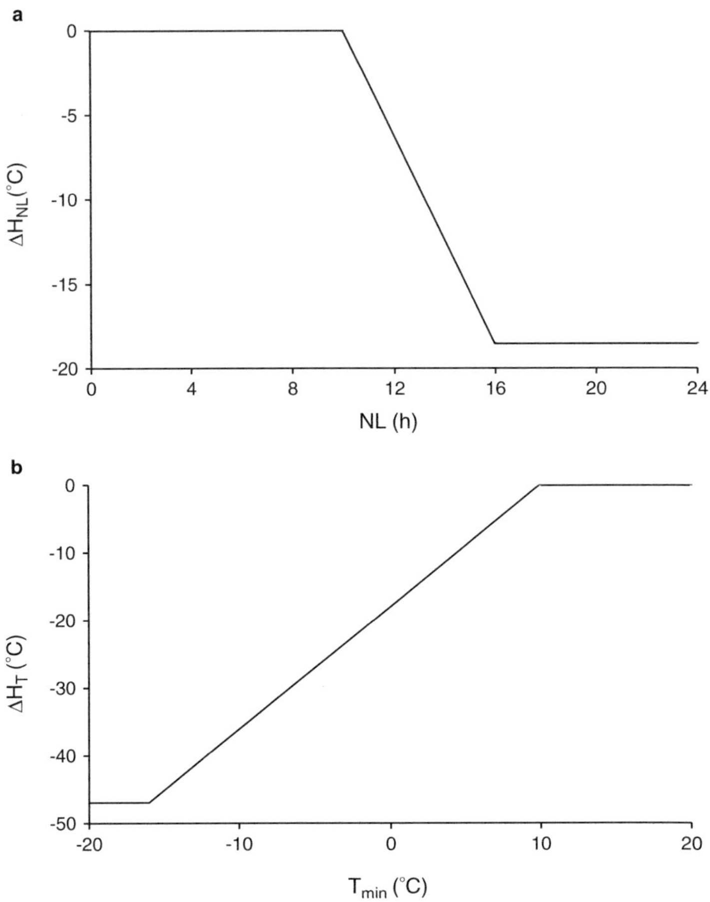

Fig. 5.15 Effects of (**a**) night length (NL) and (**b**) daily minimum air temperature, T_{min}, on tree frost hardiness in Leinonen's (1996a) model. The symbols ΔH_{NL} and ΔH_T denote the potential increase in the stationary frost hardiness caused by the two environmental factors in the additive model (Eq. 5.3). Note that due to the negative scale included in frost hardiness variables, the effect of each of the two environmental factors on the potential increase of stationary frost hardiness increases as the negative value of the corresponding variable ΔH_{NL} or ΔH_T decreases. For further explanation, see text and Figs. 5.11c and 5.12b (Modified from Leinonen et al. 1996; published with the permission of the Finnish Forest Research Institute, currently the Natural Resources Institute Finland, and the Finnish Society of Forest Science)

instant t. Thus, at any given moment, the stationary frost hardiness equals the minimal stationary frost hardiness plus the increase caused by the additive effect of night length and air temperature. The hardening competence $C_h(t)$ determines how great a part, if any, of the potential increase caused by the additive effect of air temperature and night length is realised in the stationary frost hardiness and, with the delay described in Eq. 5.1, in the frost hardiness itself (Fig. 5.11c).

During the latter half of the active growth phase, $C_h = 0$ (Fig. 5.12b).[8] The stationary frost hardiness thus equals the minimal stationary frost hardiness, $S^s_{h,min}$, regardless of the prevailing night length and air temperature (Eq. 5.3). During lignification, the hardening competence increases from zero to unity, so that the stationary frost hardiness becomes increasingly responsive to the additive hardening effects of night length and air temperature. During rest, the maximum value of $C_h = 1$ is retained, so that the trees are maximally responsive to these hardening effects. During quiescence and the first half of the active growth phase, the hardening competence decreases, attaining the value of zero at approximately the mid-point of the active growth phase (Fig. 5.12b). During this development the stationary frost hardiness becomes less responsive to the additive hardening effects of air temperature and night length. The shortening nights in winter and spring are assumed not to cause dehardening, so regardless of the prevailing night length the minimum value of ΔH_{NL} (indicating maximum effect of night length on hardiness) is assumed to occur between the winter and summer solstices (Leinonen 1996a).

The fourth new ecophysiological aspect introduced in Leinonen's (1996a) model concerns growth cessation. *Pinus sylvestris* has a fixed growth habit (Sect. 3.5.3.1), and therefore Leinonen (1996a) modelled its growth cessation in terms of temperature sum accumulation rather than in terms of Koski and Sievänen's (1985) joint effect model, which was used in Kellomäki et al.'s (1992, 1995) model of the annual cycle. Due to this difference, night length affects the annual phenological cycle in Kellomäki et al.'s (1992, 1995) model (Fig. 5.11b), but not in Leinonen's (1996a) model (Fig. 5.11c).

5.3.2.4 Predictions of Leinonen's (1996a) Model

The functioning and predictions of Leinonen's (1996a) model for the climatic conditions of Jyväskylä, central Finland, are illustrated in Fig. 5.16.[9] The simulations with the long-term air temperature data used in several chapters of the present volume were initiated from June 1883, but results are shown here for only one annual cycle, starting at the predicted height growth cessation on 9 July 1960.

In July 1960, the air temperatures generally stayed at a relatively high level (Fig. 5.16a), so that the high-temperature requirement of lignification was met on 31 July (Fig. 5.16b). During lignification, the hardening competence, C_h, increased linearly from zero to unity as the state of lignification, S_l, increased from zero to 100 % (Fig. 5.12b), so that for the lignification phase, the values of C_h and S_l are superimposed on each other in the figure. Accordingly, the full hardening

---

[8] The value of C(t) is not zero at the beginning of the active growth phase because in *Pinus sylvestris* the hardiness of the previous year's needles does not attain its minimum at the time of the onset of shoot growth (Leinonen et al. 1997). With other tree species, the hardening competence may attain the zero value at or even before the beginning of the active growth phase.

[9] The details of the model of the annual phenological cycle are explained in Sect. 3.6.2.

competence $C_h = 1$ was attained at the end of the lignification phase on 31 July, when $S_l = 100\ \%$ (Fig. 5.16b).

In August, air temperatures still remained at a relatively high level (Fig. 5.16a); therefore, as rest break requires exposure to chilling temperatures, little rest break occurred in August as indicated by the relatively low values of the state of rest break, S_r, for the time (Fig. 5.16b). Towards the end of August and later, air

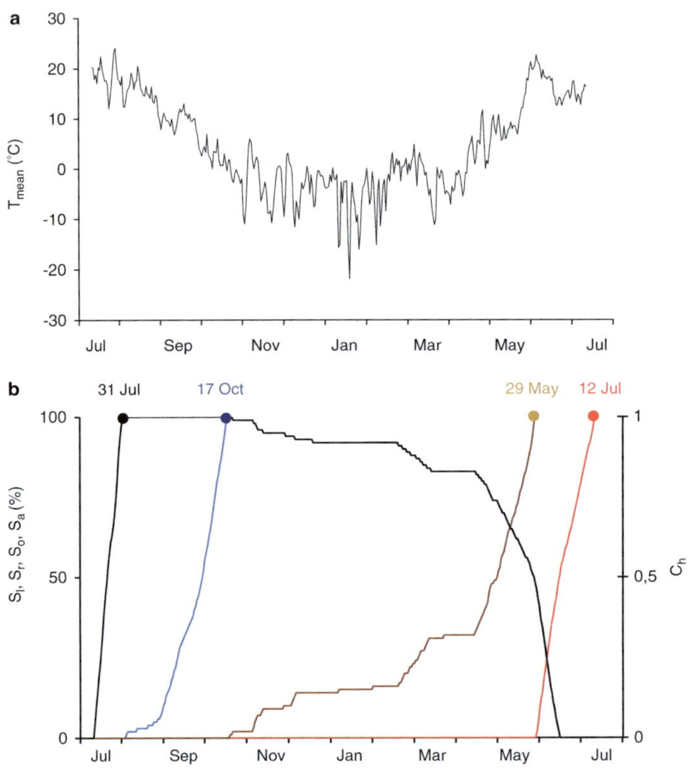

Fig. 5.16 An analysis of the functioning of Leinonen's (1996a) model in predicting needle frost hardiness in *Pinus sylvestris*. In all panels, the time scale on the horizontal axis ranges from July 1960 to July 1961. (**a**) Daily mean air temperature, T_{mean}, in Jyväskylä, central Finland. (**b**) Predicted annual phenological cycle and its effect on the annual cycle of frost hardiness, as mediated by the value of the hardening competence, C_h (*black curve* and *right-hand vertical axis*). The state variables for the four phases of the annual phenological cycle (*left-hand vertical axis*): state of lignification, S_l, (*black curve* superimposed on the *line* of C_h on the *left-hand* side of the figure till the end of lignification on 31 July), state of rest break, S_r, (*blue curve* till rest completion on 17 October), state of ontogenetic development, S_o, (*brown curve* till growth onset on 29 May), and state of active growth, S_a, (red curve till height growth cessation on 12 July). (**c**) Daily minimum air temperature, T_{min}, (*black curve* and *left-hand vertical axis*) and daily night length, NL, (*red curve* and *right-hand vertical axis*) in Jyväskylä. (**d**) Predicted effects of night length (ΔH_{NL}, *red curve*) and air temperature (ΔH_T, *black curve*) on the potential increase of stationary frost hardiness. (**e**) Predicted needle frost hardiness in *Pinus sylvestris*

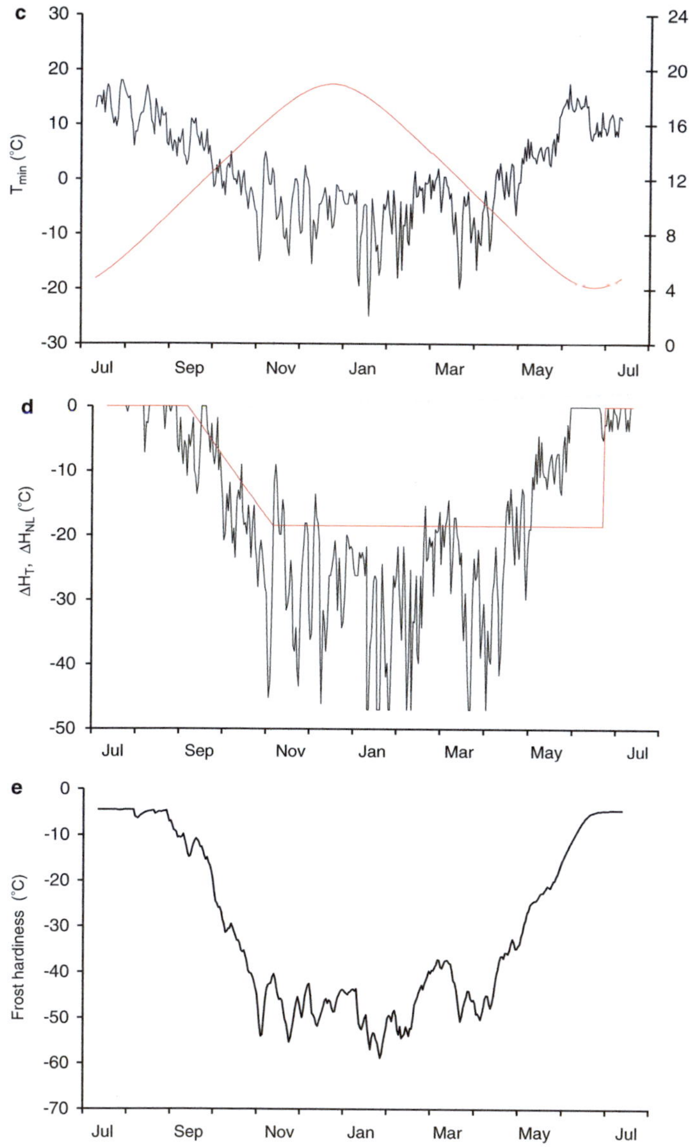

Fig. 5.16 (continued)

temperatures dropped (Fig. 5.16a) and the rate of rest break increased as indicated by the increased slope of the curve for the state of rest break, S_r, for the time (Fig. 5.16b). The chilling requirement of rest completion was met on 17 October. During rest, i.e., from 31 July to 17 October, the hardening competence remained at its maximum level $C_h = 1$ (Figs. 5.12b and 5.16b).

After rest completion on 17 October,[10] the low air temperatures (Fig. 5.16a) generally restricted the ontogenetic development, so that at any given moment in winter the state of ontogenetic development, S_o, either increased slowly or remained constant (Fig. 5.16b). However, in early November and again at the end of February and beginning of March, the air temperatures were sufficiently high (Fig. 5.16a) to cause considerable ontogenetic development, as indicated by the accelerated increase of S_o at those times (Fig. 5.16b). From mid-April onwards, the rising air temperatures (Fig. 5.16a) caused a more continuous acceleration of ontogenetic development, so that the high air temperature requirement of growth onset was met on 29 May (Fig. 5.16b).

During quiescence the hardening competence, C_h, decreases linearly with increasing S_o (Fig. 5.12b). Thus, as S_o increased relatively slowly after rest completion on 17 October, C_h remained at relatively high levels in winter but started to decrease more rapidly from mid-April onwards, when ontogenetic development was accelerated (Fig. 5.16b). At the onset of height growth on 29 May, C_h had dropped to about a half of its maximum value prevailing during rest (Fig. 5.16b). This describes the specific case addressed in the simulation, i.e., the *Pinus sylvestris* needles of the previous year are still hardy at the time of the onset of elongation growth (Leinonen et al. 1997). Despite a drop in the latter part of June 1961, the air temperatures generally remained at a sufficiently high level after the growth onset to promote active growth (Fig. 5.16a), so that the high-temperature requirement of active growth was met on 12 July 1961 (Fig. 5.16b). During the first half of the active growth phase, the hardening competence decreased to zero, where it remained until the end of this phase (Figs. 5.12b and 5.16b).

The potential direct effects of air temperature and night length on frost hardiness are represented by the effects of these environmental factors on the potential increase in stationary frost hardiness (Fig. 5.16d). According to the parameter values defined by Leinonen (1996a), the effect of night length increases linearly when night length increases from 10 to 16 h (Fig. 5.15a). Thus, at the latitude of Jyväskylä, the value of ΔH_{NL} decreases linearly from $\Delta H_{NL} = 0$ °C on 5 September to $\Delta H_{NL} = -18.5$ °C on 4 November (Fig. 5.16d).[11] As it is assumed that decreasing night length during winter and spring does not cause dehardening, the maximum effect of night length on frost hardiness, indicated by the minimum value of ΔH_{NL}, is assumed to prevail until summer solstice, after which the value of ΔH_{NL} abruptly changes to zero (Fig. 5.16d).

Regarding the effect of air temperature on the potential increase of stationary frost hardiness, Leinonen's (1996a) parameterization correspondingly implies a linear increase in the temperature range from +10 °C to −16 °C. This is seen as a

[10] Note that the model includes the simplifying assumption of abrupt rest completion when the chilling requirement is met. See the discussion in Sect. 3.6.4.

[11] This is the case in leap years, as the year 1960 was. In other years the corresponding events occur one calendar day later.

corresponding decrease of $\Delta H_T = 0$ °C at $T_{min} = +10$ °C to $\Delta H_T = -47$ °C at $T_{min} = -16$ °C (Fig. 5.15b). Accordingly, on warm summer days with $T_{min} > +10$ °C and cold winter days with $T_{min} < -16$ °C, ΔH_T stays at the constant levels of 0 °C and −47 °C, respectively; otherwise ΔH_T fluctuates according to the fluctuation of T_{min} (Figs. 5.16c, d).

Hardening competence, C_h, represents the indirect environmental regulation of frost hardiness, so that it mediates the constraints caused by the annual phenological cycle, which itself is also regulated by environmental factors, on the annual cycle of frost hardiness (Fig. 5.11c; Eq. 5.3). Thus, on any given day, the value of C_h (Fig. 5.16b) determines the proportion of potential increase in the stationary frost hardiness caused by the additive effects of NL and T_{min} (Fig. 5.16d), which is finally realised in the value of the stationary frost hardiness in the calculation of the rate of change of frost hardiness with Eq. 5.1.

In July 1960, C_h is on the increase during lignification (Fig. 5.16b), but the prevailing values of NL are too small and the values of T_{min} too high (Fig. 5.15c) to have any noticeable effect on ΔH_{NL} and ΔH_T, respectively (Fig. 5.16d). In August, C_h is at its maximal level (Fig. 5.16b) and the T_{min} values are low enough on some days (Fig. 5.16c) to have an effect on ΔH_T (Fig. 5.16d). These events of low T_{min}, however, are of such short duration that despite their effect on ΔH_T (Fig. 5.16d), they cause only slight transient hardening of the needles (Fig. 5.16e) because of the inertia represented by the time constant τ in Eq. 5.1.

From the beginning of September, the T_{min} values drop more constantly (Fig. 5.16c), and from 5 September, NL is also long enough (Figs. 5.15a and 5.16c) to cause a potential increase in the stationary frost hardiness (Fig. 5.16d). Thus, as the increasing direct hardening effect of T_{min} and NL is combined with the maximum or near-maximum value of C_h in September and early to mid-October, an almost monotonous hardening is predicted. The hardening is interrupted only by a few short episodes of slight dehardening caused by transient episodes of unusually high temperatures (Fig. 5.16e).

At the beginning of November, the predicted needle frost hardiness attains an approximately maximal level (Fig. 5.16e). From then on until mid-April, the predicted frost hardiness fluctuates (Fig. 5.16e) according to the fluctuations of ΔH_T (Fig. 5.16d) caused by fluctuations of T_{min} (Fig. 5.16c), though the overall level of frost hardiness decreases slightly as a result of a decrease in the hardening competence, C_h (Fig. 5.16b). From mid-April on, the rising air temperatures (Figs. 5.16a, c) cause both a rapid decrease of C_h (Fig. 5.16b) and increase of ΔH_T (Fig. 5.16d). Together, these two factors predict an almost monotonic dehardening in spring, so that the hardiness level of −10 °C is attained on 9 June 1961 (Fig. 5.16e). However, due to the hardening competence still remaining in early June (Fig. 5.16b) and the inertia represented by the time constant τ, the minimum level of hardiness is not attained until 3 July 1961 (Fig. 5.16e).

5.3.2.5 The Realism and Future Development of the Models

Leinonen's (1996a) model predicted the annual cycle of the frost hardiness of *Pinus sylvestris* growing in central Finland accurately (Fig. 5.14b). Unfortunately, Kellomäki et al.'s (1992, 1995), and Leinonen's (1996a) model have never been tested with the same set of data. Even so, though Kellomäki et al.'s (1992, 1995) model also predicted the frost hardiness of *Pinus sylvestris* with reasonable accuracy in a test with a smaller set of data (Fig. 5.14a), Leinonen's (1996a) model is the more realistic of the two. This is because in addition to being explicitly parameterized for the species examined, it addresses the gradual change in the environmental responses of frost hardiness that takes place during the annual phenological cycle. The relatively high accuracy obtained with the evidently less realistic model of Kellomäki et al. (1992, 1995) is once again a good demonstration of the relationship between model realism and model accuracy, i.e., high accuracy does not imply high realism.

Zhang et al. (2003) tested the additive theory of the effects of air temperature and night length on frost hardening with second-year *Pinus sylvestris* seedlings. Starting at the cessation of elongation growth, the seedlings were exposed to three combinations of long and short nights and low and high air temperatures. The frost hardiness of four tissue types, i.e., stems, needless, buds, and roots, was monitored in the different treatments. According to the additive theory, the hardiness attained in the long night and low temperature treatment should be equal to the sum of those attained in the short night and low temperature and long night and high temperature treatments minus the minimal hardiness prevailing at the beginning of the experiment (Eq. 5.3). This prediction was not realised with any of the four tissue types examined. Furthermore, the effects of the various treatments varied markedly among the four tissue types examined. For instance, the hardening of the stems took place similarly during the first 30 days of each treatment, after which it was the most rapid in the long night and high temperature treatment; with roots, this treatment caused the least hardening (Zhang et al. 2003).

Though Leinonen's (1996a) model is supported by several empirical studies specifically addressing the hardening of *Pinus sylvestris*, Zhang et al.'s (2003) experimental results call for a revision of it, as it is based on the additive theory of the effects of air temperature and night length on the hardening of trees (Eq. 5.3). In further model development, the earlier empirical evidence supporting Leinonen's (1996a) model should be compared with the contradictory evidence presented by Zhang et al. (2003). Should the latter evidence be found to be more comprehensive than the former, then the model should be revised.

More generally, Zhang et al.'s (2003) results provide an illustrative demonstration of the hypothetico-deductive (HDM) approach adopted in the present volume (Fig. 2.2 in Sect. 2.1). All models, including the most thoroughly tested ones, should ultimately be regarded as research tools for theoretical research and deduction of further research hypotheses. Accordingly, models supported by relatively sound empirical evidence, too, should be subjected to further empirical tests, and if counteracting evidence is accumulated in the tests, the model should be revised. To this end, the modelling framework adopted in the present volume facilitates a flexible introduction of novel regulation principles of the annual cycle into the models whenever this is considered necessary (Hänninen and Kramer 2007; Hänninen and Lundell 2007).

Considering Leinonen's (1996a) model, the main new principle introduced in it is the concept of hardening competence, which mediates the constraints caused by the annual phenological cycle on the annual cycle of frost hardiness more realistically than the method applied in the older model of Kellomäki et al. (1992, 1995) does (Fig. 5.12). Thus, even if Leinonen's (1996a) model were to be revised with respect to the additive effect included in it, there would probably be no need to revise the idea of hardening competence introduced in it.

Experimental evidence shows that in many tree species there is considerable variation among different provenances in the annual cycle of frost hardiness (Sect. 6.3). In order to address this variation in modelling, it may be adequate in several cases to reconsider the values of the model parameters, such as the high temperature requirement of growth onset, H_{crit}, (Sect. 3.2.1) or the time constant of hardening/dehardening, τ, (Eq. 5.1). This is the case when the main drivers and modes of action of the annual cycle of frost hardiness are similar among the different provenances, so that there are only quantitative differences in the responses among the provenances. However, even in this basically straightforward case the amount of experimental work required for the development of realistic models for the numerous provenances is considerable.

The different environmental responses found by Zhang et al. (2003) for the different tissue types suggest that ideally, a specific model should be formulated not only for each tree species but also for each tissue type of the species. This notion accords with earlier findings of differential hardening in different tree tissues (Sakai and Larcher 1987; Larcher 2003). However, practical limitations usually prevent such an ideal practice, so that compromises must be made. In doing so, it is most crucial to consider the limitations of the models applied. The main problem, especially with integrated models, is their complexity, which hampers their parameterization for different species and tree tissues. In comparison with several other models used in tree ecophysiology and climate change research, these models are not especially complicated mathematically, but their testing often requires large hierarchical experimental setups, such as the one reported in Fig. 5.10. This has led Greer et al. (2001) to suggest that more simplified models should be developed but that their ecophysiological basis should not be lost. Once again, the concept of model coverage is essential for this effort. This concept is discussed further in Chap. 8.

5.4 Year-to-Year Variation in Frost Hardiness and Damage

Year-to-year variation in the annual cycle of frost hardiness is examined in the present volume with reference to a simulation experiment with 90-year long-term data for Jyväskylä, central Finland.[12] The results show large year-to-year variation for winter months, as indicated by a frost hardiness range of approximately 30 °C for winter days (Fig. 5.17). The large variation is caused by the large variation in air temperature. In winter, high air temperatures decrease frost hardiness both directly by increasing the value of ΔH_T (Fig. 5.16d) and indirectly by lowering the value of C_h (Fig. 5.16b). Towards spring, the year-to-year variation is reduced (Fig. 5.17) because the decreasing hardening competence (Fig. 5.16b) restricts the degree to which the potential direct effects of air temperature on frost hardiness are realized. If frost hardiness were regulated by direct effects only, then the sub-model used for that response (Fig. 5.15b) would predict large variation in spring as well.

In summer there is generally little year-to-year variation in frost hardiness, and in mid-July there is practically none (Fig. 5.17). This phenomenon is caused mainly by the limited hardening competence prevailing during the active growth phase and the initial phases of lignification (Fig. 5.12b). Accordingly, the year-to-year

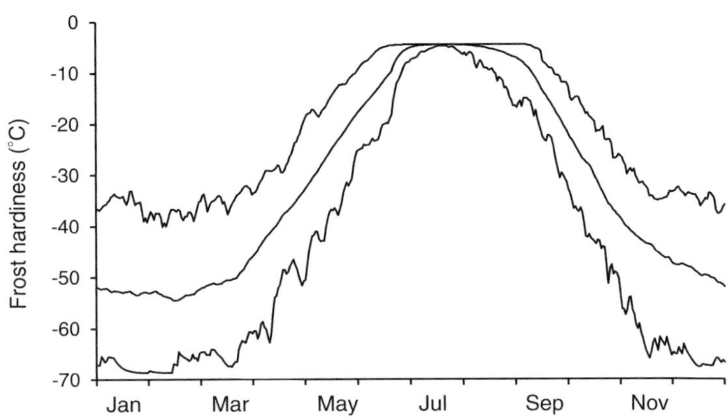

Fig. 5.17 Simulated year-to-year variation in the annual cycle of needle frost hardiness in *Pinus sylvestris* over 90 years in 1884–1979 in Jyväskylä, central Finland (the years 1911–1916 were excluded on account of missing meteorological observations). For each day, the year-to-year maximum (*lower curve*), mean (*middle curve*) and minimum (*upper curve*) hardiness, as predicted by Leinonen's (1996a) model, is presented

[12] As indicated in Chap. 1, the data cover 92 years, but as the simulation was initiated at growth cessation in July 1883 and terminated at growth cessation in July 1980, the predicted frost hardiness data for the first and the last year were incomplete and were therefore rejected from the statistics for Fig. 5.17.

variation predicted for the summer months is mainly caused by the variation in the timing of active growth and lignification (results not shown). Variation in the progress of lignification also causes variation in the increase of hardening competence, and for this reason the variation in frost hardiness increases in August (Fig. 5.17). However, full hardening competence is attained every year, at least later in autumn. Thus, as there is no year-to-year variation in the effect of night length (Fig. 5.16d), the increasing late-autumn variation in frost hardiness is caused solely by the variation in air temperature.

Despite the wide year-to-year variation in the annual cycle of frost hardiness, very little frost damage was predicted for the 90 years in the simulation. The predicted annual needle loss caused by frost damage was 2.0 % on average, ranging from a minimum of 0.5 % to a maximum of 6.5 %. In all, then, these simulation results suggest that the *Pinus sylvestris* trees can acclimate well to the climatic conditions of central Finland. This result is in accordance with the notion that native tree species and provenances are generally well adapted to the climate prevailing at their natural growing sites.

5.5 Summary

Frost hardiness is a relatively well-defined ecophysiological whole-tree attribute of the seasonality of boreal and temperate trees. When measuring it, however, one needs to specify several aspects of it, so that the results of different studies are not always directly comparable with each other. There is plenty of experimental data available on the environmental regulation of frost hardiness. Air temperature and night length are usually the most important environmental factors regulating frost hardiness. Frost hardiness is closely related to the seasonal ontogenetic development of trees, so that any comprehensive study of the annual cycle of frost hardiness should examine it in combination with the annual phenological cycle.

In recent decades, integrated models explicitly addressing the changing environmental responses of frost hardiness over the annual phenological cycle have been developed for boreal and temperate trees. The models often have high accuracy and, being based on explicit ecophysiological considerations, they evidently have considerable realism as well. However, the multitude of effects of, and interactions between, various environmental factors found in experimental studies suggests that the realism of the models is decreased by insufficient coverage of the essential factors affecting frost hardiness. Experimental studies have also revealed great genetic differences among tree species and provenances, and physiological differences among different tree tissues in the environmental regulation of frost hardiness. This necessarily implies that the generality of any one model of frost hardiness cannot be particularly high. This reasoning, then, calls for novel experimental work with different tree species and tissues, guided by the hypothetico-deductive modelling framework introduced in Sect. 2.1.

List of Symbols

t time

Time-Dependent Environmental Variables

$T(t)$ air temperature (°C)
$T_{mean}(t)$ daily mean air temperature (°C)
$T_{min}(t)$ daily minimum air temperature (°C)
$NL(t)$ daily night length (h)

Time-Dependent Rate and State Variables

$R_h(t)$ rate of change of frost hardiness (rate of hardening/dehardening) (°C day^{-1})
$S_h(t)$ state of frost hardiness (°C)
$S_r(t)$ state of rest break (%)
$S_o(t)$ state of ontogenetic development (%)
S_a state of active growth (%)
S_l state of lignification (%)

Auxiliary Frost Hardiness Variables

$S^s{}_h(t)$ stationary frost hardiness (°C)
$\Delta H_T(t)$ potential increase of the stationary frost hardiness caused by air temperature (°C)
$\Delta H_{NL}(t)$ potential increase of the stationary frost hardiness caused by night length (°C)

Variable Mediating the Effects on the Annual Phenological Cycle on Frost Hardiness

$C_h(t)$ hardening competence

Model Parameters

τ	time constant (day)
a	slope of linear dependence of stationary frost hardiness on daily minimum air temperature
b	interception of linear dependence of stationary frost hardiness on daily minimum air temperature
$S^{s}_{h,min}$	minimal stationary frost hardiness

References

Aitken, S. N., & Hannerz, M. (2001). Genecology and gene resource management strategies for conifer cold hardiness. In F. J. Bigras & S. J. Colombo (Eds.), *Conifer cold hardiness* (pp. 23–53). Dordrecht: Kluwer Academic Publishers.

Aronsson, A. (1975). Influence of photo- and thermoperiod on the initial stages of frost hardening and dehardening of phytotron-grown seedlings of Scots pine (*Pinus silvestris* L.) and Norway spruce (*Picea abies* (L.) Karst.). *Studia Forestalia Suecica, 128*, 1–20.

Aronsson, A. (1980). Frost hardiness in Scots pine (Pinus silvestris L.) II. Hardiness during winter and spring in young trees of different mineral nutrient status. *Studia Forestalia Suecica, 155*, 1–27.

Bannister, B., & Neuner, G. (2001). Frost resistance and the distribution of conifers. In F. J. Bigras & S. J. Colombo (Eds.), *Conifer cold hardiness* (pp. 3–21). Dordrecht: Kluwer Academic Publishers.

Bigras, F. J., Ryyppö, A., Lindström, A., & Stattin, E. (2001). Cold acclimation and deacclimation of shoots and roots of conifer seedlings. In F. J. Bigras & S. J. Colombo (Eds.), *Conifer cold hardiness* (pp. 57–88). Dordrecht: Kluwer Academic Publishers.

Burr, K. E., Hawkins, C. D. B., L'Hirondelle, S. J. L., Binder, W. D., George, M. F., & Repo, T. (2001). Methods for measuring cold hardiness of conifers. In F. J. Bigras & S. J. Colombo (Eds.), *Conifer cold hardiness* (pp. 369–401). Dordrecht: Kluwer Academic Publishers.

Cannell, M. G. R., Sheppard, L. J., Smith, R. I., & Murray, M. B. (1985). Autumn frost damage on young *Picea sitchensis* 2. Shoot frost hardening, and the probability of frost damage in Scotland. *Forestry, 58*, 145–166.

Cao, Y., Repo, T., Silvennoinen, R., Lehto, T., & Pelkonen, P. (2010). An appraisal of the electrical resistance method for assessing root surface area. *Journal of Experimental Botany, 61*, 2491–2497.

Cao, Y., Repo, T., Silvennoinen, R., Lehto, T., & Pelkonen, P. (2011). Analysis of the willow root system by electrical impedance spectroscopy. *Journal of Experimental Botany, 62*, 351–358.

Chen, H.-H., & Li, P. H. (1978). Interactions of low temperature, water stress, and short days in the induction of stem frost hardiness in red osier dogwood. *Plant Physiology, 62*, 833–835.

Christersson, L. (1978). The influence of photoperiod and temperature on the development of frost hardiness in seedlings of *Pinus silvestris* and *Picea abies*. *Physiologia Plantarum, 44*, 288–294.

Colombo, S. J., Menzies, M. I., & O'Reilly, C. O. (2001). Influence of nursery cultural practices on cold hardiness of coniferous forest tree seedlings. In F. J. Bigras & S. J. Colombo (Eds.), *Conifer cold hardiness* (pp. 223–252). Dordrecht: Kluwer Academic Publishers.

Edgerton, L. J. (1954). Fluctuations in the cold hardiness of peach flower buds during rest period and dormancy. *Proceedings of the American Society for Horticultural Science, 64*, 175–180.

Fuchigami, L. H., Weiser, C. J., Kobayashi, K., Timmis, R., & Gusta, L. V. (1982). A degree growth stage (°GS) model and cold acclimation in temperate woody plants. In P. H. Li & A. Sakai (Eds.), *Plant cold hardiness and freezing stress. Mechanisms and crop implications* (Vol. 2, pp. 93–116). New York: Academic Press.
Greer, D. H. (1983). Temperature regulation of the development of frost hardiness in *Pinus radiata* D. Don. *Australian Journal of Plant Physiology, 10*, 539–547.
Greer, D. H., Leinonen, I., & Repo, T. (2001). Modelling cold hardiness development and loss in conifers. In F. J. Bigras & S. J. Colombo (Eds.), *Conifer cold hardiness* (pp. 437–460). Dordrecht: Kluwer Academic Publishers.
Hänninen, H., & Kramer, K. (2007). A framework for modelling the annual cycle of trees in boreal and temperate regions. *Silva Fennica, 41*, 167–205.
Hänninen, H., & Lundell, R. (2007). Dynamic models in plant ecophysiology. In E. Taulavuori & K. Taulavuori (Eds.), *Physiology of northern plants under changing environment* (pp. 157–175). Trivandrum, Kerala: Research Signpost.
Hänninen, H., Leinonen, I., Repo, T., & Kellomäki, S. (1996). Overwintering and productivity of Scots pine in a changing climate. *Silva Fennica, 30*, 229–237.
Hänninen, H., Zhang, G., Rikala, R., Luoranen, J., Konttinen, K., & Repo, T. (2013). Frost hardening of Scots pine seedlings in relation to the climatic year-to-year variation in air temperature. *Agricultural and Forest Meteorology, 177*, 1–9.
Holliday, J. (2009). Genomics of cold hardiness in forest trees. In L. V. Gusta, M. E. Wisniewski, & K. Tanino (Eds.), *Plant cold hardiness. From the laboratory to the field* (pp. 293–304). Oxfordshire: CAB International.
Irving, R. M., & Lanphear, F. O. (1967). Dehardening and the dormant condition in *Acer* and *Viburnum*. *Proceedings of the American Society for Horticultural Science, 91*, 699–705.
Junttila, O. (2007). Regulation of annual shoot growth cycle in northern tree species. In K. Taulavuori & E. Taulavuori (Eds.), *Physiology of northern plants under changing environment* (pp. 177–210). Trivandrum, Kerala: Research Signpost.
Kellomäki, S., Väisänen, H., Hänninen, H., Kolström, T., Lauhanen, R., Mattila, U., & Pajari, B. (1992). A simulation model for the succession of the boreal forest ecosystem. *Silva Fennica, 26*, 1–18.
Kellomäki, S., Hänninen, H., & Kolström, M. (1995). Computations on frost damage to Scots pine under climatic warming in boreal conditions. *Ecological Applications, 5*, 42–52.
King, D. A., & Ball, M. C. (1998). A model of frost impacts on seasonal photosynthesis of *Eucalyptus pauciflora*. *Australian Journal of Plant Physiology, 25*, 27–37.
Kobayashi, K. D., Fuchigami, L. H., & Weiser, C. J. (1983). Modeling cold hardiness of red-osier dogwood. *Journal of the American Society for Horticultural Science, 108*, 376–381.
Koski, V., & Sievänen, R. (1985). Timing of growth cessation in relation to the variations in the growing season. In P. M. A. Tigerstedt, P. Puttonen, & V. Koski (Eds.), *Crop physiology of forest trees* (pp. 167–193). Helsinki: Helsinki University Press.
Kramer, K. (1994). A modelling analysis of the effects of climatic warming on the probability of spring frost damage to tree species in The Netherlands and Germany. *Plant, Cell and Environment, 17*, 367–377.
Larcher, W. (2003). *Physiological plant ecology. Ecophysiology and stress physiology of functional groups* (4th ed.). Berlin: Springer. 513 p.
Leinonen, I. (1996a). A simulation model for the annual frost hardiness and freeze damage of Scots pine. *Annals of Botany, 78*, 687–693.
Leinonen, I. (1996b). Dependence of dormancy release on temperature in different origins of *Pinus sylvestris* and *Betula pendula* seedlings. *Scandinavian Journal of Forest Research, 11*, 122–128.
Leinonen, I., Repo, T., Hänninen, H., & Burr, K. E. (1995). A second-order dynamic model for the frost hardiness of trees. *Annals of Botany, 76*, 89–95.
Leinonen, I., Repo, T., & Hänninen, H. (1996). Testing of frost hardiness models for *Pinus sylvestris* in natural conditions and in elevated temperature. *Silva Fennica, 30*, 159–168.

Leinonen, I., Repo, T., & Hänninen, H. (1997). Changing environmental effects on frost hardiness of Scots pine during dehardening. *Annals of Botany, 79*, 133–138.

Luoranen, J. (2000). Control of growth and frost hardening of silver birch container seedlings: Growth retardants, short day treatment and summer planting. *The Finnish Forest Research Institute, Research Papers, 777*, 1–167.

Luoranen, J., Lahti, M., & Rikala, R. (2008). Frost hardiness of nutrient-loaded two-year-old *Picea abies* seedlings in autumn and at the end of freezer storage. *New Forests, 35*, 207–220.

Mäkelä, A., Hari, P., Berninger, F., Hänninen, H., & Nikinmaa, E. (2004). Acclimation of photosynthetic capacity in Scots pine to the annual cycle of temperature. *Tree Physiology, 24*, 369–376.

Pelkonen, P., & Hari, P. (1980). The dependence of the springtime recovery of CO_2 uptake in Scots pine on temperature and internal factors. *Flora, 169*, 398–404.

Proebsting, E. L. (1963). The role of air temperatures and bud development in determining hardiness of dormant Elberta peach fruit buds. *Proceedings of the American Society for Horticultural Science, 83*, 259–269.

Quamme, H. A. (1995). Deep supercooling in buds of woody plants. In R. E. Lee Jr., G. J. Warren, & L. V. Gusta (Eds.), *Biological ice nucleation and its applications* (pp. 183–199). St. Paul: APS Press.

Räisänen, M., Repo, T., & Lehto, T. (2009). Cold acclimation of Norway spruce roots and shoots after boron fertilization. *Silva Fennica, 43*, 223–233.

Raulo, J., & Leikola, M. (1974). Tutkimuksia puiden vuotuisen pituuskasvun ajoittumisesta. Summary: Studies on the annual height growth of trees. *Communicationes Instituti Forestalis Fenniae, 81*(2), 1–19.

Repo, T. (1991). Rehardening potential of Scotch pine seedlings during dehardening. *Silva Fennica, 25*, 13–21.

Repo, T. (1992). Seasonal changes of frost hardiness in *Picea abies* and *Pinus sylvestris* in Finland. *Canadian Journal of Forest Research, 22*, 1949–1957.

Repo, T. (1993). Impedance spectroscopy and temperature acclimation of forest trees. *University of Joensuu, Faculty of Forestry. Research Notes* 9. 53 p.

Repo, T., & Pelkonen, P. (1986). Temperature step response of dehardening in *Pinus sylvestris* seedlings. *Scandinavian Journal of Forest Research, 1*, 271–284.

Repo, T., Mäkelä, A., & Hänninen, H. (1990). Modelling frost resistance of trees. In H. Jozefek (Ed.), Modelling to understand forest functions. *Silva Carelica, 15*, 61–74.

Repo, T., Zhang, M. I. N., Ryyppö, A., Vapaavuori, E., & Sutinen, S. (1994). Effects of freeze-thaw injury on parameters of distributed electrical circuits of stems and needles of Scots pine seedlings at different stages of acclimation. *Journal of Experimental Botany, 45*, 823–833.

Repo, T., Hänninen, H., & Kellomäki, S. (1996). The effects of long-term elevation of air temperature and CO_2 on the frost hardiness of Scots pine. *Plant, Cell and Environment, 19*, 209–216.

Repo, T., Zhang, G., Ryyppö, A., & Rikala, R. (2000). The electrical impedance spectroscopy of Scots pine (*Pinus sylvestris* L.) shoots in relation to cold acclimation. *Journal of Experimental Botany, 51*, 2095–2107.

Repo, T., Leinonen, I., Wang, K.-Y., & Hänninen, H. (2006). Relation between photosynthetic capacity and cold hardiness in Scots pine. *Physiologia Plantarum, 126*, 224–231.

Rikala, R., & Repo, T. (1987). Frost resistance and frost damage in *Pinus sylvestris* seedlings during shoot elongation. *Scandinavian Journal of Forest Research, 2*, 433–440.

Rousi, M. (1985). Neulastuhot osa luonnon järjestelmää. *Metsä ja Puu, 9*, 16–19.

Sakai, A., & Larcher, W. (1987). *Frost survival of plants. Responses and adaptation to freezing stress*. Berlin: Springer. 321 pp.

Sakai, A., & Weiser, C. J. (1973). Freezing resistance of trees in North America with reference to tree regions. *Ecology, 54*, 118–126.

Sutinen, M.-L. (2007). Lipid changes during cold-acclimation and de-acclimation of herbaceous and woody species. In K. Taulavuori & E. Taulavuori (Eds.), *Physiology of northern plants under changing environment* (pp. 19–37). Trivandrum, Kerala: Research Signpost.

Sutinen, M.-L., Palta, J. P., & Reich, P. B. (1992). Seasonal differences in freezing stress resistance of needles of *Pinus nigra* and *Pinus resinosa*: Evaluation of the electrolyte leakage method. *Tree Physiology, 11*, 241–254.

Sutinen, M.-L., Arora, R., Wisniewski, M., Ashworth, E., Strimbeck, R., & Palta, J. (2001). Mechanisms of frost survival and freeze-damage in nature. In F. J. Bigras & S. J. Colombo (Eds.), *Conifer cold hardiness* (pp. 89–120). Dordrecht: Kluwer Academic Publishers.

Taulavuori, K., Niinimaa, A., Laine, K., Taulavuori, E., & Lähdesmäki, P. (1997). Modelling frost resistance of Scots pine seedlings using temperature, daylength and pH of cell effusate. *Plant Ecology, 133*, 181–189.

Taulavuori, K., Prasad, M. N. V., Taulavuori, E., & Laine, K. (2005a). Metal stress consequences on frost hardiness of plants at northern high latitudes: A review and hypothesis. *Environmental Pollution, 135*, 209–220.

Taulavuori, K., Taulavuori, E., & Laine, K. (2005b). Ultraviolet radiation and plant frost hardiness in the Subarctic. *Arctic, Antarctic, and Alpine Research, 37*, 11–15.

Weiser, C. J. (1970). Cold resistance and injury in woody plants. *Science, 169*, 1269–1278.

Winter, F. (1973). Ein Simulationsmodell über die Phänologie und den Verlauf der Frostresistenz von Apfelbäumen. *Ecologia Plantarum, 8*, 141–152.

Wisniewski, M. (1995). Deep supercooling in woody plants and the role of plant structure. In R. E. Lee Jr., G. J. Warren, & L. V. Gusta (Eds.), *Biological ice nucleation and its applications* (pp. 163–181). St. Paul: APS Press.

Zhang, G. (2001). Cold acclimation in Scots pine. *The Finnish Forest Research Institute, Research Papers* 817. 52 p.

Zhang, G., Ryyppö, A., Vapaavuori, E., & Repo, T. (2003). Quantification of additive response and stationarity of frost hardiness by photoperiod and temperature in Scots pine. *Canadian Journal of Forest Research, 33*, 1772–1784.

Chapter 6
Evolutionary Aspects of the Annual Cycle

Contents

Abstract The hypothetico-deductive modelling framework introduced in Chap. 2 is applied to examining the evolutionary aspects of the annual cycle in boreal and temperate trees. For use of growth resources and competition (capacity adaptation), early onset and late cessation of growth are selected for. However, due to the risk of spring and autumn frost damage (survival adaptation), they are simultaneously selected against. This trade-off is examined by means of computer simulations with models representing various regulation principles of the annual cycle. Considerable differences among the principles are reported. When the principles are equally constrained for avoidance of frost damage, some of them allow the trees to use the growing season more comprehensively than others. Next, differences among the provenances of the tree species are examined within the framework of the modelling approach. The annual cycle of each provenance is adapted to its native climate, and this adaptation is manifested in several traits associated with the environmental regulation of the annual cycle. In the models of the annual cycle, this

H. Hänninen, *Boreal and Temperate Trees in a Changing Climate*, Biometeorology 3,
DOI 10.1007/978-94-017-7549-6_6

genetic differentiation is readily addressed via the values of the model parameters, such as critical night length of growth cessation or the chilling requirement of rest break. The possibilities of addressing the effects of the maternal environment on the annual cycle traits of the offspring are also discussed, and so is the emerging approach of combining genetic and ecophysiological modelling of the annual cycle.

Keywords Annual cycle • Boreal trees • Capacity adaptation • Chilling requirement • Climatic adaptation • Climatic variation • Computer simulations • Critical night length • Ecophysiological modelling • High temperature requirement • Photoperiodic ecotypes • Seasonality • Survival adaptation • Temperate trees

6.1 Proximate and Ultimate Causes for Tree Seasonality

In evolutionary biology, causal factors affecting organisms are usually subdivided into the categories of proximate and ultimate causes (Mayr 1961, 1982). Proximate causes denote those present-day factors that affect the organisms within an ecological time scale, i.e., during the lifetime of an individual organism. Thus, as they affect the organisms in real time, their effects can be established with experiments, at least in principle. Ultimate causes, on the contrary, denote those factors that have caused evolutionary change via natural selection during the evolutionary history of the species. Interestingly, the proximal effect can be realised by a factor different from the one that was the ultimate cause during the evolutionary history.

The concepts of proximate and ultimate causes are especially relevant for the seasonality of boreal and temperate trees. The main ultimate cause for the seasonality of these trees is the occurrence of low freezing air temperatures in winter. The trees cannot tolerate the low freezing temperatures while in the active growth phase, so that natural selection has removed those tree genotypes from the populations that had insufficient seasonal acclimation to the climatic seasonality (Heide 1985). Seasonal acclimation is realised via the environmental responses of tree development discussed in Chaps 3, 4, and 5, so that the environmental factors causing these ecophysiological responses constitute the proximate causes for tree seasonality.

In boreal and temperate trees, the proximate causes are different from the ultimate one, i.e., the occurrence of low freezing air temperatures in winter, almost without exception. Increasing night length, for instance, is an essential proximate cause for growth cessation (Sect. 3.5) and frost hardening in autumn (Sect. 5.3.2). Ultimately, however, these ecophysiological responses are adaptations to the occurrence of low freezing temperatures, not to the long nights. The increasing night length provides the tree with the environmental cue of the approaching season with low freezing temperatures. The response to the proximate factor is crucial, for if the cessation of growth and the initiation of frost hardening were to take place only upon the occurrence of the ultimate reason, i.e., the low freezing air temperatures, the tree would be late in its frost hardening process. In this way, the proximate environmental factors provide the trees not only with matter and energy but also with information concerning their future environmental conditions (Lockhart 1983; Häkkinen and Hari 1988; Cannell 1989; Hänninen et al. 1990).

6.2 The Trade-Off Between Survival and Growth

6.2.1 *Survival Adaptation and Capacity Adaptation*

Heide (1985) divided the adaptive phenomena and traits of plants into two categories. He introduced the concept of *capacity adaptation* for all traits that facilitate the plants' performance in competition and the concept of *survival adaptation* for all traits that help the plants to overcome any form of stress. In boreal and temperate conditions, the low air temperature prevailing in winter is the main stress factor requiring survival adaptation, which is also required, especially in many temperate regions, for tolerance of drought. Furthermore, biotic stresses, such as herbivores and pathogens, require survival adaptation as well.

There is often a trade-off between the traits facilitating survival adaptation and those facilitating capacity adaptation. It is especially pronounced in the annual cycle of boreal and temperate trees (Lockhart 1983; Bennie et al. 2010). With respect to capacity adaptation, a long growing season, i.e., early onset and late cessation of photosynthetic production, nutrient uptake, and growth, would be beneficial because it facilitates the use of the growth resources and thus improves the competitive ability of the individual tree. With respect to survival adaptation, however, a long growing season might be risky, as it exposes the trees to late spring and early autumn frosts. Thus the timing of the events in the annual cycle of boreal and temperate trees is subjected to a stabilising selection caused by the two selective pressures working in opposite directions.

The role of frost in the survival adaptation of boreal and temperate trees is realised in years with an exceptionally severe freezing stress. This can happen either in winter, with exceptionally low frost temperatures occurring, or in spring or autumn, with frost occurring unseasonally, either exceptionally late in spring or exceptionally early in autumn. Exceptional years are rare by definition, which is why studying the trade-off between survival and growth in boreal and temperate trees requires a long time span. Ideally, the trade-off should be studied on an evolutionary time scale of thousands of years, but that is not possible in practice, of course. In long-term provenance trials, the time span is typically decades (Kalela 1938; Beuker et al. 1998). Accordingly, these trials have been a valuable experimental source of information on the climatic adaptation of boreal and temperate forest tree species (Beuker 1994a, b). In most cases, however, only the long-term survival and growth of the trees is monitored, neglecting the role of the annual cycles of the different provenances included in the experiment. Such an approach does not enable one to examine specific ecophysiological hypotheses concerning the trade-off between survival and growth.

The modelling approach adopted in the present volume enables such an examination, though lack of meteorological data puts an evolutionary time scale of thousands of years out of the question in modelling studies, too. In such studies the research problem is essentially theoretical in the sense that the implications of different models of the regulation of the annual cycle for the trade-off are compared by means of model calculations (Häkkinen and Hari 1988). These studies belong to the discipline of evolutionary ecology and therefore apply the optimality approach typical for the discipline (Mäkelä et al. 2002; Cowan 2002). Accordingly, they use theoretical calculations for assessing the implications of different regulation principles of the annual cycle, represented by their corresponding models, for the overall adaptation of the trees. According to the optimality principle it may then be hypothesised that the model realised in the tree populations will be the one maximising overall adaptation in the theoretical calculations. It is the task of empirical studies to examine which, if any, of the regulation principles studied is realised in real trees. This method is a special case of the hypothetico-deductive research strategy, where theoretical calculations are used for guiding empirical studies (Fig. 2.2 in Sect. 2.1).

6.2.2 The Timing of Growth Onset

Häkkinen and Hari (1988) introduced a theoretical approach with the optimality principle to studies of the annual cycle of boreal and temperate trees. By means of computer simulations using long-term air temperature data from Jyväskylä, central Finland, they examined the implications of five regulation principles of the springtime timing of growth onset for the trade-off between growth and survival.

Each regulation principle was represented by its respective model. In the time model, growth onset was simply assumed to take place on a fixed calendar day each year, whereas in the other four models, the rate of ontogenetic development towards growth onset was assumed to depend on the air temperature. In three of the models the air temperature response of the rate of development remained constant. The response varied among the three models, however, being either piece-wise linear as in the day degree model, exponential, or sigmoidal as in Sarvas's (1972) PU model. Finally, Pelkonen and Hari's (1980) fluctuating model was also included in the analysis. In this model, feedback is included, i.e., the air temperature response of the rate of development changes according to the value of the state of development.[1]

[1] With the exception of the exponential response, these models are discussed in Sects. 3.2.1 and 4.2.1.1.

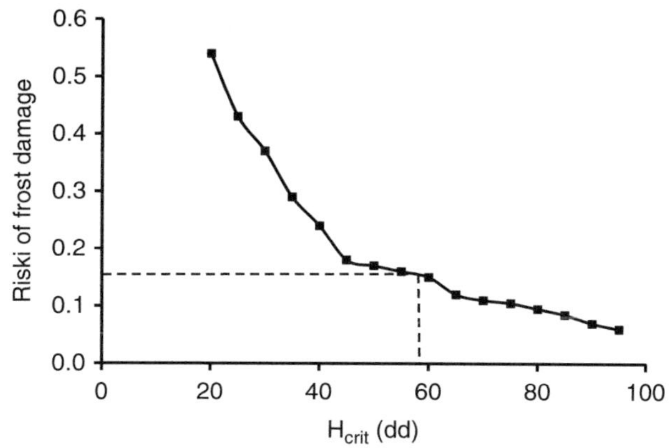

Fig. 6.1 Dependence of boreal trees' simulated risk of spring frost damage on the high temperature requirement of growth onset, H_{crit}. The simulations were carried for the climatic conditions of Jyväskylä, central Finland, for a period of 96 years with a piece-wise linear air temperature response of the rate of ontogenetic development towards growth onset. The model corresponds to the day degree model discussed in Sect. 3.2.1.1 except that the time step of the calculations was 6 h; dd = arbitrary day degree unit. The *dashed lines* illustrate the determination of the H_{crit} value for a frost damage risk of 0.15 (Häkkinen and Hari 1988; redrawn with the permission of the Finnish Forest Research Institute, currently the Natural Resources Institute Finland, and the Finnish Society of Forest Science)

With any model or regulation principle, the trade-off dilemma remains, i.e., the risk of frost damage decreases with delayed springtime growth onset, but this delay simultaneously shortens the growing season. Häkkinen and Hari (1988) examined this dilemma in terms of the five models of growth onset. To take the day degree model as an example, their sequence of calculations with each model was as follows.

First, the timing of the growth onset was calculated for each of the 96 years represented in the meteorological data. Spring frost damage was assumed to occur in the years when the daily minimum air temperature, T_{min}, dropped below a threshold value of −2 °C, representing the minimal frost hardiness of the newly emerging tissue after growth onset, between the simulated growth onset and 1 July. Second, the risk of frost damage was calculated as the proportion of years, out of the 96 years examined, when simulated frost damage occurred. This entire set of calculations was carried out by using a range of values for the high temperature requirement of growth onset, H_{crit}. Finally, the risk of frost damage obtained for each value of H_{crit} was plotted against the value of H_{crit} used in the simulation (Fig. 6.1).

In order to compare the implications of all the five models for the trade-off between survival and growth, Häkkinen and Hari (1988) carried out the calculations summarised in Fig. 6.1 for the day degree model with the other four models, too. After that, a critical parameter value corresponding to a fixed risk of frost damage was determined for each of the five models in the way indicated for the

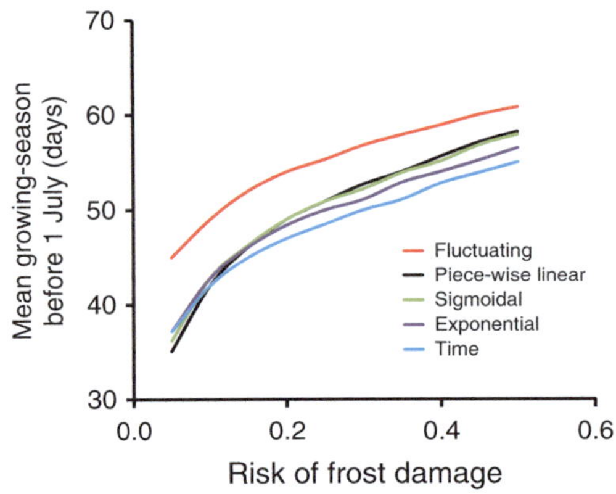

Fig. 6.2 Simulated dependence of the degree of capacity adaptation on the degree of survival adaptation in boreal trees. Capacity adaptation is represented on the *vertical axis* by the mean number of growing-season days before 1 July. The degree of survival adaptation decreases with the increase of the risk of spring frost damage represented on the *horizontal axis*. The portrayed relationship is based on simulations with the climatic conditions of Jyväskylä, central Finland, for a period of 96 years. Five different principles of regulating growth onset, each represented by its respective model indicated on the right-hand side of the figure, were used in the simulations (Häkkinen and Hari 1988; redrawn with the permission of the Finnish Forest Research Institute, currently the Natural Resources Institute Finland, and the Finnish Society of Forest Science)

day degree model in Fig. 6.1. Subsequently, the timing of growth onset was calculated for the 96 years with each model by using the critical parameter value thus determined. In this way it was possible to compare the degree of capacity adaptation implied by each of the models with a fixed degree of survival adaptation, i.e., a fixed risk of spring frost damage. Finally, the operations were repeated with different fixed values of the risk of frost damage, and the degree of capacity adaptation (i.e., the length of the growing season before 1 July) was plotted against the degree of survival adaptation (i.e., the risk of spring frost damage, with an increasing value of the risk indicating decreasing survival adaptation) (Fig. 6.2).

The results show that with all levels of the risk of spring frost damage, Pelkonen and Hari's (1980) fluctuating model implies a higher degree of capacity adaptation than any of the other four models (Fig. 6.2). Mathematically speaking, as discussed by Häkkinen and Hari (1988), this is caused jointly by the temporal autocorrelation appearing in the air temperature data and the feedback principle involved in the fluctuating model (Sect. 4.2.1.1). Due to the autocorrelation, air temperatures drop into the above-zero range well before the sub-zero temperatures are reached. The feedback involved in the fluctuating model enables a reversal of the development whenever the air temperature drops. Due to this reversal, the simulated development in any spring with exceptionally late frosts does not reach the state of growth onset before the occurrence of damaging frosts has stopped. Thus, to avoid frost damage even during those critical years, it is not necessary in this model to increase the value of the parameter affecting growth cessation, i.e., the parameter corresponding to H_{crit} (Fig. 6.1), by as much as in the other models. Accordingly, for more normal springs, when the last frosts occur earlier, the fluctuating model predicts an earlier growth onset than the other models do.

Following the optimality principle, Häkkinen and Hari (1988) concluded on the basis of their computational results (Fig. 6.2) that if the regulation principles described in the five models have undergone natural selection, then it is probably the principle described by the fluctuating model that has been selected for. This conclusion may be criticised, however, because the reversal included in the fluctuating development model is not possible in the irreversible ontogenetic development towards growth onset addressed in Häkkinen and Hari's (1988) study. In regard to this phenomenon, then, Pelkonen and Hari's (1980) fluctuating model is unrealistic. This evident shortcoming should not, however, obscure the merits of Häkkinen and Hari's (1988) pioneering study in introducing the optimality approach to quantitative simulation studies of the annual cycle of boreal and temperate trees.

6.2.3 *The Timing of Growth Cessation*

Hänninen et al. (1990) applied the optimality approach introduced by Häkkinen and Hari (1988) to a study of the trade-off between survival and growth in the case of growth cessation in boreal and temperate trees. With computer simulations, they examined the implications of three regulation principles of growth cessation for this trade-off. Growth cessation was assumed to be induced by the attainment of the critical night length, NL_{crit}, (Ekberg et al. 1979; Junttila 1980), by the accumulation of the critical temperature sum, TS_{crit}, (Raulo and Leikola 1974), or by the attainment of the critical value, JE_{crit}, of a variable representing the joint effect of these two factors (Koski and Selkäinaho 1982; Koski and Sievänen 1985).[2] For the joint effect principle, the model variable was defined as follows:

$$JE(t) = NL(t) + b \cdot S_{dd}(t) \qquad (6.1)$$

where $S_{dd}(t)$ = accumulated temperature sum, NL(t) = night length, and JE(t) = the value of the joint effect variable, all three variables at time instant t. The joint effect variable is expressed in units of night length (h), so that the value of the temperature sum is converted into units of night length by means of the parameter b. In the calculations, the value of $b = 0.0088 \text{ h (dd)}^{-1}$ was used. This value

[2] The models corresponding to the three principles are discussed in Sect. 3.5.

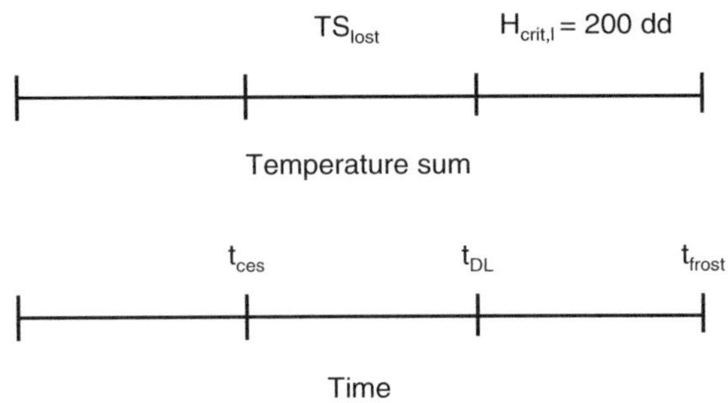

Fig. 6.3 The principle and the variables used in a theoretical examination of the growth cessation trade-off in boreal trees. The examination is based on the degree of capacity adaptation implied by a fixed degree of survival adaptation. t_{ces} = date of growth cessation, t_{DL} = deadline day when growth must cease at the latest in order to avoid autumn frost damage, t_{frost} = first day when the daily minimum air temperature, T_{min}, drops below −10 °C, TS_{lost} = lost temperature sum, i.e., the temperature sum available but not used for growth, $H_{crit,l}$ = the high temperature requirement of lignification, and dd = day degree unit. With each of the growth cessation principles examined, the TS_{lost} is minimised on the constraint that growth cessation occurs no later than the t_{DL} each year. The principles are then compared on the basis of the distributions of TS_{lost} they imply (Modified from Hänninen et al. 1990; published with the permission of Oxford University Press)

represents the slope of the fitted line in Fig. 3.23c in Sect. 3.5.3.1. The simulations were carried out by using the same 96-year air temperature data from Jyväskylä as were used in Häkkinen and Hari's (1988) study of the springtime trade-off (Figs. 6.1 and 6.2).

In Hänninen et al.'s (1990) study, the calculations were based on the assumption that after the active growth phase comes a phase of lignification and that there is a high temperature requirement of lignification, $H_{crit,l} = 200$ dd.[3] It was further assumed that if the air temperature drops below −10 °C during lignification, then frost damage occurs. On the basis of these assumptions the following three dates were determined for each year: the day of growth cessation, t_{ces}, the day when the daily minimum air temperature T_{min} drops below −10 °C for the first time, t_{frost}, and the deadline day, t_{DL}, after which exactly 200 dd-units are accumulated before t_{frost} (Fig. 6.3).

Optimally, growth cessation would take place on day t_{DL} each year, so that the tree would combine sufficient survival adaptation with maximal capacity adaptation. This is not possible, however, because t_{DL} varies from year to year according to the early-autumn air temperature conditions. In years with cool autumns, t_{DL} comes earlier than in years with warm autumns. Thus, in Hänninen et al.'s (1990)

[3] The phase of lignification is discussed in Sects. 3.6.2 and 5.3.2. Since the proper value of $H_{crit,l}$ was not known, a sensitivity analysis with a range of values from 50 to 300 dd was carried out. The conclusions of the study were not affected by the value of $H_{crit,l}$ applied.

study, the deadline day t_{DL} varied by more than a month, from 14 August to 15 September, during the 96 years. Therefore, with growth cessation occurring earlier than day t_{DL} in most years, the trees necessarily lose part of the temperature sum accumulation (TS_{lost} in Fig. 6.3). This reasoning prompts the evolutionary problem addressed by Hänninen et al. (1990): among the three regulation principles studied, which one minimises the lost temperature sum, TS_{lost}?

With each of the three regulation principles, the value of the corresponding model parameter, i.e., NL_{crit}, TS_{crit}, or JE_{crit}, was determined first, so that growth cessation takes place each year as late as possible but not later than the deadline day, t_{DL} (Fig. 6.3). The determination was based on the regularity in the development of the model variables, i.e., after summer solstice the value of each of the model variables NL(t), S_{dd}(t), and JE(t) increases monotonically. This implies that the higher the value of the model parameter, the later the timing of the simulated growth cessation. Accordingly, in order to minimise TS_{lost}, the value of the corresponding model parameter is maximised within the constraint that during each year, growth cessation takes place no later than on t_{DL} (Fig. 6.3).

With the night length principle, the value of the model parameter NL_{crit} was simply determined as the value of NL(t) = 7.95 h prevailing on the earliest deadline day (t_{DL} = 14 August) among the 96 years examined. With that value of NL_{crit}, growth cessation takes place exactly on t_{DL} in the critical year (i.e., the year with the earliest t_{DL}) and earlier than t_{DL} in other years (Fig. 6.4a).

With the temperature sum principle, the critical year is determined on the basis of temperature sum accumulation, i.e., the critical year is the one when the temperature sum accumulated on t_{DL} is the lowest among the 96 years examined.[4] This value of S_{dd}(t) was taken as the value of the model parameter, and in this way the value of TS_{crit} = 718 dd was obtained. When growth cessation is simulated with this TS_{crit} over the 96 years, then analogously to the night length model, growth cessation takes place exactly on t_{DL} in the critical year and earlier than t_{DL} in other years (Fig. 6.4b).

With the joint effect principle, the model parameter JE_{crit} was determined otherwise similarly to the temperature sum principle, but instead of the accumulated temperature sum TS(t), the joint effect variable JE(t) was used in the calculations (Eq. 6.1). Accordingly, the model parameter JE_{crit} = 14.45 h was obtained as the lowest value of JE(t) prevailing on t_{DL} among the 96 years examined (Fig. 6.4c).

For the night length principle, the critical year causing the constraint was 1973, when the autumn was exceptionally cool, so that the accumulation of temperature sum ceased exceptionally early (Fig. 6.4a, red curve). In all other years, part of the

[4] Due to year-to-year variation in the temperature conditions, the critical year for the temperature sum principle is not necessarily the same as the critical year for the night length principle, where the critical year is simply the year with the earliest t_{DL}.

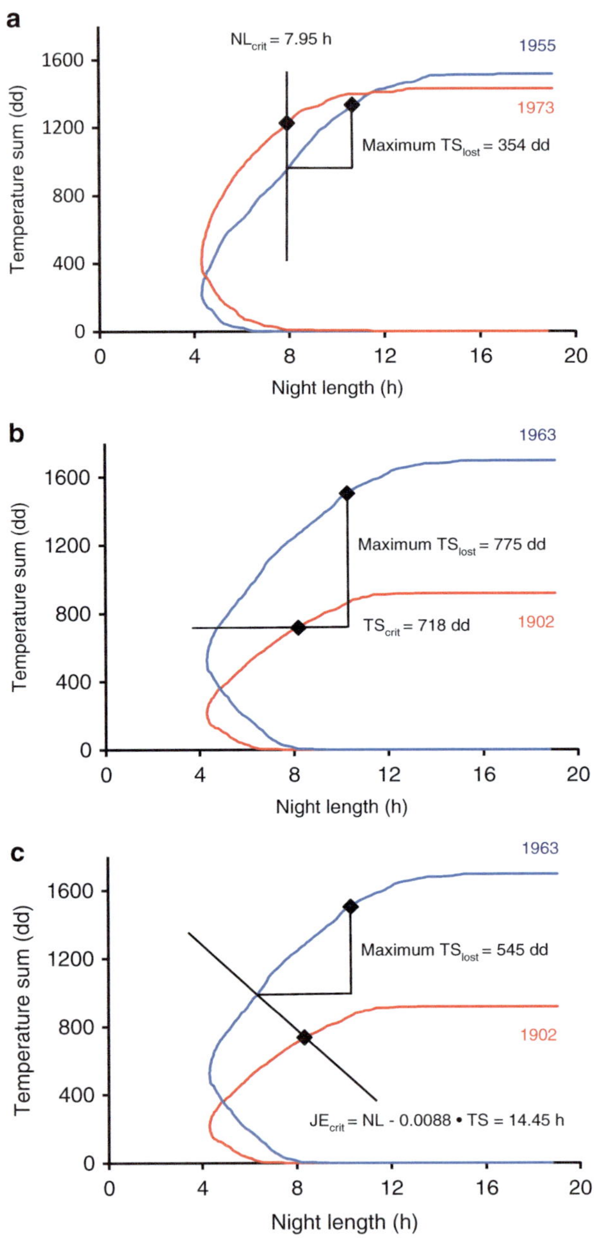

Fig. 6.4 Determination of the growth cessation-inducing critical values of the model parameters (*red curves*) and the maximum lost temperature sum (maximum TS_{lost}, *blue curves*) for the three models in a theoretical examination of the growth-cessation trade-off. In each curve, the accumulation of temperature sum (*vertical axis*) is plotted against the night length (*horizontal axis*), so that the turning point of the curve at the night length of 4.15 h represents the summer solstice situation.

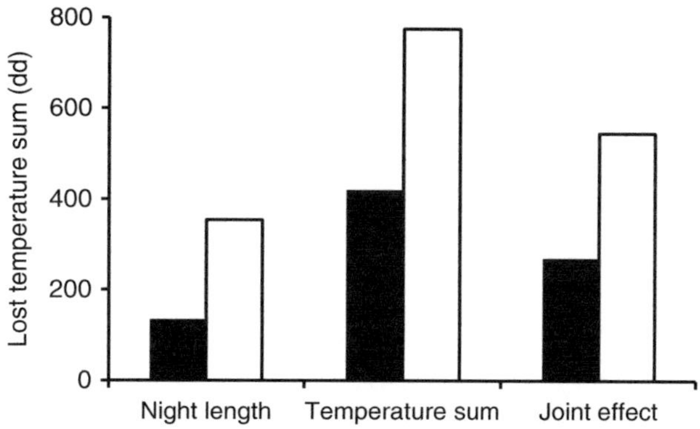

Fig. 6.5 Simulated mean (*black bars*) and maximum (*white bars*) values of lost temperature sum, TS_{lost}, over the 96 years, in a theoretical examination of the growth cessation trade-off in boreal trees under the climatic conditions of Jyväskylä, central Finland. The value of TS_{lost} indicates the temperature sum available but not used for growth, so that low values of TS_{lost} indicate a high degree of capacity adaptation. Simulations were carried out with three models representing the regulation of growth cessation by night length, temperature sum, and the joint effect of the two environmental factors, respectively (Figure drawn on the basis of data from Hänninen et al. (1990))

temperature sum available for growth was lost because growth cessation was predicted to occur before t_{DL}. The mean and maximum values of the lost temperature sum, TS_{lost}, were 133 and 354 dd, respectively (Fig. 6.5). The maximum value of TS_{lost} was obtained in 1955, when the autumn was quite warm, so that the accumulation of temperature sum continued exceptionally long (Fig. 6.4a, blue curve).

For the temperature sum principle, the value of $TS_{crit} = 718$ dd was obtained in the exceptionally cool summer of the critical year 1902 (Fig. 6.4b, red curve). In all other years, part of the temperature sum available for growth was lost, the mean and maximum values of the lost temperature sum, TS_{lost}, being 419 and 775 dd, respectively (Fig. 6.5). The maximum value of TS_{lost} was obtained in the

Fig. 6.4 (continued) Growth cessation is predicted to occur when the coloured curve representing the year meets the *black line* representing the critical value of the model parameter. For each of the two years, the *black diamond* indicates the deadline day, t_{DL}. **a** The night length model, where $NL_{crit} = 7.95$ h is the critical night length determined on the basis of the critical year 1973. **b** The temperature sum model, where $TS_{crit} = 718$ dd is the critical temperature sum determined on the basis of the critical year 1902. **c** The joint effect model, where $JE_{crit} = 14.45$ h is the critical value of the joint effect variable determined on the basis of the critical year 1902. The simulations were carried out for the climatic conditions of Jyväskylä, central Finland, for a period of 96 years, but results are shown only for the years determining the values of NL_{crit}, TS_{crit}, JE_{crit}, and the maximum values of TS_{lost} (Hänninen et al. 1990; redrawn with the permission of Oxford University Press)

exceptionally warm summer of 1963, when the total temperature sum accumulation exceeded 1,600 dd (Fig. 6.4b, blue curve).[5]

For the joint effect principle, the value of $JE_{crit} = 14.45$ h was obtained in the exceptionally cool summer of 1902 (Fig. 6.4c, red curve), which was the critical year for the temperature sum principle, too (Fig. 6.4b, red curve). As with the other models, part of the temperature sum available for growth was lost in other years, the mean and maximum values of the lost temperature sum, TS_{lost}, with the JE model being 268 dd and 545 dd, respectively (Fig. 6.5). As with the temperature sum model, the maximum value of TS_{lost} was obtained in the exceptionally warm summer of 1963 (Fig. 6.4c, blue curve), but in this case the value of TS_{lost} was lower than it was with the temperature sum model (Fig. 6.4b, blue curve).

Considered together, the results show that among the three regulation principles for the cessation of growth, the one based on night length provides the best evolutionary solution for the dilemma of the trade-off between survival and growth. The night length principle implies the smallest loss of temperature sum potentially available for growth, followed by the joint effect principle, while the principle based on temperature sum alone provides the largest loss of temperature sum (Fig. 6.5). As discussed in Hänninen et al. (1990), this order is caused by the large year-to-year variation in air temperature (Fig. 6.4b, c). Because of that variation, any given value of the accumulated temperature sum does a poor job of predicting the subsequent accumulation of temperature sum during the rest of the growing season. Thus, as the value of TS_{crit} needs to be determined on the basis of the coolest year (Fig. 6.4b, red curve), it implies a great loss of temperature sum in most other years (Fig. 6.4b, blue curve; Fig. 6.5). In contrast, the prevailing night length provides more information about the subsequent accumulation of temperature sum during the rest of the growing season (Fig. 6.4a), so that it implies a smaller loss of temperature sum (Fig. 6.5). The principle of joint effect takes an intermediate position in this trade-off comparison, being inferior to the night length principle and an improvement over the temperature sum principle (Figs. 6.4 and 6.5).

Thus, though one might have hypothesised a priori that a regulation principle based on two environmental factors would provide a better and safer solution for the dilemma of trade-off between survival and growth than any principle based on one factor only (Koski and Sievänen 1985), that notion did not hold true in this case. Introducing new environmental factors with poor predictive properties into the

---

[5] During the writing of the present volume, a small systematic error was revealed in the daily mean air temperatures of the long-term meteorological data from Jyväskylä. In the new calculations carried out for the present volume, the error was corrected. Accordingly, the maximum total accumulation of temperature sum in Jyväskylä is about 1,500 dd (Fig. 1.2b in Sect. 1.1.2) rather than over 1,600 dd as presented in Fig. 6.4b, c. This unfortunate shortcoming affected the details of the results not only in the study by Hänninen et al. (1990) reviewed in this section but also in several other original publications reviewed in different chapters of the present volume. However, as the systematic error did not imply erroneous conclusions in any of these publications, the slightly erroneous details of the results were not corrected in the reviews of the published papers in the present volume.

regulation principle of the annual cycle does not improve it but hampers it with respect to solving the trade-off dilemma (Hänninen et al. 1990).

Following the optimality principle, it can be concluded that if the three regulation principles examined have undergone natural selection, then it is probably the principle based on night length regulation that has been selected for (Hänninen et al. 1990). This conclusion accords with the mainline literature, which emphasises the role of night length in the growth cessation of boreal and temperate trees.[6]

However, as discussed in Sect. 3.5.3, a closer look at the literature reveals that the case is more complicated than that. Several interactions with air temperature have been found, and more specifically, the joint effect principle of Koski and Sievänen (1985) has received considerable support in experimental studies. Thus the observations deviate considerably from the prediction of the optimality principle. As is common in studies related to the optimality principle, this discrepancy, too, may come from the theoretical principle addressing only natural selection and ignoring other phenomena that shape the genetic structure of tree populations, such as a gene flow among the populations (Savolainen et al. 2007). In their discussion of the natural selection addressed by the optimality principle, Hänninen et al. (1990) suggest three potential and partly overlapping explanations for the discrepancy.

First, natural selection always works only on traits existing in the tree population, not on any theoretical traits, no matter how optimal they might be for the trees. Thus it is possible that no trees with pure night length regulation have ever existed in those developmental lines where joint effect regulation is seen at present. Second, natural selection works on entire trees, not on separate sets of traits. It is possible, then, that a sub-optimal ecophysiological trait is somehow linked to other traits in the overall physiology of the tree, so that this linking constrains the variation in the first trait seen in the present populations. Third, a sub-optimal trait may be a manifestation of adaptation to previous environmental factors different from the current ones. The last explanation does not appear plausible, however, in the case of the trade-off examined by Hänninen et al. (1990), because it is unlikely that the large year-to-year variation in air temperature has ever been absent in the evolutionary history of boreal trees.

---

[6] In species with a fixed growth habit, height growth cessation is regulated by the accumulation of temperature sum and occurs early in the growing season (Sect. 3.5.3.1). In these species, the cessation of height growth is not a manifestation of an overall cessation of growth and the onset of frost hardening, so that they were outside the scope of Hänninen et al.'s (1990) study.

6.3 Differences Among Tree Provenances

6.3.1 An Ecophysiological Approach to Genetic Differences

Most boreal and temperate tree species have wide geographical ranges, and the different provenances of a given tree species are therefore adapted to different climatic conditions. There are considerable differences among the provenances in several traits, such as growth rate, productivity, and resistance to various biotic and abiotic stresses. These differences have been a classic research theme in the closely related fields of quantitative genetics, ecological genetics, genecology, plant breeding, and forest genetics (Langlet 1971; Flint 1974; Eriksson et al. 1978; Rehfeldt et al. 1999). The research has addressed the heritability of the various traits and has focused on the interaction of environmental and genetic factors in determining the various tree traits (Billington and Pelham 1991; Savolainen et al. 2004).

Comparative research on various tree provenances is often relevant for practical forestry and horticulture. In forestry, provenance transfers have been used on a routine basis for a long time. When provenance transfers are used in order to increase wood production, special attention should be paid to the survival adaptation of the transferred trees in the new conditions (Viherä-Aarnio and Heikkilä 2006; Viherä-Aarnio and Velling 2008).

Similarly, the adaptation of different cultivars of fruit trees to the conditions prevailing at the orchard needs to be considered in practical horticulture (Säkö 1985; Egea et al. 2003).

Detailed discussion of hereditary phenomena, whether at the molecular or the population level, is outside the scope of the present volume. The present volume mainly addresses the outcome of the hereditary mechanisms, i.e., the differences in seasonality among the various tree provenances, and this is done within the ecophysiological whole-tree modelling framework applied throughout the volume. Accordingly, when genetic differences are addressed in this chapter, the approach is ecological here, too, so that the focus is on the relationship between the tree and its environment rather than on the inheritance of this relationship. The genetic differences among provenances can be readily addressed in the modelling approach adopted in the present volume by means of the genotype-specific values of the model parameters, such as the high temperature requirement of growth onset, H_{crit}, or the critical night length of growth cessation, NL_{crit}.

6.3.2 Tree Development in Spring and Early Summer

6.3.2.1 Linsser's Principle

Differences in the spring phenology among provenances of boreal and temperate trees have been studied for a long time by means of common garden experiments. In these experiments, provenances originating from cool regions have often commenced growth earlier than those originating from warm regions (Beuker 1994b; Aitken and Hannerz 2001). This phenomenon has also been documented in several growth chamber and greenhouse experiments with seedlings (Eriksson et al. 1978; Hänninen 1990; Søgaard et al. 2008). These findings show that the rate of development in spring is generally higher in the provenances from cool regions than those from warm regions. This regularity accords with the general notion that as an adaptation to the shortness of the growing season, plants growing in cool and cold climates generally have comparatively high rates of metabolic activity (Crawford 2008).

According to the traditional explanation, phenological differences among tree provenances in spring are caused by differences in the high temperature requirement of growth onset, H_{crit}. Accordingly, H_{crit} is generally lower for northern and high-altitude provenances than for southern and low-altitude ones (Table 6.1). This regularity was addressed quantitatively by Linsser (1867) early on. According to the so-called Linsser's principle, the value of H_{crit} in each provenance of a given tree species is a certain fixed proportion of the mean total annual accumulation of high temperature units during the growing season in the climate of the natural growing site of the provenance, H_{tot}:

$$100 \cdot \frac{H_{crit}}{H_{tot}} = c \tag{6.2}$$

where c = constant, indicating the *relative high temperature requirement*, which is identical across the different populations of any given species according to Linsser's principle.

Although there are plenty of empirical data supporting the qualitative generalisation that H_{crit} is lower in northern and high-altitude provenances than in southern and low-altitude ones, the exact principle presented by Linsser (1867) has only rarely been tested empirically. However, as part of his large-scale studies addressing the development of the generative meristems of trees, Sarvas (1967) tested the principle. In addition to his own observations on the timing of male flowering in different tree species and provenances, he collected additional data

Table 6.1 Differences in the annual cycle traits among provenances of boreal and temperate trees. The differences are summarised with the values of parameters belonging to the ecophysiological modelling approach applied in the present volume

Provenance	C_{crit}	H_{crit}	NL_{crit}
High latitude/altitude		(Low)[a]	Low
Low latitude/altitude		(High)[a]	High
Maritime	High		
Continental	Low		

C_{crit} chilling requirement of rest completion, H_{crit} high temperature requirement of growth onset, NL_{crit} critical night length of growth cessation

[a]See the text and Alberto et al. (2013)

Table 6.2 A test of Linsser's (1867) principle with observations on the timing of male flowering in five tree species in Finland (Tuusula and Punkaharju) and in central Europe (Eberswalde, Bědovice, and Krakow), together with calculations of temperature sum based on air temperature records made in or near the experimental stands

Species	Location	Year	H_{crit} (dd)	H_{tot} (dd)	c (%)	Original reference
Alnus glutinosa	Tuusula	1964	12.5	1333	1.1	Sarvas (1967)
	Eberswalde	1934	25	2013	1.2	Scamoni (1955)
	Eberswalde	1935	27	2013	1.3	Scamoni (1955)
Betula pendula	Punkaharju	1964	51	1248	4.1	Sarvas (1967)
	Bědovice	1958	95	1772	5.3	Chalupa (1964)
	Bědovice	1960	90	1772	5.1	Chalupa (1964)
	Bědovice	1961	97	1772	5.4	Chalupa (1964)
	Eberswalde	1933	85	2013	4.2	Scamoni (1955)
	Eberswalde	1934	87	2013	4.3	Scamoni (1955)
	Eberswalde	1935	82	2013	4.1	Scamoni (1955)
	Krakow	1935	99	2131	4.6	Jentys-Szaferowa (1938)
Quercus robur	Punkaharju	1964	172	1367	12.6	Sarvas (1967)
	Bědovice	1958	195	1772	11.0	Chalupa (1964)
	Bědovice	1960	205	1772	11.6	Chalupa (1964)
	Bědovice	1961	204	1772	11.5	Chalupa (1964)
	Eberswalde	1934–1936	243	2013	12.1	Scamoni (1955)
Picea abies	Tuusula	1964	139	1333	10.4	Sarvas (1967)
	Bědovice	1958	180	1772	10.1	Chalupa (1964)
	Eberswalde	1936	199	2013	9.9	Scamoni (1955)
Pinus sylvestris	Tuusula	1964	230	1333	17.2	Sarvas (1967)
	Bědovice	1959	265	1772	15.0	Chalupa (1964)
	Eberswalde	1933	330	2013	16.4	Scamoni (1955)
	Eberswalde	1936	344	2013	17.1	Scamoni (1955)

H_{crit} high temperature requirement of flowering, here specified with the day degree model as the temperature sum at the time of flowering, H_{tot} mean total annual accumulation of temperature sum during the growing season in the experimental stand, *c* relative high temperature requirement of flowering; *dd* day degree unit (Redrawn from Sarvas (1967), with the permission of the Finnish Academy of Science and Letters)

from the literature. He found that the relative high temperature requirement for male flowering turned out to be quite constant across the provenances in each of the species studied despite the great variation in the absolute high temperature requirement among the provenances (Table 6.2). Thus Sarvas's (1967) findings support Linsser's principle.

Sarvas (1969) also developed Linsser's (1867) idea further, stating that there is a lower limit for the high temperature requirement of growth onset, H_{crit}. Therefore, in provenances further north, the decreasing value of the mean total annual accumulation of high temperature units at the growing sites of the trees, H_{tot}, attains a limit after which their H_{crit} cannot go as low as the fixed proportion of H_{tot} would call for (Eq. 6.2). Thus Linsser's principle does not hold for the northernmost provenances:

$$100 \cdot \frac{H_{crit}}{H_{tot}} > c \tag{6.3}$$

According to Sarvas (1969), this is an indication of incomplete adaptation to the cool climatic conditions in the northernmost parts of the geographical range of the tree species.

Due to its clarity and simplicity, Linsser's principle is an appealing hypothesis for quantitatively explaining large-scale differences among various geographical provenances of boreal and temperate trees. If the hypothesis is found to be valid in empirical tests, the relationship embodied in it can be put to multiple uses in studies of the annual cycle of boreal and temperate trees (Hänninen et al. 2009).

However, in a recent review Alberto et al. (2013) found results partially contradictory to Linsser's principle (see also Aitken and Hannerz 2001). Rather than examining the quantitative relationship suggested in the principle (Eq. 6.2), they did a general review of the clines observed in the timing of growth onset in provenance trials of boreal and temperate trees. In addition to the cline discussed in this section, which gave qualitative support to Linsser's principle, i.e., the cline where growth onset takes place earlier in high than in low latitude or altitude provenances, Alberto et al. (2013) also found cases where no cline was observed, and even clines where the relationship was the other way round. These findings not only contradict Linsser's principle (Eq. 6.2) in quantitative terms but also question the phenomenon suggested in the principle more profoundly and conceptually.

These findings of Alberto et al. (2013) show, once again, that even established theories addressing the seasonality of boreal and temperate trees should be interpreted with care. However, rather than falsifying the concept of high latitude and altitude provenances being earlier in their spring phenology than low latitude and altitude ones, the results of Alberto et al. (2013) call for further studies examining the circumstances under which Linsser's principle is realised, at least qualitatively, and under which it is not.

6.3.2.2 Computational Examination of the Effect of Natural Selection on H_{crit}

Using the theoretical computational approach introduced by Häkkinen and Hari (1988) and discussed earlier in this chapter, Leinonen and Hänninen (2002) examined two hypotheses about the way natural selection operates on the high temperature requirement of growth onset, H_{crit}, in different tree provenances. They did a theoretical study of the trade-off between survival and growth in the growth onset of three *Picea abies* provenances covering a latitude range from southern Russia to northern Finland. Here the three provenances are referred to as the southern, the central, and the northern one according to their respective latitudes (Fig. 6.6). Using the day degree model with the threshold air temperature of $T_{thr} = +5$ °C, Leinonen

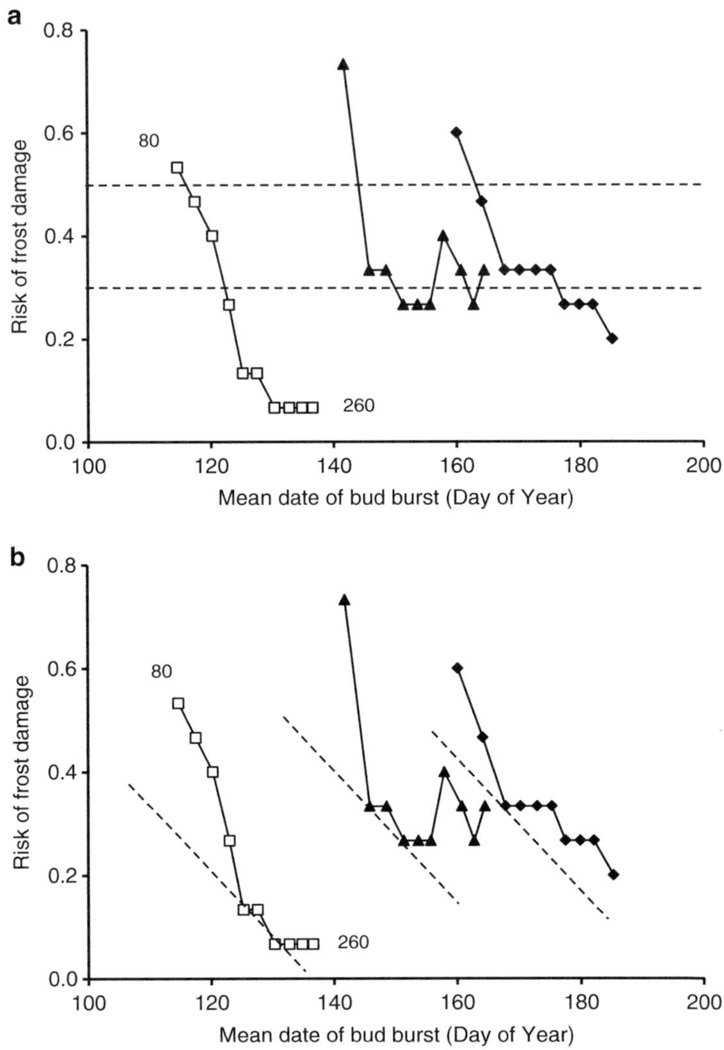

Fig. 6.6 Test of two hypotheses for the determination of the high temperature requirement of growth onset, H_{crit}, in boreal and temperate trees by natural selection. The dependence of the risk of spring frost damage (*vertical axis*) on the mean date of bud burst (*horizontal axis*) in *Picea abies* trees was calculated by means of the day degree model by using ten values of H_{crit} so that they ranged from 80 dd (the *left-hand point* of each curve) to 260 dd (the *right-hand point* of each curve) with steps of 20 dd. The calculations were carried out with 15-year air temperature data from a southern (Voronez, 51°40′N; *white squares*), a central (Tampere, 61°28′N; *black triangles*), and a northern location (Muonio, 67°58′N; *black diamonds*) (**a**) The selection is hypothesised to operate by allowing a fixed maximum risk of spring frost damage. The *horizontal dashed lines* indicate two possible risk levels. (**b**) The selection is hypothesised to operate according to the principle of gradual trade-off between survival and capacity adaptation, represented here by the *sloping dashed lines*. The amount of increase in the risk of spring frost damage allowed by natural selection depends on the lengthening of the growing season corresponding to the increase of the risk (Leinonen and Hänninen 2002; redrawn with the permission of the Finnish Forest Research Institute, currently the Natural Resources Institute Finland, and the Finnish Society of Forest Science)

and Hänninen (2002) examined the effect of H_{crit} on the survival and capacity adaptation of the three provenances. As expected, a higher H_{crit} delayed the date of bud burst and decreased the risk of spring frost damage, i.e., the degree of capacity adaptation decreased and that of survival adaptation increased in all three provenances (Fig. 6.6).

According to the first hypothesis in Leinonen and Hänninen's (2002) study, natural selection results in a fixed minimum level of survival adaptation, i.e., in a fixed maximum risk of frost damage (Häkkinen and Hari 1988; Hänninen and Hari 1996). With the high-risk level of 0.5, this hypothesis predicted the optimal value of H_{crit} to be in the range of 80–100 dd in all three provenances (Fig. 6.6a). Correspondingly, with the lower risk of 0.3, the optimal value was found to be in the range of 120–140 dd in the southern and the central provenance and considerably higher, i.e., in the range of 180–200 dd in the northern provenance (Fig. 6.6a). Thus the only difference predicted by the hypothesis of fixed maximum risk of spring frost damage is contradicted by the pattern found for *Picea abies*, where H_{crit} is lower in northern than in southern provenances (Table 6.2; Hänninen 1990; Søgaard et al. 2008).

The second hypothesis presented by Leinonen and Hänninen (2002) assumed that the selection operating on the value of H_{crit} is based on a gradual trade-off between survival and capacity adaptation (Fig. 6.6b). Accordingly, a lengthening of the growing season caused by decreasing H_{crit} is selected for as long as the concomitant increase in spring frost damage is not too great in comparison. Thus, as described by the sloping dashed lines in Fig. 6.6b, the maximum permitted increase in frost damage is assessed with reference to the corresponding amount of lengthening of the growing season. This hypothesis predicted the values of $H_{crit} =$ 190, 130, and 120 dd for the southern, the central, and the northern provenance, respectively (Fig. 6.6b). Thus, as this hypothesis predicted a pattern similar to the one observed for *Picea abies* (Table 6.2; Hänninen 1990; Søgaard et al. 2008), Leinonen and Hänninen (2002) concluded that natural selection has probably shaped the high temperature requirement of growth cessation in the different *Picea abies* provenances on the basis of a gradual trade-off between survival and capacity adaptation rather than a fixed degree of minimum survival adaptation.

6.3.3 Tree Development in Late Summer and Early Autumn

The critical night length of growth cessation, NL_{crit}, has been examined in numerous studies with various boreal and temperate tree species. A vast body of literature has accumulated in this research tradition, and in practically all the publications, a similar conclusion has been drawn: the value of NL_{crit} is lower in northern than in southern provenances (Fig. 6.7; Table 6.1; Dormling et al. 1968; Heide 1974; Eriksson et al. 1978; Håbjørg 1978; Junttila 1980; Junttila and Skaret 1990; Howe et al. 1995; Viherä-Aarnio et al. 2005, 2006). Classically, these genetically differentiated tree provenances have been referred to as *photoperiodic*

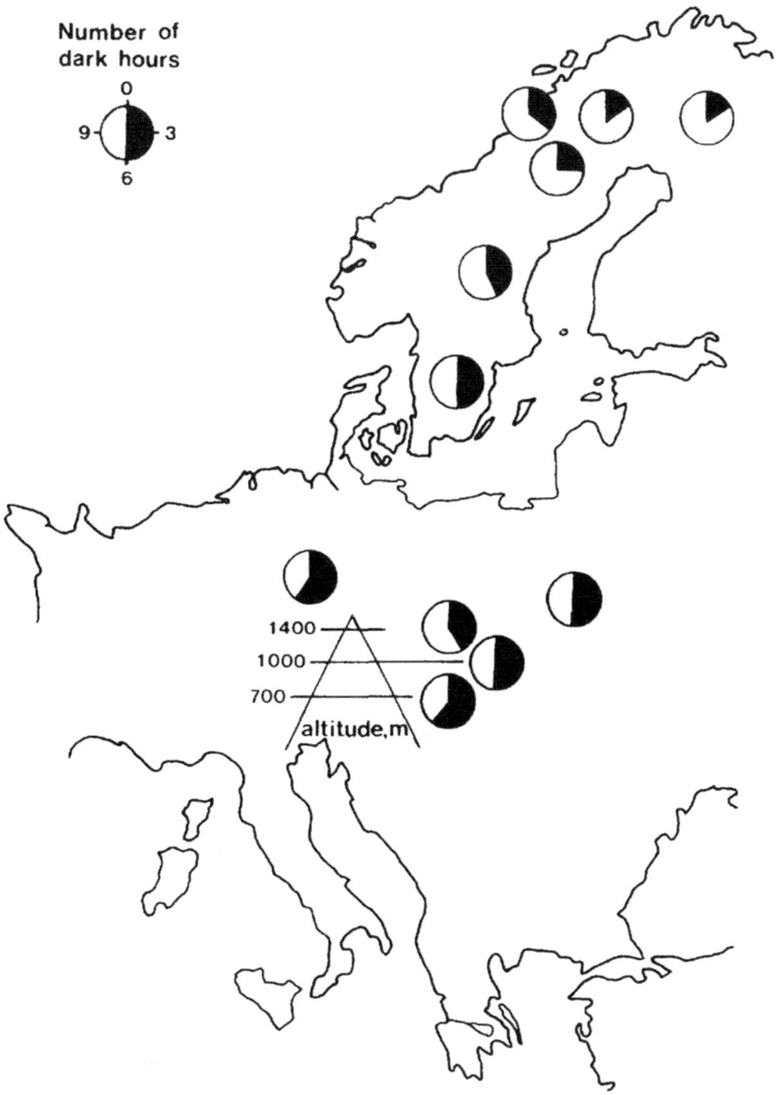

Fig. 6.7 Geographical variation in the critical night length for bud set in *Picea abies* seedlings. The dark parts of the *circles* indicate the number of dark hours in the diurnal cycle required for bud set in 50 % of the seedlings (Ekberg et al. 1979; redrawn with the permission of John Wiley & Sons, Inc)

ecotypes of growth cessation among boreal and temperate trees. The concept of ecotype entails the idea of strict borders between the genetically differentiated provenances. Therefore, as the present view is that the genetic structure of tree populations changes gradually across the geographical range of the species, one should, strictly speaking, use the concept *ecocline*, i.e., one referring to gradual change (Junttila 2007). However, as the concept of ecotype continues to be generally used in this context, it is also used in the present volume.

From the evolutionary point of view, there are two obvious reasons for the existence of photoperiodic ecotypes of growth cessation in boreal and temperate trees. First, on a given date in late summer before autumn equinox, the nights are shorter in northern than in southern locations (Fig. 1.1d in Sect. 1.2). Thus, even if growth cessation took place on the same date in the North as in the South, it would take place with shorter nights in the North than in the South. Second, due to the earlier arrival of winter in the North, growth cessation has to take place earlier in the northern than in the southern locations. Due to these astronomical and climatic phenomena, shorter night lengths are selected for in the North than in the South.

In addition to the latitude gradient, in mountainous regions there may be genetic variation in the critical night length, NL_{crit}, on an altitude gradient also: the values of NL_{crit} are lower in high- than in low-altitude populations (Table 6.1; Fig. 6.7; Heide 1974; Håbjørg 1978). At a given latitude, the photoperiodic conditions are practically identical among sites located at different altitudes. Therefore, the earlier arrival of winter at high than at low altitudes is the only evolutionary reason for the genetic differentiation of the value of NL_{crit} along the altitude gradient.

Due to their lower NL_{crit}, northern trees cease growth earlier than southern ones do when grown together in a common garden experiment or in common experimental conditions. Therefore, as growth cessation is closely related to the early phases of frost hardening in boreal and temperate trees (Sect. 5.3), northern trees in general also harden earlier than southern ones do when grown in the same conditions (Junttila and Kaurin 1990; Hurme et al. 1997; Beuker et al. 1998; Repo et al. 2000).

6.3.4 Tree Development in Late Autumn and Winter

Using the theoretical computational approach introduced by Häkkinen and Hari (1988), Hänninen and Hari (1996) examined the effects of the chilling requirement of rest completion, C_{crit}, and the high temperature requirement of growth onset, H_{crit}, on the survival and capacity adaptation of boreal and temperate trees growing at the four locations presented in Fig. 1.1 in Sect. 1.1.1. They simulated the timing of growth onset with the sequential model of it, using slightly modified versions of Sarvas's (1972, 1974) air temperature responses for the rate of rest break and that of ontogenetic development.[7] For each of the four locations, they used daily air temperature data for 14 years as input in the simulations.

[7] The three sub-models forming the overall model are described in detail in Sects. 3.2.1.4, 3.3.2.2, and 3.4.1.1, respectively. The functioning of the model is illustrated in Fig. 3.18 in Sect. 3.4.1.1.

In Hänninen and Hari's (1996) examination, the theoretical tree genotype was characterised by a combination of the two requirements (C_{crit}, H_{crit}). For each theoretical tree genotype, the criterion for survival adaptation was a fixed maximal risk of spring frost damage, so that after the simulated growth onset, the probability of killing frost ($T_{min} \leq -10$ °C, T_{min} = daily minimum temperature) was zero and the probability of damaging frost (-10 °C $< T_{min} \leq -5$ °C) less than 0.1. The criterion for capacity adaptation was formulated so that the accumulation of high temperature units after growth onset was at least 90 %, on average, of the theoretical maximum calculated for the optimal genotype, which maximises the accumulation of high temperature units within the constraints set by the criterion for survival adaptation.

The results for two locations, i.e., Eskdalemuir and Jyväskylä, are shown in Fig. 6.8. As expected, combinations of low values of C_{crit} and H_{crit} predicted insufficient survival adaptation caused by frost damage due to premature growth onset at every location, whereas combinations of high values of C_{crit} and H_{crit} predicted insufficient capacity adaptation caused by loss of growing season due to an excessively delayed growth onset. Optimal genotypes were those with intermediate values of C_{crit} and H_{crit}, and there was also an interaction, so that both the lower and the upper limit for the optimality of H_{crit} decreased with increasing values of C_{crit} (Fig. 6.8). This is because in the sequential model an increase in C_{crit} delays growth onset by delaying the onset of ontogenetic development towards it (Figs. 3.14 and 3.18 in Sects. 3.3.2.2 and 3.4.1.1, respectively).

Most importantly, the results of Hänninen and Hari's (1996) simulation study showed quantitatively that in order to avoid spring frost damage caused by premature growth onset, the chilling requirement of trees growing in the climatic conditions of Eskdalemuir, Scotland, needs to be much higher (Fig. 6.8a) than that of trees growing in the other locations examined (Fig. 6.8b, results shown for Jyväskylä only). This is caused by the maritime climate of Eskdalemuir, where air temperatures between 0 and 10 °C prevail during a large part of winter. Thus, as these temperatures cause both rest break and, after partial or full ontogenetic competence has been resumed, ontogenetic development towards growth onset, there is a risk that rest completion takes place too early in genotypes with a low chilling requirement, leading to subsequent ontogenetic development and premature growth onset. Accordingly, Hänninen and Hari's (1996) theoretical simulation study predicted that if rest break is regulated entirely by accumulated chilling, then the chilling requirement is higher in maritime than in more continental climates (Table 6.1).

Leinonen (1996) did an experimental study of the chilling requirement in seedlings of various provenances of *Betula pendula* and *Pinus sylvestris*. His results agreed with Hänninen and Hari's (1996) prediction (Fig. 6.8) in that the chilling requirement was found to be higher in maritime than in continental provenances. Unlike Leinonen (1996), most other comparative studies have examined the variation in the chilling requirement along a latitudinal gradient. In these studies the chilling requirement has been found both to increase (Perry and Wang 1960; Kriebel and Wang 1962; Farmer 1968; Myking and Heide 1995) and to decrease

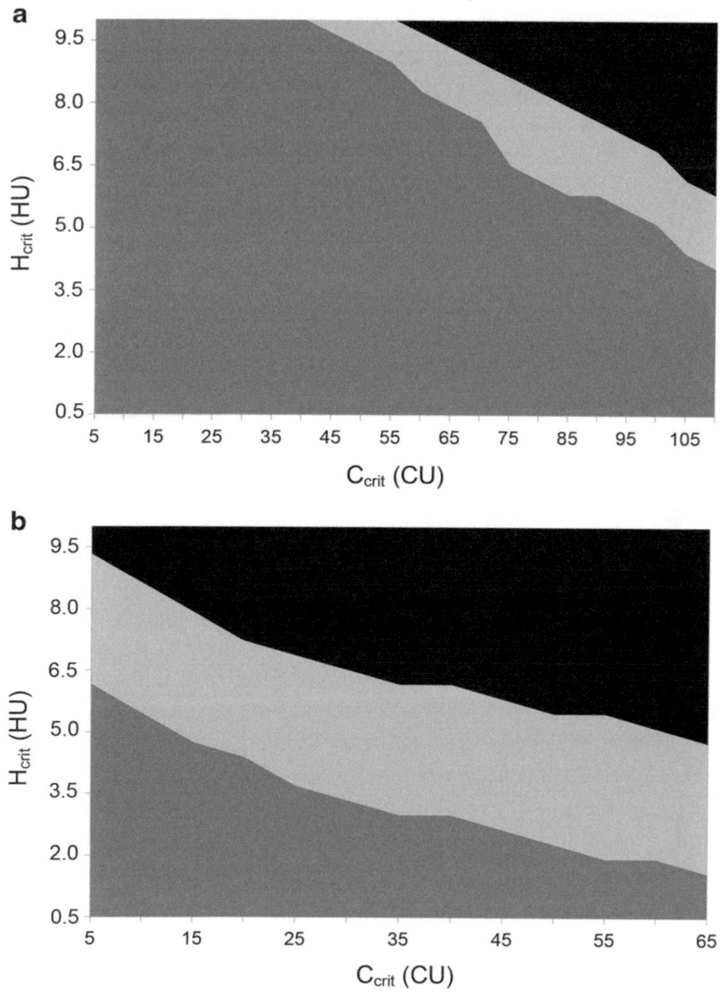

Fig. 6.8 A theoretical examination of the survival adaptation and the capacity adaptation of different tree genotypes in the climatic conditions of (**a**) Eskdalemuir, Scotland, and (**b**) Jyväskylä, central Finland. Each point on the planes represents a theoretical tree genotype with the chilling requirement of rest completion, C_{crit}, indicated on the *horizontal axis* and the high temperature requirement of growth onset, H_{crit}, on the *vertical axis*. CU and HU indicate the chilling and high temperature units of the model applied. In the figures, the *dark grey areas* betoken genotypes with insufficient survival adaptation, the *light grey areas* stand for optimal genotypes, and the black areas signify genotypes with insufficient capacity adaptation. Note the different scales of the *horizontal axes* in the two figures. This examination of survival and capacity adaptation was based on simulated timing of growth onset and the related risk of frost damage in each of the theoretical genotypes over 14 years. Note that in comparison with the original study, the scale of the *vertical axis* has been changed as explained in Sect. 3.2.1.4 (Hänninen and Hari 1996; redrawn with the permission of the Finnish Forest Research Institute, currently the Natural Resources Institute Finland, and the Finnish Society of Forest Science)

(Boyer and South 1989; Hänninen 1990; Hannerz et al. 2003) with increasing latitude of the provenance, and in some of these studies the differences among the provenances were relatively small. These findings suggest that unlike the gradient between maritime and continental climates (Fig. 6.8; Hänninen and Hari 1996; Leinonen 1996), the latitudinal gradient as such is not essential for the natural selection operating on the chilling requirement of rest completion.

6.4 Differences Among Tree Species Under the Same Climatic Conditions

The studies discussed in Sect. 6.3 show that the seasonality of tree provenances is adapted to the climatic conditions prevailing at their respective native growing sites. As a result of this adaptation there are great differences among provenances in their responses to the environmental cues, such as night length or chilling, during the annual cycle. This does not imply, however, that under any given climatic condition one specific way of the timing of seasonality were to be superior to all others. This is readily shown by the well-known phenomenon where, e.g., the phenological events in spring do not take place simultaneously in all tree species growing at the site (Lechowicz 1984; Vitasse 2013). A thorough study of this evolutionarily interesting phenomenon is outside the scope of the present volume, so that only a brief discussion of some of the related theories is offered below.

The phenological differences among species growing in similar climatic conditions suggest that there are several alternative ways to solve the trade-off between survival and growth discussed in Sect. 6.2. It has been suggested that there is a major difference in this trade-off between early and late successional tree species (Körner and Basler 2010; Caffarra and Donnelly 2011). Accordingly, early successional species are opportunistic, so that their strategy is based on a combination of efficient use of the growth resources by early growth onset (high capacity adaptation) and a relatively high risk of frost damage (low survival adaptation). In contrast, the late successional species follow a more conservative strategy of late growth onset (low capacity adaptation) with a concomitant low risk of frost damage (high survival adaptation).

In temperate deciduous trees, bud burst usually occurs earlier in understory trees than in canopy trees. Using a sophisticated experimental set-up, Vitasse (2013) demonstrated that this difference is caused by ontogenetic changes related to the ageing of the trees. This reasoning is plausible from the evolutionary point of view, for regardless of the species, the earlier bud burst of understory trees helps them to avoid the shading caused by canopy trees in spring.

However, a difference between two species in the timing of growth onset does not necessarily imply a difference in their corresponding strategies in regard to capacity vs. survival adaptation. This is because the occurrence of growth onset does not entail the same degree of dehardening in all tree species (Larcher 2003). It may be hypothesised that a trade-off is involved in this case, too, but that it occurs at the physiological level rather than the whole-tree level (growth onset, frost damage). According to this hypothesis, the delayed dehardening before and at the onset of visible growth implies some kind of metabolic cost to the tree. This is a plausible notion, for otherwise delayed dehardening should have been selected for similarly in all tree species.

Lechowicz (1984) attributed the differences in the timing of bud burst among temperate deciduous trees to their wood anatomy. Species with a wood anatomy allowing high water conduction capacity run a high risk of cavitation of xylem water columns in winter. In these species, cambial activity has to take place prior to bud burst, so that bud burst occurs in these species (e.g., *Fraxinus* sp., *Quercus* sp.) relatively late as a rule (Lechowicz 1984). This explanation is analogous to the discussion presented in the previous paragraph in the sense that to understand the evolutionary background of phenological differences between species at the whole-tree level, one also needs to examine the phenomena at the anatomical and the physiological level.

Finally, besides climatic seasonality, the selection on the phenological timing in plants is also affected by various biotic interactions, which, however, are outside the scope of the present volume. Among the most important ones are interactions with pollinators and seed predators (Elzinga et al. 2007).

6.5 Effects of the Maternal Environment[8]

In the past two decades it has been found that the maternal environment, i.e., the environment in which the tree seeds develop, affects the properties of the annual cycle in the offspring of *Picea abies* considerably. This was a ground-breaking notion at the time it was introduced. Norwegian scientists took the lead in putting it forward, and they have also provided considerable empirical support for it over the years (Johnsen 1989a, b; Johnsen et al. 1989, 2005a, b; Kvaalen and Johnsen 2008; for a review, see Johnsen et al. 2009). The molecular basis of the phenomenon remains unknown although several hypotheses, such as that of epigenetic effects, have been presented to explain the phenomenon (Rohde and Junttila 2008). For *Picea abies,* the empirical evidence supporting the notion of the effects of the maternal environment is conclusive, but for other species there are evidently only very few, if any, studies addressing the phenomenon (Rohde and Junttila 2008).

Due to the effects of the maternal environment, northern trees grown at more southern locations in provenance trials or in seed orchards produce offspring with properties normally seen in southern provenances, such as later onset and later cessation of growth (Johnsen et al. 2009). Correspondingly, properties typical for

[8] This concept, referring to the conditions where the seeds develop, should not be confused with the concept of maternal effects sensu stricto (Rohde and Junttila 2008).

northern provenances appear in the offspring of southern provenances grown at northern locations. These findings show that traits such as the high temperature requirement of growth onset, H_{crit}, or the critical night length of growth cessation, NL_{crit}, are not only genetically controlled. In this way, the effects of the maternal environment provide an additional mechanism for retaining genetic variation in the tree population. The effects of the maternal environment are also important in practical forestry, in which seed orchards are often founded at non-native locations in order to improve the growing conditions and avoid background pollination. The implications of the effects of the maternal environment for the projected climate change are discussed in Chap. 8.

In principle, the effects of the maternal environment can be readily addressed in the modelling approach adopted in the present volume: the values of the model parameters, such as H_{crit} and NL_{crit}, are not constant for a given tree genotype but are affected by the environmental conditions prevailing during seed development. The empirical results available at present may already allow the formulation of the first approximate models for this relationship. If such a model can be developed, then each seedling cohort will have its own specific value for the model parameter, depending on the environmental conditions during the year when the development of the seeds was in the responsive phase. Furthermore, as in the case of any other phenomenon of the annual cycle, in the absence of sufficient data for developing a model for the effects of the maternal environment on the value of the model parameter, the implications of those effects can be examined theoretically by means of sensitivity analyses based on different assumptions about the effects.

6.6 Modelling Evolutionary Change in Tree Populations

The ecophysiological modelling approach adopted in the present volume readily facilitates studies addressing the effects of natural selection on different tree genotypes, represented by their respective value(s) of the model parameter(s) (Table 6.1). This approach is exemplified in Fig. 6.8, where the performance of different tree genotypes, defined by the combination of the chilling requirement of rest completion, C_{crit}, and the high temperature requirement of growth onset, H_{crit}, is examined at two locations. As discussed in Sect. 6.3.4, the results of this theoretical examination predict which genotypes will be selected for and which ones selected against in the environments considered.

This approach, however, only sorts the theoretical population assumed at the beginning of the simulation by eliminating some and saving other genotypes. In order to address evolutionary change, i.e., change in the gene frequencies in the population from one generation to the next, one should also consider the genetic aspect inherent in the sexual reproduction within the population. This has been the classic research theme in the discipline of quantitative genetics (Savolainen et al. 2007). In the models of quantitative genetics, however, the ecophysiological phenomena causing the selection, such as the one addressed in Fig. 6.8, are not explicitly considered.

Kramer et al. (2008) introduced a modelling approach in which both the ecophysiological and the genetic aspects of evolutionary change in tree populations are considered simultaneously. Their genetic-ecophysiological model, ForGEM, simulates the birth, growth, and death of individual trees, explicitly addressing such ecophysiological phenomena as light interception, photosynthesis, allocation of photosynthetic products, and growth in different parts of the tree. The genetic phenomena they addressed included the gene flow due to the dispersal of pollen, the probability of mating among different genotypes depending on the distance between the plants and the overlap of their flowering periods, and the gene flow due to the dispersal of seeds. The ForGEM model was parameterized for *Fagus sylvatica* on the basis of large sets of data for different European populations of that species.

Kramer et al. (2008) examined the adaptive potential of *Fagus sylvatica* to environmental changes by means of model simulations with the ForGEM, addressing, among other issues, the dominant role of forest management practices in determining the population size and regeneration events. On the basis of their simulations, Kramer et al. (2008) concluded that the European *Fagus sylvatica* has high potential for adapting to environmental changes, provided that the recruitment intervals are short and that many mother trees contribute to the next generation, averting the selection-based loss of genetic variation within the population. Kramer et al. (2008) also concluded that climatic change might cause evolutionary change, i.e., change in the gene frequencies of adaptive traits in the populations of European *Fagus sylvatica*, within a time frame as short as a few tree generations.

Though Kramer et al.'s (2008) ForGEM model also addressed the timing of bud burst and the related frost risk, it did not address the ecophysiological phenomena behind the timing of growth onset, discussed in Chap. 3 of the present volume. However, as ForGEM has a modular structure, it facilitates the introduction of novel sub-models for different genetic and ecophysiological processes. Thus, Kramer and Hänninen (2009) introduced an ecophysiological sub-model to simulate the onset of growth in ForGEM. By means of scenario simulations with the ForGEM thus refined, they then examined the evolutionary change caused by climatic warming in the ecophysiological phenomena related to growth onset. The results indicate that climatic warming lowers the chilling requirement of rest completion and thereby accelerates bud burst in the examined populations of *Fagus sylvatica*.

Though limited in scope, Kramer and Hänninen's (2009) preliminary modelling exercise shows how the models of the annual cycle discussed in various chapters of the present volume can also be applied to studies addressing the evolutionary changes likely to be caused by the projected climate change. By allowing populations in different parts of the geographic area to adapt to the local conditions, i.e., to attain local parameter values for the models of the annual cycle, this approach opens new possibilities for modelling the impacts of climate change. The potential of the approach was demonstrated later in more extensive studies by Kramer et al. (2010, 2015).

6.7 Summary

Dobzhansky's (1973) famous phrase "Nothing in biology makes sense except in the light of evolution" is very much true in studies concerning the seasonality of boreal and temperate trees, as their annual cycle is a major adaptive strategy in the seasonal climate prevailing at their growing sites. Several traits of the annual cycle have undergone a stabilising selection because the traits causing early onset and late cessation of growth are selected for by capacity adaptation but simultaneously selected against by survival adaptation. This trade-off was examined in this chapter through the ecophysiological modelling approach adopted throughout the present volume. In the studies reviewed, the theoretical optimality principle typical for evolutionary ecology was applied. Though the predictions of the optimality principle may only rarely be realised as such in real tree populations, simulation studies provide additional quantitative insights into the climatic adaptation of boreal and temperate tree species. Such studies also prompt research questions for further empirical work, as is expected in the hypothetico-deductive research strategy adopted in the present volume.

As boreal and temperate tree species have wide geographical ranges, the different provenances of these tree species are adapted to different climatic conditions. The variation among the provenances in several traits, many of them related to seasonality, has been studied for a long time, and the differences among the provenances are relatively well understood at present. To understand the climatic adaptation of boreal and temperate tree species, both genetic and ecophysiological processes need to be addressed. In the present chapter, however, the discussion was mainly delimited to the ecophysiological aspects. In this way, rather than considering the evolutionary process itself, the focus was on its outcome, i.e., the differences among the provenances in relation to their native climates. The modelling approach is well suited for this purpose, as the differences among the tree provenances can be readily addressed by means of the parameter values of the models. In this way, the modelling approach also provided additional quantitative insights into the adaptive significance of the differences. Furthermore, the recently found effects of the maternal environment on the annual cycle of the offspring of *Picea abies* can also be addressed in terms of the modelling approach.

Finally, a modelling approach synthesising the genetic and the ecophysiological aspects of tree adaptation was briefly discussed. Though this approach is in its initial stages as yet, it provides a promising vision of mechanistic research on the evolutionary change taking place in tree populations by explicitly considering the ecophysiological and other ecological phenomena behind the differences in fitness among the different tree genotypes.

List of Symbols

t time

Time-Dependent Environmental Variables

$NL(t)$ daily night length (h)

Other Environmental Variables

H_{tot} mean total annual accumulation of high temperature units during the growing season (HU)

Time-Dependent Model Variables

$S_{dd}(t)$ accumulated temperature sum (dd)
$JE(t)$ variable for the joint effect of night length and temperature sum (h)

Other Model Variables

t_{frost} the day when the daily minimum air temperature T_{min} drops below $-10\,°C$ for the first time
t_{ces} the day of growth cessation
t_{DL} the deadline day of growth cessation
TS_{lost} the lost temperature sum (dd)

Model Parameters

H_{crit} high temperature requirement of growth onset (dd)
$H_{crit,l}$ high temperature requirement of lignification (dd)
NL_{crit} critical night length of growth cessation (h)
TS_{crit} critical temperature sum of growth cessation (dd)
b parameter converting accumulated temperature sum into hours in the joint effect model ($h\ dd^{-1}$)

JE_{crit} critical value of JE(t) for growth cessation (h)
C_{crit} chilling requirement of rest completion (CU)
H_{crit} high temperature requirement of growth onset (HU)

Other Symbols

dd day degree unit
CU chilling unit
HU high temperature unit (general term covering all specific units)
c the relative high temperature requirement of growth onset

References

Aitken, S. N., & Hannerz, M. (2001). Genecology and gene resource management strategies for conifer cold hardiness. In F. J. Bigras & S. J. Colombo (Eds.), *Conifer cold hardiness* (pp. 23–53). Dordrecht: Kluwer Academic Publishers.

Alberto, F. J., Aitken, S. N., Alía, R., Gonzáles-Martínez, S. C., Hänninen, H., Kremer, A., Lefèvre, F., Lenormand, T., Yeaman, S., Whetten, R., & Savolainen, O. (2013). Potential for evolutionary responses to climate change – Evidence from tree populations. *Global Change Biology, 19*, 1645–1661.

Bennie, J., Kubin, E., Wiltshire, A., Huntley, B., & Baxter, R. (2010). Predicting spatial and temporal patterns of bud-burst and spring frost risk in north-west Europe: The implications of local adaptation to climate. *Global Change Biology, 16*, 1503–1514.

Beuker, E. (1994a). Long-term effects of temperature on the wood production of *Pinus sylvestris* L. and *Picea abies* (L.) Karst. in old provenance experiments. *Scandinavian Journal of Forest Research, 9*, 35–45.

Beuker, E. (1994b). Adaptation to climatic changes of the timing of bud burst in populations of *Pinus sylvestris* L. and *Picea abies* (L.) Karst. *Tree Physiology, 14*, 961–970.

Beuker, E., Valtonen, E., & Repo, T. (1998). Seasonal variation in the frost hardiness of Scots pine and Norway spruce in old provenance experiments in Finland. *Forest Ecology and Management, 107*, 87–98.

Billington, H. L., & Pelham, J. (1991). Genetic variation in the date of bud burst in Scottish birch populations: Implications for climate change. *Functional Ecology, 5*, 403–409.

Boyer, J. N., & South, D. B. (1989). Seasonal changes in intensity of bud dormancy in loblolly pine seedlings. *Tree Physiology, 5*, 379–385.

Caffarra, A., & Donnelly, A. (2011). The ecological significance of phenology in four different tree species: Effects of light and temperature on bud burst. *International Journal of Biometeorology, 55*, 711–721.

Cannell, M. G. R. (1989). Chilling, thermal time and the date of flowering of trees. In C. J. Wright (Ed.), *Manipulation of fruiting* (pp. 99–113). London: Butterworths.

Chalupa, V. (1964). Dynamika kveteni lesních dřevin. Summary: The flowering of forest trees. *Práce výzkumných ústavů lesnických ČSSR, 28*, 141–173.

Cowan, I. (2002). Fit, fitter, fittest; Where does optimisation fit in? *Silva Fennica, 36*, 745–754.

Crawford, R. M. M. (2008). *Plants at the margin. Ecological limits and climate change*. Cambridge: Cambridge University Press. 478 p.

Dobzhansky, T. (1973). Nothing in biology makes sense except in the light of evolution. *American Biology Teacher, 35*, 125–129.

Dormling, I., Gustafsson, Å., & von Wettstein, D. (1968). The experimental control of the life cycle in Picea abies (L.) Karst. I. Some basic experiments on the vegetative cycle. *Silvae Genetica, 17*, 44–64.

Egea, J., Ortega, E., Martínez-Gómez, P., & Dicenta, F. (2003). Chilling and heat requirements of almond cultivars for flowering. *Environmental and Experimental Botany, 50*, 79–85.

Ekberg, I., Eriksson, G., & Dormling, I. (1979). Photoperiodic reactions in conifer species. *Holarctic Ecology, 2*, 255–263.

Elzinga, J. A., Atlan, A., Biere, A., Gigord, L., Weis, A. E., & Bernasconi, G. (2007). Time after time: Flowering phenology and biotic interactions. *Trends in Ecology and Evolution, 22*, 432–439.

Eriksson, G., Egberg, I., Dormling, I., Matérn, B., & von Wettstein, D. (1978). Inheritance of bud-set and bud-flushing in *Picea Abies* (L.) Karst. *Theoretical and Applied Genetics, 52*, 3–19.

Farmer, R. E., Jr. (1968). Sweetgum dormancy release: Effects of chilling, photoperiod, and genotype. *Physiologia Plantarum, 21*, 1241–1248.

Flint, H. L. (1974). Phenology and genecology of woody plants. In H. Lieth (Ed.), *Phenology and seasonality modeling* (pp. 83–97). Berlin: Springer-Verlag.

Håbjørg, A. (1978). Photoperiodic ecotypes in Scandinavian trees and shrubs. *Scientific Reports of the Agricultural University of Norway, 57*(33), 1–20.

Häkkinen, R., & Hari, P. (1988). The efficiency of time and temperature driven regulation principles in plants at the beginning of the active period. *Silva Fennica, 22*, 163–170.

Hannerz, M., Ekberg, I., & Norell, L. (2003). Variation in chilling requirements for completing bud rest between provenances of Norway spruce. *Silvae Genetica, 52*, 161–168.

Hänninen, H. (1990). Modelling bud dormancy release in trees from cool and temperate regions. *Acta Forestalia Fennica, 213*, 1–47.

Hänninen, H., & Hari, P. (1996). The implications of geographical variation in climate for differentiation of bud dormancy ecotypes in Scots pine. In P. Hari, J. Ross, M. Mecke (Eds.), Production process of Scots pine; Geographical variation and models. *Acta Forestalia Fennica, 254*, 11–21.

Hänninen, H., Häkkinen, R., Hari, P., & Koski, V. (1990). Timing of growth cessation in relation to climatic adaptation of northern woody plants. *Tree Physiology, 6*, 29–39.

Hänninen, H., Luoranen, J., Rikala, R., & Smolander, H. (2009). Late termination of freezer storage increases the risk of autumn frost damage to Norway spruce seedlings. *Silva Fennica, 43*, 817–830.

Heide, O. M. (1974). Growth and dormancy in Norway spruce ecotypes (*Picea abies*) I. Interaction of photoperiod and temperature. *Physiologia Plantarum, 30*, 1–12.

Heide, O. M. (1985). Physiological aspects of climatic adaptation in plants with special reference to high-latitude environments. In Å. Kaurin, O. Junttila, & J. Nilsen (Eds.), *Plant production in the north* (pp. 1–22). Tromsø: Norwegian University Press.

Howe, G. T., Hackett, W. P., Furnier, G. R., & Klevorn, R. E. (1995). Photoperiodic responses of a northern and southern ecotype of black cottonwood. *Physiologia Plantarum, 93*, 695–708.

Hurme, P., Repo, T., Savolainen, O., & Pääkkönen, T. (1997). Climatic adaptation of bud set and frost hardiness in Scots pine (*Pinus sylvestris*). *Canadian Journal of Forest Research, 27*, 716–723.

Jentys-Szaferowa, J. (1938). Z badań biometrycznych nad zbiorowym gatunkiem Betula alba L. II. O możliwości krzyżowania sie gatunków Betula verrucosa Ehrh. i Betula pubescens Ehrh. Summary: Biometrical studies on the collective species Betula alba L. II. The possibility of hybridization between species Betula verrucosa Ehrh. and Betula pubescens Ehrh. *Instytut badawcsy lasów państwowych. Seria A*, nr. 40. Warszawa.

Johnsen, Ø. (1989a). Phenotypic changes in progenies of northern clones of *Picea abies* (L.) Karst. grown in a southern seed orchard. I. Frost hardiness in a phytotron experiment. *Scandinavian Journal of Forest Research, 4*, 317–330.

Johnsen, Ø. (1989b). Phenotypic changes in progenies of northern clones of *Picea abies* (L.) Karst. grown in a southern seed orchard. II. Seasonal growth rhythm and height in field trials. *Scandinavian Journal of Forest Research, 4*, 331–341.

Johnsen, Ø., Dietrichson, J., & Skaret, G. (1989). Phenotypic changes in progenies of northern clones of *Picea abies* (L.) Karst. grown in a southern seed orchard. III. Climatic damage and growth in a progeny trial. *Scandinavian Journal of Forest Research, 4*, 343–350.

Johnsen, Ø., Dæhlen, O. G., Østreng, G., & Skøppa, T. (2005a). Daylength and temperature during seed production interactively affect adaptive performance of *Picea abies* progenies. *New Phytologist, 168*, 589–596.

Johnsen, Ø., Fossdal, C. G., Nagy, N., Mølmann, J., Dæhlen, O. G., & Skøppa, T. (2005b). Climatic adaptation in *Picea abies* progenies is affected by the temperature during zygotic embryogenesis and seed maturation. *Plant, Cell and Environment, 28*, 1090–1102.

Johnsen, Ø., Kvaalen, H., Yakovlev, I., Dæhlen, O. G., Fossdal, C. G., & Skøppa, T. (2009). An epigenetic memory from time of embryo development affects climatic adaptation in Norway spruce. In L. V. Gusta, M. E. Wisniewski, & K. Tanino (Eds.), *Plant cold hardiness. From the laboratory to the field* (pp. 99–107). Oxfordshire: CAB International.

Junttila, O. (1980). Effect of photoperiod and temperature on apical growth cessation in two ecotypes of Salix and Betula. *Physiologia Plantarum, 48*, 347–352.

Junttila, O. (2007). Regulation of annual shoot growth cycle in northern tree species. In E. Taulavuori & K. Taulavuori (Eds.), *Physiology of northern plants under changing environment* (pp. 177–210). Trivandrum, Kerala: Research Signpost.

Junttila, O., & Kaurin, Å. (1990). Environmental control of cold acclimation in *Salix pentandra. Scandinavian Journal of Forest Science, 5*, 195–204.

Junttila, O., & Skaret, G. (1990). Growth and survival of seedlings of various *Picea* species under northern climatic conditions. *Scandinavian Journal of Forest Science, 5*, 69–81.

Kalela, A. (1938). Zur Synthese der experimentellen Untersuchungen über Klimarassen der Holzarten. *Communicationes Instituti Forestalis Fenniae, 26*(1), 1–445.

Körner, C., & Basler, D. (2010). Phenology under global warming. *Science, 327*, 1461–1462.

Koski, V., & Selkäinaho, J. (1982). Experiments on the joint effect of heat sum and photoperiod on seedlings of Betula pendula. *Communicationes Instituti Forestalis Fenniae, 105*, 1–34.

Koski, V., & Sievänen, R. (1985). Timing of growth cessation in relation to the variations in the growing season. In P. M. A. Tigerstedt, P. Puttonen, & V. Koski (Eds.), *Crop physiology of forest trees* (pp. 167–193). Helsinki: Helsinki University Press.

Kramer, K., & Hänninen, H. (2009). The annual cycle of development of trees and process-based modelling of growth to scale up from the tree to the stand. In A. Noormets (Ed.), *Phenology of ecosystem processes* (pp. 201–227). Dordrecht: Springer Science+Business Media, LLC.

Kramer, K., Buiteveld, J., Forstreuter, M., Geburek, T., Leonardi, S., Menozzi, P., Povillon, F., Schelhaas, M. J., Teissier du Cros, E., Vendramin, G. G., & van der Werf, D. C. (2008). Bridging the gap between ecophysiological and genetic knowledge to assess the adaptive potential of European beech. *Ecological Modelling, 216*, 333–353.

Kramer, K., Degen, B., Buschbom, J., Hickler, T., Thuiller, W., Sykes, M. T., & de Winter, W. (2010). Modelling exploration of the future of European beech (*Fagus sylvatica* L.) under climate change – Range, abundance, genetic diversity and adaptive response. *Forest Ecology and Management, 259*, 2213–2222.

Kramer, K., van der Werf, B., & Schelhaas, M.-J. (2015). Bring in the genes: Genetic-ecophysiological modeling of the adaptive response of trees to environmental change. With application to the annual cycle. *Frontiers in Plant Science 5*: Article 742, doi:10.3389/fpls.2014.00742.

Kriebel, H. B., & Wang, C. W. (1962). The interaction between provenance and degree of chilling in bud-break of sugar maple. *Silvae Genetica, 11*, 125–130.

Kvaalen, H., & Johnsen, Ø. (2008). Timing of bud set in *Picea abies* is regulated by a memory of temperature during zygotic and somatic embryogenesis. *New Phytologist, 177*, 49–59.
Langlet, O. (1971). Two hundred years genecology. *Taxon, 20*, 653–721.
Larcher, W. (2003). *Physiological plant ecology. Ecophysiology and stress physiology of functional groups* (4th ed.). Berlin: Springer-Verlag. 513 p.
Lechowicz, M. J. (1984). Why do temperate deciduous trees leaf out at different times? Adaptation and ecology of forest communities. *The American Naturalist, 124*, 821–842.
Leinonen, I. (1996). Dependence of dormancy release on temperature in different origins of *Pinus sylvestris* and *Betula pendula* seedlings. *Scandinavian Journal of Forest Research, 11*, 122–128.
Leinonen, I., & Hänninen, H. (2002). Adaptation of the timing of bud burst of Norway spruce to temperate and boreal climates. *Silva Fennica, 36*, 695–701.
Linsser, C. (1867). Die periodischen Erscheinungen des Planzenlebens in ihrem Verhältniss zu den Wärmeerscheinungen. *Mémoires de l'Académie impériale des sciences de St. Pétersbourg*. XI, Nr. 7. 44 p.
Lockhart, J. A. (1983). Optimum growth initiation time for shoot buds of deciduous plants in a temperate climate. *Oecologia, 60*, 34–37.
Mäkelä, A., Givnish, T. J., Berninger, F., Buckley, T. N., Farquhar, G. D., & Hari, P. (2002). Challenges and opportunities of the optimality approach in plant ecology. *Silva Fennica, 36*, 605–614.
Mayr, E. (1961). Cause and effect in biology. *Science, 134*, 1501–1506.
Mayr, E. (1982). *The growth of biological thought: Diversity, evolution, and inheritance*. Cambridge, MA: The Belknap Press of Harvard University Press. 974 p.
Myking, T., & Heide, O. M. (1995). Dormancy release and chilling requirement of buds of latitudinal ecotypes of *Betula pendula* and *B. pubescens*. *Tree Physiology, 15*, 697–704.
Pelkonen, P., & Hari, P. (1980). The dependence of the springtime recovery of CO_2 uptake in Scots pine on temperature and internal factors. *Flora, 169*, 398–404.
Perry, T. O., & Wang, C. W. (1960). Genetic variation in the winter chilling requirement for date of dormancy break for *Acer rubrum*. *Ecology, 41*, 790–794.
Raulo, J., & Leikola, M. (1974). Tutkimuksia puiden vuotuisen pituuskasvun ajoittumisesta. Summary: Studies on the annual height growth of trees. *Communicationes Instituti Forestalis Fenniae, 81*(2), 1–19.
Rehfeldt, G. E., Ying, C. C., Spittlehouse, D. L., & Hamilton, D. A., Jr. (1999). Genetic responses to climate in *Pinus contorta*: Niche breadth, climate change, and reforestation. *Ecological Monographs, 69*, 375–407.
Repo, T., Zhang, G., Ryyppö, A., Rikala, R., & Vuorinen, M. (2000). The relation between growth cessation and frost hardening in Scots pines of different origins. *Trees, 14*, 456–464.
Rohde, A., & Junttila, O. (2008). Remembrances of an embryo: Long-term effects on phenology traits in spruce. *New Phytologist, 177*, 2–5.
Säkö, J. (1985). Winter hardiness problems in fruit growing in Finland. *Acta Horticulturae, 168*, 31–36.
Sarvas, R. (1967). The annual period of development of forest trees. *Proceedings of the Finnish Academy of Science and Letters, 1965*, 211–231.
Sarvas, R. (1969). *Genetical adaptation of forest trees to the heat factor of the climate*. Second world consultation on forest tree breeding. Washington, DC, 7–16 Aug 1969. FAO/IUFRO, FO-FTB-69-2/15. 11 p.
Sarvas, R. (1972). Investigations on the annual cycle of development of forest trees. Active period. *Communicationes Instituti Forestalis Fenniae, 76*(3), 1–110.
Sarvas, R. (1974). Investigations on the annual cycle of development of forest trees. II. Autumn dormancy and winter dormancy. *Communicationes Instituti Forestalis Fenniae, 84*(1), 1–101.
Savolainen, O., Bokma, F., García-Gil, R., Komulainen, P., & Repo, T. (2004). Genetic variation in cessation of growth and frost hardiness and consequences for adaptation of *Pinus sylvestris* to climatic changes. *Forest Ecology and Management, 197*, 79–89.

Savolainen, O., Pyhäjärvi, T., & Knürr, T. (2007). Gene flow and local adaptation in trees. *Annual Review of Ecology, Evolution, and Systematics, 38*, 595–619.

Scamoni, A. (1955). Beobachtungen über den Pollenflug der Waldbäume in Eberswalde. Summary: Observations on pollen dispersal by forest trees at Eberswalde. *Forstgenetik und Forstpflanzenzüchtung, 4*, 113–122.

Søgaard, G., Johnsen, Ø., Nilsen, J., & Junttila, O. (2008). Climatic control of bud burst in young seedlings of nine provenances of Norway spruce. *Tree Physiology, 28*, 311–320.

Viherä-Aarnio, A., & Heikkilä, R. (2006). Effect of latitude of seed origin on moose (*Alces alces*) browsing on silver birch (*Betula pendula*). *Forest Ecology and Management, 229*, 325–332.

Viherä-Aarnio, A., & Velling, P. (2008). Seed transfers of silver birch (*Betula pendula*) from the Baltic to Finland – Effect on growth and stem quality. *Silva Fennica, 42*, 735–751.

Viherä-Aarnio, A., Häkkinen, R., Partanen, J., Luomajoki, A., & Koski, V. (2005). Effects of seed origin and sowing time on timing of height growth cessation of *Betula pendula* seedlings. *Tree Physiology, 25*, 101–108.

Viherä-Aarnio, A., Häkkinen, R., & Junttila, O. (2006). Critical night length for bud set and its variation in two photoperiodic ecotypes of *Betula pendula*. *Tree Physiology, 26*, 1013–1018.

Vitasse, Y. (2013). Ontogenic changes rather than difference in temperature cause understory trees to leaf out earlier. *New Phytologist, 198*, 149–155.

Chapter 7
Upscaling to Higher Levels of Organisation

Contents

Abstract Upscaling of the models of the annual cycle of boreal and temperate trees to higher levels of organisation is briefly discussed. The models of the annual phenological cycle (Chap. 3), the annual cycle of photosynthesis in evergreen conifers (Chap. 4), and the annual cycle of frost hardiness (Chap. 5) address the modelled phenomena at the level of individual trees. Over the last decades, these models have been increasingly introduced as sub-models into large-scale process-based ecosystem models. In particular, models for the annual cycle of photosynthesis have often been used as sub-models. A simulation study reviewed here shows that unless the boreal restrictions described by the annual cycle model are taken into consideration, a drastic overestimate of the Gross Primary Production of the tree stand is obtained in simulations. About 15 years ago the models of the annual cycle were upscaled to the continental level in modelling the geographical ranges of tree species. The large-scale models thus refined predicted the geographical ranges of tree species accurately, suggesting that the seasonal phenomena of trees are major determinants of the geographical ranges of the tree species. While this approach opens an entirely new promising avenue for research, the annual cycle models applied in these studies still require further development for better realism. These models have also been addressed to a limited extent in the Dynamic Global Vegetation Models, but in this case, too, the annual cycle models applied evidently require further development.

H. Hänninen, *Boreal and Temperate Trees in a Changing Climate*, Biometeorology 3,
DOI 10.1007/978-94-017-7549-6_7

Keywords Annual cycle • BIOMASS • Boreal trees • Computer simulations • DGVM • Ecophysiological modelling • Ecosystem modelling • Frost damage • PHENOFIT • Photosynthetic capacity • Seasonality • Temperate trees • Upscaling

7.1 Upscaling to the Stand and the Ecosystem Level

7.1.1 The Principle of Upscaling

During the last few decades, plant ecophysiological information has been used to develop process-based dynamic models, or process models for short, for various terrestrial ecosystems (Fig. 7.1; Kramer et al. 2002; Zaehle et al. 2014). These studies adopt a bottom-up approach, starting from modelling the rate of a physiological phenomenon during a short time interval within a limited space unit in the stand. For instance, the rate of photosynthesis of one leaf or shoot during 1 min is first modelled on the basis of the amount of light intercepted by the leaf or shoot. Subsequently, the photosynthetic production of the stand over the growing season is obtained by integrating the rate of photosynthesis with respect to space and time (Kolari and Hari 2008).

In addition to light interception and photosynthesis, other essential ecophysiological phenomena, such as the uptake and transpiration of water, the uptake of nutrients, the allocation of photosynthates to the different parts of the plant, and finally, the growth of the plants, i.e., the construction of plant tissues, are often addressed also. In the most comprehensive process-based ecosystem models, the death of plant tissues, the flow of litter into the soil, and the decomposition of dead organic matter in the soil are also addressed (Zaehle et al. 2014). In this way, these ecosystem models describe the entire cycles of various nutrients. Accordingly, the most comprehensive process-based models address the ecophysiology of the entire ecosystem, and for this reason they have also become important tools in ecological research related to the climate change (Hari et al. 2008; Kramer and Hänninen 2009).

Process-based models are generally modular constructions, so that they allow the introduction of new sub-models, describing additional ecophysiological phenomena not previously addressed. From the 1980s onwards, models of the annual cycle of boreal and temperate trees have also been introduced into the process-based stand and ecosystem models as sub-models (Kramer and Hänninen 2009). The importance of the annual cycle for modelling at the stand or ecosystem level is illustrated by the fact that most process-based models designed for boreal and temperate forests presently include models of the annual cycle as sub-models. Kramer et al. (2002), for instance, compared process-based models with results of eddy covariance measurements in a large-scale study at six sites in Europe. Five out of the six process-based models compared had at least one annual cycle model as a sub-model.

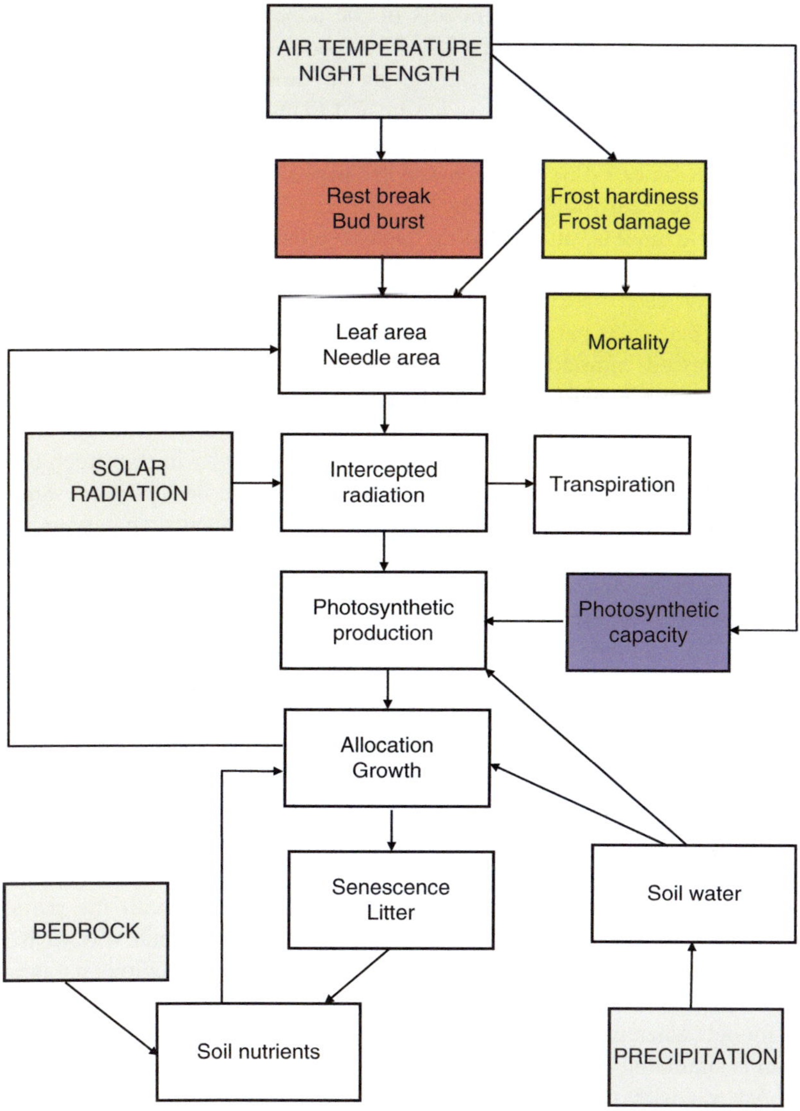

Fig. 7.1 A simplified schematic diagram of process-based ecosystem models. The *grey boxes* and *upper case* letters indicate meteorological and geological factors that ultimately drive the functioning of the ecosystem by setting limits to biotic phenomena. The *colours* indicate sub-models for the annual cycle of boreal and temperate trees discussed in earlier chapters: the annual phenological cycle (*red*, Chap. 3), the annual cycle of photosynthesis in evergreen conifers (*blue*, Chap. 4), and the annual cycle of frost hardiness (*yellow*, Chap. 5)

The principle of introducing models of the annual cycle into larger process-based models is illustrated in Fig. 7.1. The models of rest break and bud burst discussed in Chap. 3 are used for calculating the time when leaves develop in the canopy and start intercepting solar radiation and fixing CO_2 from the atmosphere in photosynthesis. Additional sub-models for the growth of the leaf area are also applied here (Kramer 1996). As discussed in Chaps. 3 and 6, the genetic differences among plants are readily addressed by the genotype-specific parameters involved in the annual cycle models, such as the high temperature requirement of growth onset, H_{crit}. Thus, community dynamics and interspecific competition can be addressed by designating for each species its own sub-model of rest break and bud burst, including the particular parameter values for the species. Intraspecific competition can be addressed similarly by dividing the modelled tree population into sub-populations, each characterised by its specific sub-model and particular parameter values (Kramer and Hänninen 2009; Kramer et al. 2015).

The models of the annual cycle of photosynthetic capacity in evergreen conifers discussed in Chap. 4 are used as sub-models (Fig. 7.1), in basically the same way they are used when modelling the seasonality of photosynthetic capacity in a single shoot. The light response of the rate of CO_2 exchange for each shoot changes in accordance with varying air temperature, as depicted in Fig. 4.2 (Sect. 4.1.2.1), but now the calculations carried out for a given moment are repeated with all small-space units, i.e., all shoots, in the canopy. In these calculations, light attenuation down the canopy is addressed so that the amount of light reaching a given shoot decreases as a function of the needle area above the shoot (Mäkelä et al. 2004; Kolari and Hari 2008). When the sub-model for photosynthetic capacity is introduced into the main model, then under any given environmental conditions the rate of photosynthesis of the entire canopy is seen to be higher after the rising of air temperatures in late spring than after being exposed to the low air temperatures generally prevailing earlier in the spring, similarly to that of individual shoots. Intra- and inter-specific genetic variation can be addressed with the parameter values of the sub-model for photosynthetic capacity in the same way as is done with the models of rest break and bud burst discussed in the previous paragraph.

The models of the annual cycle of frost hardiness discussed in Chap. 5 are used as sub-models when the frost damage potentially occurring in the tree population in question is simulated (Fig. 7.1). If the daily minimum air temperature, T_{min}, is low in comparison with the predicted daily frost hardiness, damage is predicted to occur. The effects of slight to medium frost damage are modelled by reducing the photosynthesizing leaf or needle area (Hänninen et al. 2005), whereas more serious damage increases the mortality of the trees in the stand (Kellomäki et al. 1992, 1995). Intra- and inter-specific genetic variation in frost hardiness and damage can be readily addressed with the parameter values of the frost hardiness model in the same way as is done with the other annual-cycle models presented in Fig. 7.1 (Kramer et al. 2015).

7.1.2 The Seasonality of Photosynthesis and Forest Productivity

Even though models of the annual cycle of boreal and temperate trees are commonly used nowadays as sub-models in process-based stand and ecosystem models, the implications of the phenomena associated with the annual cycle of trees for the stand level have only rarely been explicitly examined in the current climate by means of process-based models. Rather, process-based models with annual cycle models as sub-models have been used mainly in studies addressing the effects of the climate change. These studies were reviewed in detail by Kramer and Hänninen (2009), and two cases illustrating the approach are discussed in Sect. 8.5.

Bergh et al. (1998), however, examined the importance of the springtime recovery of photosynthetic capacity in northern conifers at the stand level in the current climatic conditions. Using the process-based model BIOMASS (McMurtrie et al. 1994), they studied the photosynthetic production of *Picea abies* stands in northern Sweden. BIOMASS is a generic growth model, so that with proper parameterisation it can be used for simulating the growth of stands of various tree species. Bergh et al. (1998) parameterised BIOMASS for *Picea abies* growing in northern Sweden, and they also introduced into it additional modifications describing the restrictions to photosynthetic production specific for boreal conditions. These consist of the effects of low air temperature, including those of frost (Sect. 4.2.2), and the effects of frozen soil on the photosynthetic capacity of the trees.

Using the version of BIOMASS parameterised for *Picea abies* growing in northern Sweden, Bergh et al. (1998) simulated the gross primary production, GPP, of tree stands both with and without the boreal restrictions. The results show that the boreal restrictions have a drastic effect on the predicted GPP (Fig. 7.2). If there were no restrictions caused by low air temperatures, frost, and frozen soil, the ambient climatic conditions would allow considerable photosynthetic production as early as February and March, and the GPP would almost attain its maximum value in late April and May. However, due to the restrictions, very little photosynthesis occurs before May, and the maximum level of GPP is not attained until the end of June (Fig. 7.2). In autumn, a similar but smaller reduction of GPP is caused by the restrictions specific to boreal conditions.

Overall, the reduction of GPP in northern Sweden stands of *Picea abies* caused by the restrictions specific to boreal conditions varied between 34 and 44 % in Bergh et al.'s (1998) simulations, depending on the simulation year and whether the soil was treated with irrigation and fertilisation. These results clearly demonstrate the importance of the boreal restrictions for the photosynthetic production of boreal coniferous stands. If the GPP is simulated without explicitly addressing the effect of low air temperature, frost, and frozen soil on photosynthesis when the global carbon cycle is modelled, for instance, seriously erroneous results will be obtained (Bergh et al. 1998).

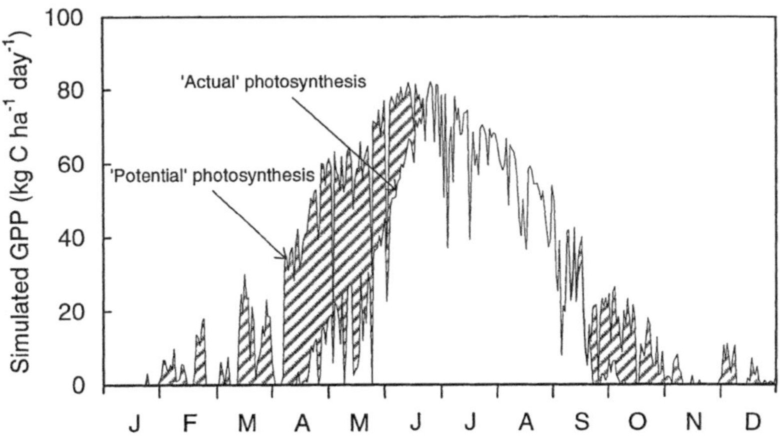

Fig. 7.2 Simulated gross primary production, GPP, of a northern Swedish stand of *Picea abies*. The GPP was simulated for the climatic conditions of 1991 by means of the process-based growth model BIOMASS, parameterised for *Picea abies* growing in northern Sweden. "Actual photosynthesis" and "Potential photosynthesis" indicate simulations with and without, respectively, modifications taking into account the restrictions to photosynthesis specific for the boreal conditions. The restrictions are caused by the effects of low air temperatures, including frost, on the photosynthetic capacity of the needles and the effects of frozen soil on photosynthetic production. Traits such as the generally low rates of photosynthesis in boreal conifers are included in both simulations (Redrawn from Bergh et al. 1998; with the permission of Elsevier Limited)

7.2 Upscaling to the Continental and the Global Level

7.2.1 Modelling the Geographical Ranges of Tree Species

Climatic factors are among the most important ones determining the geographical ranges of species: trees, other plants, and animals. Accordingly, the ranges of different species have often been explained by climatic indices such as the mean temperature of the coldest month or the temperature sum accumulation during the growing season. These indices are also used in the climatic envelope, or niche-based, models, which are often used for examining the determination of the geographical ranges of various species and also for projecting the effects of climate change on the shifts of these ranges (Bakkenes et al. 2002; Thomas et al. 2004; Thuiller et al. 2005; Bradley 2009; Sect. 8.2.3.2).

However, models of the annual cycle have only rarely been used for examining the effects of climatic factors on the geographical ranges of boreal and temperate trees. One obvious reason for this is that comprehensive models of the annual cycle specifically parameterised for a given tree species are available for a few species only. Models restricted to a single attribute and part of the annual cycle, such as the timing of bud burst and flowering in spring, have been parameterised for a number of species (Chuine et al. 1998; Morin et al. 2009), but models of the essential attribute of frost hardiness in particular are available for a few species only (Leinonen 1996).

Chuine and Beaubien (2001) introduced a model called PHENOFIT for examining the effects of climate factors on the geographical ranges of tree species. In PHENOFIT, several attributes of the annual cycle are addressed explicitly. These include aspects of the phenological annual cycle, such as the dates of bud burst, flowering, and fruiting, and the annual cycle of frost hardiness. In addition to the effects of frost, those of drought on the survival of the trees are also examined. PHENOFIT also addresses the reproductive success of the species examined, thus linking the ecophysiology of the annual cycle of the trees to their population ecology.

Chuine and Beaubien (2001) parameterised PHENOFIT for two North American species, i.e., *Acer saccharum* and *Populus tremuloides*. Most of the phenological events were modelled by fitting Chuine's (2000) unifying model to empirical data on the two species. Chuine and Beaubien (2001) found that PHENOFIT was able, with a few exceptions, to predict the geographical range of both species well. Accordingly, they concluded that phenology is a major determinant of the geographical ranges of tree species. Later on, PHENOFIT was similarly applied to a larger set of 17 North American species (Morin et al. 2007), and it has also been used for assessing various issues related to the climate change (Morin and Chuine 2005; Morin et al. 2008).

Despite its accuracy in predicting the geographical ranges of tree species in the current climate, the realism of PHENOFIT is compromised for two major reasons. First, the uncertainty related to the parameterisation of ecophysiological phenological models with data gathered in natural conditions (Sect. 3.4.6.1) pertains to PHENOFIT, too. Second, PHENOFIT's modelling of frost damage is on an exceptionally weak basis. This is because the authors applied the frost hardiness model of Leinonen (1996), mainly using its original parameter values (Chuine and Beaubien 2001). These parameter values were originally determined for *Pinus sylvestris* growing in central Finland, so that they hardly enable Leinonen's (1996) model to realistically describe the annual cycle of frost hardiness in the various provenances of *Acer saccharum* and *Populus tremuloides* in North America.

To take the most obvious example, Chuine and Beaubien (2001) followed Leinonen (1996) in setting the lower and the upper threshold for the range of night length causing hardening in autumn at 10 and 16 h, respectively (see Fig. 5. 15a in Sect. 5.3.2.3). In central Finland, this range is passed some time between early September and early November, depending on the exact latitude. At latitudes below N50°, however, the night length does not reach 16 h even during winter solstice. Most of the geographical ranges of both *Acer saccharum* and *Populus tremuloides* in North America are located below N50°, so that Leinonen's (1996) original parameter values cannot be valid for these two species. In later studies concerning PHENOFIT (Morin and Chuine 2005; Morin et al. 2007, 2008), less information has been given on the sub-model of frost hardiness, but the problem of using Leinonen's (1996) frost hardiness model as such evidently prevails in them, too.

The model PHENOFIT (Chuine and Beaubien 2001), including the frost hardiness and damage model of Leinonen (1996) as a sub-model, is an illustrative example of the concepts of model realism and model accuracy (Sects. 2.4 and 8.3.2.2). The high accuracy demonstrated by PHENOFIT in predicting the geographical ranges of tree species does not guarantee high realism.

However, as repeatedly pointed out in several chapters of the present volume, experimentally based species-specific information is often not available on all aspects included in detailed models of the annual cycle of boreal and temperate trees. It is necessary, then, to resort to various generalisations and assumptions; so the problems generated by including Leinonen's (1996) frost hardiness model in PHENOFIT are not exceptional as such. This situation calls for explicit and transparent discussion of the degree of empirical support for the models whenever inferences are to be made on the basis of model simulations.

The criticism presented above should not obscure the obvious merits of the modelling studies with PHENOFIT (Chuine and Beaubien 2001; Morin and Chuine 2005; Morin et al. 2007, 2008; Chuine 2010; Gritti et al. 2013). In these studies, the annual cycle of boreal and temperate trees was introduced into large-scale examinations of the way the entire geographical ranges of tree species are determined in nature. Therefore, PHENOFIT also has great potential for climate change studies. Because of its shortcomings, however, PHENOFIT should be mainly regarded as a framework for further research at present, and the projections provided by it should be interpreted with extra care. Hopefully, the above brief discussion of PHENOFIT's shortcomings will be addressed in further development of the model towards realising its potential for climate change research. It goes without saying that such an improvement in model realism requires large-scale experimental studies.

7.2.2 *Dynamic Global Vegetation Models*

Among the dynamic process-based models, the Dynamic Global Vegetation Models (DGVMs) address ecological phenomena on the largest spatial scale, i.e., that of the entire globe. Several ecophysiological processes of plants, such as photosynthesis, uptake of water and nutrients, and growth are modelled with detailed sub-models in DGVMs. They also include sub-models for soil processes, such as decomposition of litter and humus. In this way, the DGVMs simulate large-scale ecosystem processes such as the hydrological cycle, the cycles of nutrients and carbon, energy transfer, and the productivity of the ecosystems (Cramer et al. 2001; Sitch et al. 2003; Krinner et al. 2005; Hickler et al. 2006; Sato et al. 2007).

Because of practical limitations, the simulations with Dynamic Global Vegetation Models are not carried out with real plant species but a limited number of plant functional types (PFTs) that represent groups of ecologically similar plant species. Typically, PFTs such as "Tropical broad-leaved raingreen tree" and "Boreal needle-leaved evergreen tree" are used in the DGVMs (Sato et al. 2007). By means of process-based sub-models, DGVMs simulate the competition of the different PFTs, so that in addition to predicting material cycles and energy transfer, they also predict the biome to be formed in each region under the climatic conditions addressed. Accordingly, DGVMs are tested not only by comparing their predictions for dynamic phenomena, such as fluxes of carbon dioxide and water, with corresponding observations but also by comparing the predicted biomes with the observed ones over the entire globe (Cramer et al. 2001; Hickler et al. 2006). Currently, the DGVMs are an important tool in studies projecting the ecological implications of the climate change at the global level.

In comparison with the great detail of the current DGVMs in addressing various ecophysiological phenomena, such as the momentary rate of photosynthesis in the different Plant Functional Types, the phenomena related to the annual cycle of perennial plants, such as boreal and temperate trees, remain relatively poorly represented in them. Among the latter phenomena, it is usually only the annual phenological cycle, as discussed in Chap. 3, that has been addressed in the DGVMs, which mainly use simplified sub-models for bud burst and leaf senescence (Sitch et al. 2003; Sato et al. 2007). However, as demonstrated in Sect. 7.1.1, models of the annual cycle are readily introduced as sub-models to stand and ecosystem models. Since a basically similar approach with modular dynamic sub-models is also taken in DGVMs, sub-models for various aspects of the annual cycle can be readily introduced into DGVMs as well. In further development of DGVMs, then, the annual cycles of photosynthetic capacity and frost hardiness in boreal and temperate trees can also be addressed if deemed necessary.

7.3 Summary

The information obtained from ecophysiological studies at the whole-plant level is often upscaled nowadays to the stand and the ecosystem level by introducing dynamic ecophysiological plant-level models as sub-models into larger stand and ecosystem-level dynamic models. With these large-scale models, the functioning of entire ecosystems is simulated by explicitly addressing processes such as plant photosynthesis, respiration, uptake of water and nutrients, and growth. These models are usually referred to as process-based models or process models in order to differentiate them from other large-scale models, where the phenomena of lower levels of organisation are not explicitly addressed. In the most comprehensive process-based ecosystem models, the decomposition of organic matter in the soil is also simulated, so that nutrient cycles are addressed explicitly.

Over the last decades, the upscaling approach has been increasingly applied to ecophysiological models of the annual cycle of boreal and temperate trees, too. This is of essential importance, for simulation studies of forest productivity show that without explicit consideration of the annual cycle of photosynthetic capacity, a drastic overestimate of forest productivity is obtained. In addition to models of the annual cycle of photosynthesis, models of the annual phenological cycle and ones of the annual frost hardiness cycle have also been introduced into larger-scale models as sub-models.

In most cases the upscaling of models of the annual cycle in boreal and temperate trees has been carried out to the stand and ecosystem levels. Recently, however, it has been extended to the continental level in examinations of the factors determining the geographical ranges of tree species. In these studies it has been concluded that phenomena of the annual cycle are major determinants of the ranges. While opening an entirely new promising avenue of research, the models applied in these studies still require further development towards better realism. Finally, the annual cycle of boreal and temperate trees has also been addressed to a limited extent in the Dynamic Global Vegetation Models, but these models evidently require further development, too.

References

Bakkenes, M., Alkemade, J. R. M., Ihle, F., Leemans, R., & Latour, J. B. (2002). Assessing effects of forecasted climate change on the diversity and distribution of European higher plants for 2050. *Global Change Biology, 8*, 390–407.

Bergh, J., McMurtrie, R. E., & Linder, S. (1998). Climatic factors controlling the productivity of Norway spruce: A model-based analysis. *Forest Ecology and Management, 110*, 127–139.

Bradley, B. A. (2009). Regional analysis of the impacts of climate change on cheatgrass invasion shows potential risk and opportunity. *Global Change Biology, 15*, 196–208.

Chuine, I. (2000). A unified model for budburst of trees. *Journal of Theoretical Biology, 207*, 337–347.

Chuine, I. (2010). Why does phenology drive species distribution? *Philosophical Transactions of the Royal Society B, 365*, 3149–3160.

Chuine, I., & Beaubien, E. G. (2001). Phenology is a major determinant of tree species range. *Ecology Letters, 4*, 500–510.

Chuine, I., Cour, P., & Rousseau, D. D. (1998). Fitting models predicting dates of flowering of temperate-zone trees using simulated annealing. *Plant, Cell and Environment, 21*, 455–466.

Cramer, W., Bondeau, A., Woodward, F. I., Prentice, I. C., Betts, R. A., Brovkin, V., Cox, P. M., Fisher, V., Foley, J. A., Friend, A. D., Kucharic, C., Lomas, M. R., Ramankutty, N., Sitch, S., Smith, B., White, A., & Young-Molling, C. (2001). Global response of terrestrial ecosystem structure and function to CO_2 and climate change: Results from six dynamic global vegetation models. *Global Change Biology, 7*, 357–373.

Gritti, E. S., Gaucherel, C., Crespo-Perez, M.-V., & Chuine, I. (2013). How can model comparison help improving species distribution models? *PLoS ONE, 8*(7), e68823. doi:10.1371/journal.pone.0068823.

Hänninen, H., Kolari, P., & Hari, P. (2005). Seasonal development of Scots pine under climatic warming: Effects on photosynthetic production. *Canadian Journal of Forest Research, 35*, 2092–2099.

Hari, P., Salkinoja-Salonen, M., Liski, J., Simojoki, A., Kolari, P., Pumpanen, J., Kähkönen, M., Aakala, T., Havimo, M., Kivekäs, R., & Nikinmaa, E. (2008). Growth and development of forest ecosystems; the MicroForest model. In P. Hari & L. Kulmala (Eds.), *Boreal forest and climate change* (pp. 433–461). Dordrecht: Springer.

Hickler, T., Prentice, I. C., Smith, B., Sykes, M. T., & Zaehle, S. (2006). Implementing plant hydraulic architecture within the LPJ Dynamic Global Vegetation Model. *Global Ecology and Biogeography, 15*, 567–577.

Kellomäki, S., Väisänen, H., Hänninen, H., Kolström, T., Lauhanen, R., Mattila, U., & Pajari, B. (1992). A simulation model for the succession of the boreal forest ecosystem. *Silva Fennica, 26*, 1–18.

Kellomäki, S., Hänninen, H., & Kolström, M. (1995). Computations on frost damage to Scots pine under climatic warming in boreal conditions. *Ecological Applications, 5*, 42–52.

Kolari, P., & Hari, P. (2008). Photosynthetic production of a tree. In P. Hari & L. Kulmala (Eds.), *Boreal forest and climate change* (pp. 313–325). Dordrecht: Springer.

Kramer, K. (1996). Phenology and growth of European trees in relation to climate change. Thesis, Landbouw Universiteit Wageningen, the Netherlands, 210 p.

Kramer, K., & Hänninen, H. (2009). The annual cycle of development of trees and process-based modelling of growth to scale up from the tree to the stand. In A. Noormets (Ed.), *Phenology of ecosystem processes* (pp. 201–227). Dordrecht: Springer.

Kramer, K., Leinonen, I., Bartelink, H. H., Berbigier, P., Borghetti, M., Bernhofer, C., Cienciala, E., Dolman, A. J., Froer, O., Gracia, C. A., Granier, A., Grünwald, T., Hari, P., Jans, W., Kellomäki, S., Loustau, D., Magnani, F., Markkanen, T., Matteucci, G., Mohren, G. M. J., Moors, E., Nissinen, A., Peltola, H., Sabaté, S., Sanchez, A., Sontag, M., Valentini, R., & Vesala, T. (2002). Evaluation of six process-based forest growth models using eddy-covariance measurements of CO_2 and H_2O fluxes at six forest sites in Europe. *Global Change Biology, 8*, 213–230.

Kramer, K., van der Werf, B., & Schelhaas, M.-J. (2015). Bring in the genes: Genetic-ecophysiological modeling of the adaptive response of trees to environmental change. With application to the annual cycle. *Frontiers in Plant Science, 5*, 742. doi:10.3389/fpls.2014.00742.

Krinner, G., Viovy, N., de Noblet-Ducoudré, N., Ogée, J., Polcher, J., Friedlingstein, P., Ciais, P., Sitch, S., & Prentice, I. C. (2005). A dynamic global vegetation model for studies of the coupled atmosphere-biosphere system. *Global Biogeochemical Cycles, 19*, GB1015. doi:10.1029/2003GB002199.

Leinonen, I. (1996). A simulation model for the annual frost hardiness and freeze damage of Scots pine. *Annals of Botany, 78*, 687–693.

Mäkelä, A., Hari, P., Berninger, F., Hänninen, H., & Nikinmaa, E. (2004). Acclimation of photosynthetic capacity in Scots pine to the annual cycle of temperature. *Tree Physiology, 24*, 369–376.

McMurtrie, R. E., Gholz, H. L., Linder, S., & Gower, S. T. (1994). Climatic factors controlling the productivity of pine stands: A model-based analysis. *Ecological Bulletins, 43*, 173–188.

Morin, X., & Chuine, I. (2005). Sensitivity analysis of the tree distribution model PHENOFIT to climatic input characteristics: Implications for climate impact assessment. *Global Change Biology, 11*, 1493–1503.

Morin, X., Augspurger, C., & Chuine, I. (2007). Process-based modeling of species' distributions: What limits temperate tree species' range boundaries? *Ecology, 88*, 2280–2291.

Morin, X., Viner, D., & Chuine, I. (2008). Tree species range shifts at a continental scale: New predictive insights from a process-based model. *Journal of Ecology, 96*, 784–794.

Morin, X., Lechowicz, M. J., Augspurger, C., O'Keefe, J., Viner, D., & Chuine, I. (2009). Leaf phenology in 22 North American tree species during the 21st century. *Global Change Biology, 15*, 961–975.

Sato, H., Itoh, A., & Kohyama, T. (2007). SEIB–DGVM: A new Dynamic Global Vegetation Model using a spatially explicit individual-based approach. *Ecological Modelling, 200*, 279–307.

Sitch, S., Smith, B., Prentice, I. C., Arneth, A., Bondeau, A., Cramer, W., Kaplan, J. O., Levis, S., Lucht, W., Sykes, M. T., Thonicke, K., & Venevsky, S. (2003). Evaluation of ecosystem dynamics, plant geography and terrestrial carbon cycling in the LPJ dynamic global vegetation model. *Global Change Biology, 9*, 161–185.

Thomas, C. D., Cameron, A., Green, R. E., Bakkenes, M., Beaumont, L. J., Collingham, Y. C., Erasmus, B. F. N., de Siqueira, M. F., Grainger, A., Hannah, L., Hughes, L., Huntley, B., van Jaarsveld, A. S., Midgley, G. F., Miles, L., Ortega-Huerta, M. A., Peterson, A. T., Phillips, O. L., & Williams, S. E. (2004). Extinction risk from climate change. *Nature, 427*, 145–148.

Thuiller, W., Lavorel, S., Araújo, M. B., Sykes, M. T., & Prentice, I. C. (2005). Climate change threats to plant diversity in Europe. *PNAS, 102*, 8245–8250.

Zaehle, S., Medlyn, B. E., De Kauwe, M. G., Walker, A. P., Dietze, M. C., Hickler, T., Luo, Y., Wang, Y.-P., El-Masri, B., Thornton, P., Jain, A., Wang, S., Warlind, D., Weng, E., Parton, W., Iversen, C. M., Gallet-Budynek, A., McCarthy, H., Finzi, A., Hanson, P. J., Prentice, I. C., Oren, R., & Norby, R. J. (2014). Evaluation of 11 terrestrial carbon–nitrogen cycle models against observations from two temperate free-air CO_2 enrichment studies. *New Phytologist, 202*, 803–822.

Chapter 8
The Annual Cycle Under Changing Climatic Conditions

Contents

H. Hänninen, *Boreal and Temperate Trees in a Changing Climate*, Biometeorology 3,
DOI 10.1007/978-94-017-7549-6_8

Abstract The hypothetico-deductive modelling framework introduced in Chap. 2 is applied to examining the effects of climatic change on the annual cycle of boreal and temperate trees. Most emphasis is devoted to the paradoxical hypothesis that climatic warming will increase the incidence of frost damage in these trees. According to early computer simulations, trees in boreal conditions in particular would deharden and even start to grow during such mild spells in winter as are commonly projected to prevail in the future climate, so that serious damage would result during subsequent periods of frost. Empirical tests of the frost damage hypothesis suggest that the catastrophic frost damage projected in the early computer simulations will not be realised. Even so, the frost damage hypothesis cannot be ruled out. Available experimental evidence remains limited, and theoretical work with computer simulations has shown that relatively small changes in the ecophysiological traits of trees may cause premature dehardening and growth onset during mild spells in the scenario climate. There have also been several reports of considerable frost damage to boreal and temperate trees and other plants in natural conditions after unseasonally warm spells in winter even in the present climate. For these reasons, nothing conclusive can be said about the frost damage hypothesis. However, the research discussed in this and other chapters of the present volume has pointed out not only the ecophysiological traits of the trees that are critical for the frost damage hypothesis but also the experimental designs that facilitate the determining of those traits in any tree population. Overall, the importance of ecophysiological realism and continuous critical testing of the models are emphasised. Finally, the implications of the effects of climatic change on tree seasonality to the stand and ecosystem level are briefly discussed.

Keywords Boreal trees • Computer simulations • Ecological scenario • Frost hardiness • Frost damage • Model realism • Ontogenetic competence • Premature dehardening • Premature growth onset • Scenario simulation • Seasonality • Temperate trees • Unseasonally warm spells • Whole-tree chamber experiments • Winter warming

8.1 Climatic Scenarios for Ecologists

The concentration of atmospheric carbon dioxide has increased by 40 % since the pre-industrial time. At the same time, the concentrations of other greenhouse gases, such as methane, have also increased. These changes in atmospheric chemistry caused by human activities have caused a global warming of 0.85 °C over the period from 1880 to 2012 (IPCC 2013). Several changes in the functioning of ecosystems as a result of the past climatic change have been documented in scientific literature (Myneni et al. 1997; Menzel and Fabian 1999; Parmesan and Yohe 2003; Menzel et al. 2006; Linkosalo et al. 2009; IPCC 2014a, b).

The General Circulation Models (GCMs) and the Regional Climate Models (RCMs) used by meteorologists project a more pronounced warming in the future (Figs. 8.1 and 8.2). These models are based on principles of atmospheric physics and geophysics, but because of insufficient knowledge of the complicated phenomena involved, the projections of the models involve several uncertainties. For this reason, the projections of climate change relating to any given scenario of concentrations of carbon dioxide and other greenhouse gases in the atmosphere vary among the models used (Fig. 8.2; IPCC 2013).

Furthermore, the future development of the concentrations of carbon dioxide and other greenhouse gases in the atmosphere is also uncertain, for it is heavily dependent on the future development of the emissions of greenhouse gases and the land use practices affecting the carbon fluxes between organic pools and the atmosphere. Thus the scenario simulations carried out by means of the GCMs and RCMs involve a double uncertainty, i.e., that of the input and that of the models themselves (Fig. 8.1). Accordingly, the scenarios differ in the degree of global warming projected, but they all agree on the increasing trend of the global mean temperature. In addition to air temperature, the precipitation and wind conditions are also projected to change dramatically in various areas (IPPC 2013).

According to recent climate scenarios, the warming will be more pronounced in the North than at lower latitudes (IPCC 2013). Thus, especially boreal but also temperate trees will be subjected to more warming than plant and animal species living in more southern biomes. Furthermore, the warming is not projected to be uniform throughout the year at the high latitudes but to have a greater effect on the winter than the summer months (Fig. 8.2). In addition to air temperature, changes are also projected in precipitation in the boreal and temperate regions (IPCC 2013). These climate scenarios call for studies of the effects of the climate change on the annual cycle, especially the overwintering, of boreal and temperate trees.

Because of the uncertainty of the climate change scenarios, the optimal approach in assessments of the ecological effects of the projected climate change is to address a range of scenarios rather than any specific one (Murray et al. 1989; Kramer 1994).

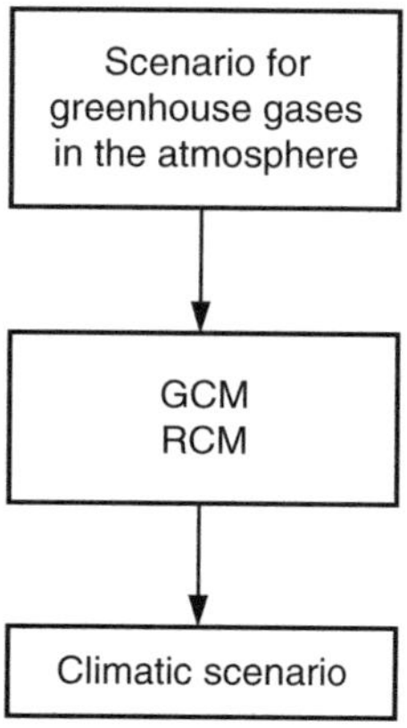

Fig. 8.1 The principle of deducing climatic scenarios with the General Circulation Models (GCMs) and Regional Circulation Models (GCMs)

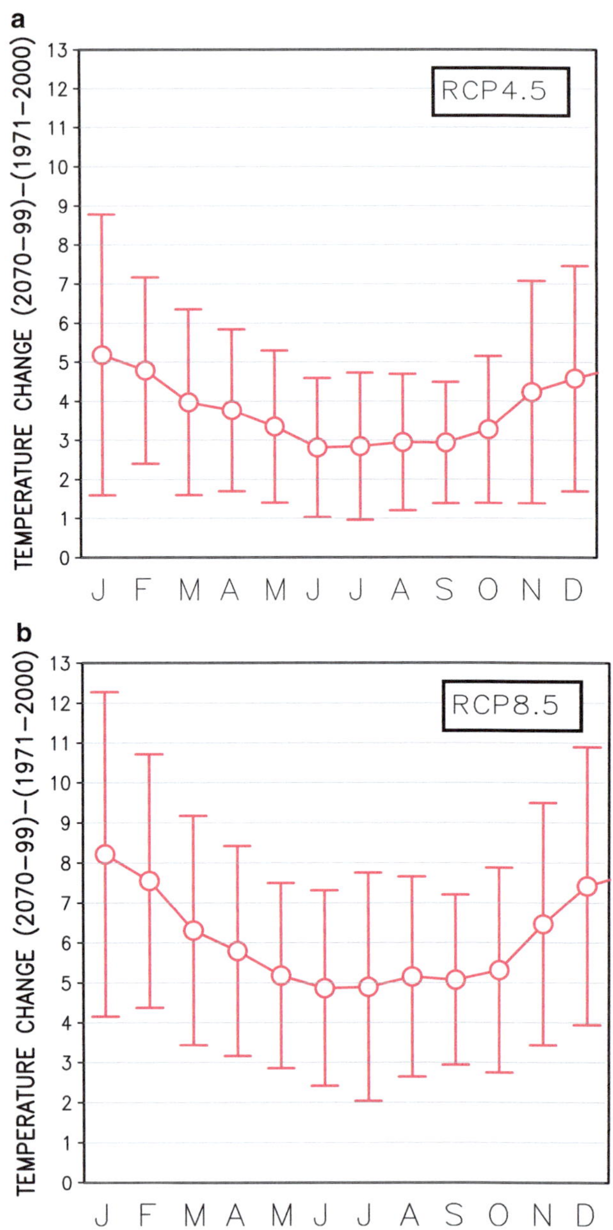

Fig. 8.2 Projected changes in monthly mean temperature for Finland from the period 1971–2000 to 2070–2099 under (**a**) the RCP4.5 and (**b**) the RCP8.5 scenarios. RCP4.5 denotes a stabilisation scenario for greenhouse gas emissions and RCP8.5 a scenario for very high emissions. The *open circles* and the curve stand for the mean of the changes simulated by 28 global climate models used in the 5th Assessment report of IPCC (2013). The *vertical bars*, derived from the scatter among the model simulations, denote the 90 % uncertainty interval of the change. The Finnish Meteorological Institute; the unpublished figures were provided by Kimmo Ruosteenoja

For the same reason, older ecological studies of the ecological effects of the climate change retain their relevance even though the climatic scenarios used as their input may appear outdated. The improvements in the GCMs and RCMs bring changes in the scenarios all the time, so that in the future an older scenario may once again seem more likely than the current one.

8.2 Methods for Assessing the Ecological Effects of Climate Change

8.2.1 Paleoecological and Dendrochronological Methods

The pollen of plants and parts of some plants and invertebrate animals remain identifiable for thousands of years in the anaerobic conditions of lake bottoms and sediments and in the peat accumulated in bogs and fens. These biological archives are addressed in paleoecological research (Berglund 1986), which has made it possible to study the former geographical ranges of several species. One research theme in paleoecological studies has been the construction of past climates on the basis of the past occurrence of different species. A somewhat similar approach is taken in the discipline of dendrochronology, where past climatic fluctuations are inferred from variations in the widths of the annual rings in trees. Both of these methods provide so-called proxy indicators, or proxies, about past climate changes (McCarroll et al. 2011). As both of these disciplines examine the past global changes documented in the archives of nature, then, they provide useful analogies for current studies related to the projected climate change (Willis and MacDonald 2011). However, as these important lessons from the past can only rarely be interpreted with reference to the annual cycle of boreal and temperate trees, they will not be discussed in this chapter.

8.2.2 Empirical Methods

8.2.2.1 An Ideal Empirical Method for Reference Purposes

An ideal empirical method for studying the ecological effects of the projected climate change would be to subject entire ecosystems for decades, if not centuries, to the projected gradual increase in the concentration of atmospheric carbon dioxide, to the gradual rise in air temperature, and to the regional changes projected to occur in the precipitation and wind conditions. In research work, such an ideal experiment is impossible to organise, of course, but since the beginning of the industrial era, humans have been organising a huge experiment of the kind on a global scale. Despite its impossibility, though, the concept of the ideal empirical method is useful because it provides a point of reference for evaluating any method used for assessing the ecological effects of the projected climate change.

8.2.2.2 Field Observations in Natural Conditions

As discussed in Sect. 8.1, the anthropogenic warming has already started according to most meteorologists. This means that the huge global experiment has also started with regard to the air temperature treatment. Thus the ideal empirical method for assessing the ecological effects of the climate change can be applied to the limited extent allowed by the changes in the concentration of atmospheric carbon dioxide and in air temperature realised up to now. It can be hypothesised, then, that the ecological effects of warming are somehow to be seen in nature already.

Fortunately, the addressing of this hypothesis has been made possible by the long-term time series available for many phenological events in both plants and animals. Collecting long-term observational time series of various phenological events is not usually considered to be scientifically the most ambitious and fashionable task. Even so, it is to the benefit of current global change research that such phenological time series have been collected for decades, with the longest of them dating back for centuries. In many cases, the original observations have come from interested laymen (Menzel and Sparks 2006).

The construction of long-term time series on the basis of various historical documents involves several methodological problems requiring the development and use of various mathematical and statistical methods (Häkkinen et al. 1995; Linkosalo et al. 1996). Furthermore, due to the large and stochastic year-to-year variation in the air temperature and the timing of phenological events, it is often hard to examine whether any possible trend in the phenological timing is caused by a corresponding trend in the air temperature; therefore, the testing of the pertinent hypothesis also requires special methodological skills (Menzel et al. 2006; Fu et al. 2015a, b). Thus, the long empirical tradition of collecting phenological time series has recently been followed by a more analytical school, where long-term phenological records are examined with sophisticated tools in the framework of global change studies. In this line of research, often carried out within large-scale international frameworks, it has repeatedly been concluded that the phenological events of many plant and animal species at the mid and high latitudes nowadays occur earlier than they did before (Menzel et al. 2006; Linkosalo et al. 2009).

Although field observations provide the ultimate test for any projection of the ecological effects of the projected climate change, this method has one essential drawback: an analysis of past changes does not alone enable any projection of future changes. Thus, despite the undisputed value of observational studies, additional methods are needed to make projections for the future. As discussed below, the development of tools and approaches for projecting future ecological effects of the climate change always takes place at the cost of losing some aspects of the ideal empirical method, which allows only observations on the past and present changes.

8.2.2.3 Long-Term Field Trials with Transferred Trees

Over a long time, arboreta with exotic tree species have been established in many countries of the boreal and temperate zone. The purpose has often been more ornamental, educational, or cultural than strictly scientific. Still, the long-term survival and growth of exotic tree species in the arboreta has revealed that individuals of many tree species have great potential for acclimation to climatic conditions new to them. Similar transfers with various provenances of native tree species have been carried out in long-term field trials (Rousi et al. 2012; Robson et al. 2013). Their purpose has usually been scientific, often with the ultimate practical aim of improving tree productivity by the use of provenances non-native to the site (Kalela 1938; Viherä-Aarnio and Velling 2008). Similarly to the arboreta where exotic tree species are grown, the provenance trials have revealed that individuals of different provenances of a given tree species have great potential for acclimation to climatic conditions new to them. Accordingly, it is standard practice in many countries to use either exotic tree species (Moore 2011) or non-native provenances of a native species (Nikkanen et al. 1999) for production of timber and wood in managed forests.

Though originally established for different purposes, long-term field trials with exotic tree species and non-native provenances provide a valuable source of information for studies addressing the ecological effects of the climate change. This is the case especially when the trees have been transferred to a warmer climate than the one prevailing at their native growing site. The field trial then provides a model system for examining the effects of climatic warming on the trees, i.e., the performance of the transferred trees in the trial is assumed to describe the response of those specific tree genotypes to a climate warming corresponding to the one caused by the transfer (Beuker 1994; Persson and Beuker 1997).

However, the approach of using provenance transfers as a model system for climate change has several limitations. First, in addition to the annual cycle of air temperature, the annual cycle of night length is also changed in the transfer to the south, which does not happen at the original site in climatic warming. Second, no change in the concentration of atmospheric carbon dioxide occurs in the transfer, whereas an increase in the concentration is projected for the original site. Third, even if the rise of the mean annual temperature involved in the transfer were identical to the one projected for the original site, the annual cycle of air temperature at the trial site would not necessarily be similar to the one projected for the original site. This problem will be discussed in more detail in Sect. 8.3.2.1. Fourth, the change of temperature conditions in the trial is abrupt, not gradual as projected for the original site. This may be a minor problem, though, because there is great year-to-year variation in the climate in natural conditions, too.

However, as discussed throughout Sect. 8.2, the ideal empirical method is possible only in observational studies. Therefore, despite the problems discussed in the preceding paragraph, the examination of existing provenance trials provides one cost-efficient way to project the effects of climate change on boreal and temperate trees. Field transfer trials come as near as possible to the ideal empirical method, i.e., observations made in completely natural conditions. In both methods the survival, growth, and reproduction of the trees are monitored in field conditions on a long-term basis, but instead of the completely natural conditions involved in the observational approach, the conditions in the field transfer trials are semi-natural. This is the cost one must pay for introducing an experimental approach that also enables projections to the future, which are not possible in the observational method.

8.2.2.4 Experiments in Controlled Conditions

The development of techniques for controlling environmental factors was essential for the emergence of many fields of experimental biology, plant ecophysiology included. For instance, the photoperiodic ecotypes of trees were found quite soon after lamps allowing artificial illumination become commonly available (Koski and Selkäinaho 1982). Nowadays, various types of growth chambers are available with versatile possibilities for controlling such environmental factors as air temperature, soil temperature, night length, amount and quality of light, humidity, and concentrations of atmospheric gases. When a lower level of control over the environmental factors is sufficient for the purposes of the experimental design, greenhouses can also be used.

Experiments with greenhouses and growth chambers located in laboratories enable a direct examination of the effects of climate change on trees. For instance, the effects of intermittent warm periods in winter on tree seedlings can be examined by transferring the seedlings into a greenhouse or growth chamber with air temperature conditions raised to correspond to the assumed mild period. There are, however, practical limitations in this approach, too. To start with, only seedlings or relatively small saplings can be used in the experiments, whereas experiments addressing mature trees are possible only by the use of detached twigs. Furthermore, several artifacts, i.e., unintended additional changes in the environmental factors, are involved in the greenhouses and growth chambers. For instance, the roots of the seedlings and saplings are usually grown in small pots in these experiments as opposed to growing freely at their natural growing site.

Clearly, then, many aspects of the ideal empirical method for addressing the ecological effects of the projected climate change are lost in experiments carried out in greenhouses and growth chambers located in laboratories. Therefore this approach is used mainly for increasing general knowledge of the basic ecophysiology of trees rather than directly addressing the effects of the projected climate change. Of course, it also serves the purposes of assessing the ecological effects of the climate change indirectly: as discussed in several earlier chapters, the experimental approach with greenhouses and growth chambers in laboratories can also be used in the development and testing of ecophysiological models, which can subsequently be used in scenario simulations addressing the ecological effects of the projected climate change.

Over the last few decades, the increased awareness of climate change and the increased need to assess its ecological effects have spurred the development of various experimental techniques for controlling the environmental factors in the field. In this approach, the purpose in most cases is to control the air temperature and/or the concentration of atmospheric carbon dioxide while keeping the growing environment of the plants otherwise as natural as possible. As the aim is to study the ecological effects of the climate change with a direct empirical approach, the methods developed come close to the ideal empirical method.

Within this overall approach, a wide range of different techniques have been used. Experimental studies carried out with tundra plants have made use of small open-top (OT) chambers, where raised air temperatures have been effected by

Fig. 8.3 A whole-tree chamber (WTC) experiment in Flakaliden, northern Sweden. On the *right* is seen the building that houses the devices controlling the environmental factors inside the WTCs. For details of the experiment, see Medhurst et al. (2006) and Wallin et al. (2013)

means of passive heating through solar radiation (Henry and Molau 1997; Marion et al. 1997). Though it is a simple and cost-effective method for studying the effects of elevated air and soil temperatures on the small plants growing in the tundra, the OT chamber method does not enable an examination of the effects of elevated concentrations of carbon dioxide. This situation is reversed in the Free Air Carbon Enrichment (FACE) experiments, where the concentration of atmospheric carbon dioxide is elevated by injecting carbon dioxide into the air at the experimental plot but the air temperature is not affected either actively or passively (Ainsworth and Long 2005; Zaehle et al. 2014).

During the last decades, relatively large whole-tree chambers (WTCs) have been used for the examination of elevated air temperature and elevated concentration of atmospheric carbon dioxide on tree saplings growing at their natural site (Kellomäki et al. 2000; Medhurst et al. 2006; Fig. 8.3). In experimental designs with WTCs, two levels, i.e., an ambient and an elevated level, of both environmental factors are normally used, and these are combined in a 2×2 factorial design. The treatment with elevated levels of both environmental factors comes quite close to the ideal empirical method, but several artifacts, such as the removal of the wind, still prevail in this approach, too. To address the artifacts, a control treatment, i.e., a chamber where both environmental factors are kept at their ambient level, is included, and the effects of the other treatments are inferred by comparing the measurement results obtained from them with the corresponding results obtained from the control chambers. Treatments where one of the two environmental factors is elevated and the other kept at the ambient level enable the examination of the relative importance of the two factors to the examined ecophysiological phenomena (Kellomäki et al. 2000; Medhurst et al. 2006).

Relatively few field experiments with WTCs have been launched in the boreal and temperate zones, obviously because of their high cost. Even so, the experiments that have been done have provided crucial new information for research into the effects of the climate change on boreal and temperate trees (Kilpeläinen et al. 2006; Barton et al. 2010; Wallin et al. 2013). A discussion of the results of the WTC experiments is therefore essential in the present volume also, but it is mainly restricted to the results concerning the effects of the environmental factors on the seasonal phenomena of the trees.

8.2.3 Numerical Methods

8.2.3.1 Computer Modelling in Ecological Scenario Studies

Computer modelling provides a powerful tool for examining the ecological effects of various climatic scenarios because the practical limitations restricting any empirical study addressing the ecological effects of the climate change are readily overcome in numerical studies carried out with computers. So, with the calculation power of modern computers, the scenario calculations can easily be extended to cover several centuries, and there is practically no limit to the number of climatic scenarios that can be addressed.

However, as emphasised throughout the present volume, numerical studies carried out with computer modelling are deductive, i.e., their results are pure mathematical implications of the assumptions of the model (Fig. 2.1 in Sect. 2.1). Ecological scenario studies carried out with computer modelling are no exception to this rule. Thus, the ecological scenario projected by the model is realised only if the assumptions written in the formulae of the computer model provide a sufficiently realistic description of the ecological phenomena addressed (Fig. 8.4). This is said, of course, assuming that the climatic scenario taken as the input for the calculations is realised in the first place.

Researchers who do solely empirical work themselves sometimes regard the use of computer modelling in ecological scenario studies as a monolithic approach. That is definitely a false notion. Saying that somebody works with computer modelling gives approximately as much information as saying that somebody else works in the laboratory doing chemical analyses. The studies using ecological computer models to assess the ecological effects of the climate change can and do take many different approaches to the research problem indicated by Fig. 8.4. The ecological models applied form a continuum, including several hybrid models

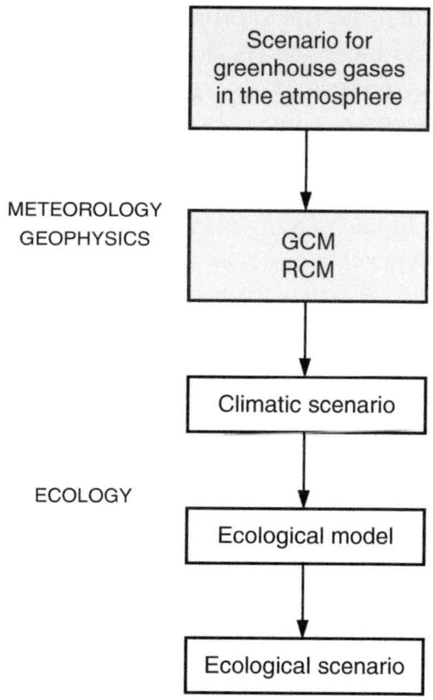

Fig. 8.4 The universal principle of ecological scenario studies addressing the effects of the projected climate change. The principle can be regarded as universal because the same scheme is ultimately used in any study assessing the ecological effects of climate change, regardless of the specific methods used. GCM and RCM stand for General Circulation Models and Regional Circulation Models, respectively

synthesising aspects from different approaches, but they can be roughly divided into two main categories, which are discussed below.

8.2.3.2 Climate Envelope Modelling

Climatic envelope, or niche-based, models follow the classic approach of biogeography in that they explain the current occurrence of a given plant or animal species by relatively rough climatic indices such as the mean annual temperature, mean temperature sum of the growing season, or average monthly precipitation prevailing at the examined site (Bakkenes et al. 2002; Bradley 2009). To develop a climate envelope model for a given species, long-term meteorological data are gathered from stations representing the geographical range of the species, and the data are used for the construction of a correlative model explaining the present occurrence of the species. The concept of climatic envelope denotes the overall climatic conditions, described by a combination of the values of the various climatic indices addressed, that allow the occurrence of the examined species. Any given species has its special requirements for each climatic index included in the model. The occurrence of the species might require, for instance, a certain minimum mean temperature sum of the growing season or a certain minimum average precipitation in July.

Climate envelope models have in general been quite successful in explaining the current geographical ranges of various plant and animal species (Bakkenes et al. 2002; Hijmans and Graham 2006). This has encouraged their users to apply their models to projecting the effects of the climate change on the future ranges of the modelled species. Once a climate envelope model has been constructed for a given species, its use in scenario studies is straightforward. To this end, data for the scenario climate are needed both for the present geographical range of the species and for locations outside the present range. By means of the scenario climate data for this extended geographical area, it is then determined whether the scenario climate of a given location meets the climate envelope requirements of the species. After this has been done for each location where the scenario climate is available, two areas are formed on the map: the areas where the scenario climate meets and the areas where it does not meet the climatic requirements of the species represented by its climatic envelope. The former areas make up the new geographical range of the species projected by the climate envelope model applied (Bakkenes et al. 2002; Hijmans and Graham 2006).

Finally, the projected and the current geographical ranges of the species are compared to determine the extent of the projected ecological effect on the species (Thuiller et al. 2005). As a result of warming scenarios, the species are typically projected to disappear from the southern parts of their present range and to gain new area north of it. Corresponding changes are projected to occur on the moisture gradient whenever sufficiently great changes in precipitation and potential evaporation are included in the climatic scenario. The threat of extinction of the species increases if the area of its projected range approaches a critical lower limit. Along this line of reasoning, Thomas et al. (2004) projected that unless powerful measures are taken to reduce the emissions of greenhouse gases, by the year 2050 up to one third of the plant and animal species of the globe will become extinct because of the climate change.

Despite its wide use in climate change research, the climate envelope method fails to account for several processes and phenomena that affect the performance of the different species under the scenario conditions. This is mainly because the method is based entirely on the current static correlations between the occurrence of the species and the climatic conditions and therefore fails to address the completely novel combinations of environmental factors that will appear in the future. For instance, unlike air temperature, night length conditions will not change as a result of the climate change. Therefore, as a result of warming, any given air temperature conditions will appear in combination with novel night length conditions in the future. As discussed in Sects. 3.5 and 6.2.3, night length is a major environmental factor affecting the annual cycle of boreal and temperate trees. Thus, though the air temperature conditions facilitating the occurrence of a given southern tree species, i.e., the climatic envelope of the species, move north as a result of the climate change, it remains unclear whether individuals of the southern tree species acclimate to the novel night length conditions prevailing in the more northern location where the climate envelope is projected to move.

For another example of novel combinations of future environmental factors, the water use efficiency of many C_3 plants increases under increased concentrations of atmospheric carbon dioxide, i.e., they transpire less water per unit of carbon dioxide fixed in photosynthesis (Mooney and Ehleringer 1997). This phenomenon may somewhat counteract the adverse effects of drought, which are commonly projected to hamper the growth of several C_3 plant species in several regions as a result of the climate change. As the climatic envelope method is based on phenomena that have taken place in the current (or past) concentrations of atmospheric carbon dioxide, it is inherently unable to account for the potential compensating effect of increased concentrations of atmospheric carbon dioxide. Furthermore, soil conditions are often not addressed in the climatic envelope methods. Thus, according to a line of reasoning similar to that presented for night length in the previous paragraph, unsuitable soil conditions may prevent the colonisation of more southern trees even though the air temperature conditions alone would facilitate such colonisation as a result of the climate change.

In conclusion, being based on the static current correlations between the occurrence of the species and climatic indices, the climate envelope models fail to address the processes that determine the geographical ranges of different species. In addition to the ecophysiological processes discussed in the present volume, other processes such as migration, interspecific competition and plant-herbivore interactions should also be addressed. It goes without saying, however, that this is possible for a limited number of species only. Therefore, despite the criticism presented above, as the climate envelope method readily enables scenario studies with a high number of species, it provides a valuable tool for deducing the first approximations of the effects of the climate change on the geographical ranges of different plant and animal species.

8.2.3.3 Process-Based Modelling

Process-based models do not address the static patterns observed in nature but the phenomena that shape those patterns. As indicated by the concept of "process", these models take a dynamic approach, i.e., they examine the phenomena over time. Process-based models belong to the category of dynamic models, then, adopting the rate-state approach introduced in Sect. 2.2. As stated before, the models of the annual cycle of boreal and temperate trees discussed and used in the present volume are dynamic ecophysiological models. Further examples of dynamic ecological models are briefly discussed in Chap. 7.

Not only ecophysiological phenomena but also physical phenomena, such as changes in the thickness and structure of the snow cover, are addressed by means of process-based models (Bartelt and Lehning 2002; Rasmus et al. 2015). Thus, as the process-based dynamic models of different natural phenomena share the same approach and the same mathematical methodology, models representing different disciplines are readily combined. This is the case particularly with the Dynamic Global Vegetation Models (DGVMs), which include sub-models for both biological and physical phenomena. As discussed in Chap. 7, DGVMs are used to simulate, among other things, the global carbon cycle. This enables a feedback loop from the DGVMs to the concentration of atmospheric carbon dioxide and to the General Circulation Models, GCMs, used by meteorologists to develop climatic scenarios.

In this chapter, process-based ecophysiological models of the annual cycle of boreal and temperate trees are applied to scenario studies mainly as such, i.e., by carrying out the model simulations at the whole-tree level particularly addressed by these models. For instance, the effects of climate change on the needle frost hardiness of individual trees are examined. In the discussion of upscaling, only two cases of scaling up the implications of whole-tree ecophysiological phenomena to the higher organisational levels are addressed. Generally, however, most of the scenario studies using ecophysiological models are carried out at the ecosystem level and use aggregated ecosystem models consisting of several sub-models for the ecophysiological and physical phenomena addressed (Kramer 1996; Kellomäki and Väisänen 1997; Kramer et al. 2002; Kolari and Hari 2008).

With process-based models, it is possible in principle to deal with all the problems hampering the use of climate envelope models discussed in the previous Sect. 8.2.3.2: all essential processes affecting the survival and growth of trees (or plants in general) can be explicitly addressed, and so can the novel combinations of environmental factors prevailing in the scenario conditions. This does not mean, however, that process-based modelling provides a shortcut to solving all problems met in assessing the ecological effects of various climatic scenarios. As already mentioned, it is impossible to carry out all the experimental studies with several species that are needed for the definition and parameterization of the numerous sub-models included in large process-based ecosystem models. Several compromises must therefore be made and, as seen later in this chapter, this is the case even with the annual cycle models discussed in the present volume, though these models are not particularly complicated.

8.2.4 *The Universal Principle of Ecological Scenario Studies*

As discussed in Sect. 8.1, the projecting of climatic scenarios involves two types of uncertainties occurring at their two respective phases (Fig. 8.1). A third uncertainty is added when ecological scenarios are deduced by means of ecological models for any of the climatic scenarios (Fig. 8.4). This is because ecological models are rarely, if ever, so advanced that their projections could be taken for granted. It cannot be overemphasised, however, that this notion, which is thoroughly substantiated in the remaining sections of this chapter, is equally valid for all ecological models, whether mathematical, graphical, or conceptual. The only essential difference among these model types is that mathematical models provide exact projections and are therefore more susceptible to falsification than the other model types, which provide more vague projections. Therefore, the exactness of the projections of mathematical models does not imply greater trustworthiness, but it enables critical testing of the models and thus helps to develop as trustworthy ecological projections as possible.

Ultimately, an ecological model of some kind is used every time the ecological effects of the climate change are assessed. This notion, referred to as *the universal principle of ecological scenario studies* in the present volume, is substantiated by the need to generalise even when carrying out strictly empirical research on the ecological effects of climate change. For instance, when studying the effects of elevated temperatures on the timing of the growth onset of trees in a whole-tree chamber experiment, an acceleration of growth onset is usually observed (Hänninen 1995a; Slaney et al. 2007). However, the empirical observation "growth onset accelerated by x days as a result of y °C warming" applies, strictly speaking, only to the individual trees observed. Therefore, when the observation is generalised to other trees belonging to the same species and provenance, an empirical model is actually formed. This model is then used according to the principle described in Fig. 8.4 when the observed acceleration of growth onset is generalised to other similar trees.

In most cases the situation is more complicated, with many other theoretical assumptions besides the generalisation needed in the model that is finally used to project the ecological effects of the climate change. Regardless of the specific research methods used, then, the principal method in ecological scenario studies is always the same: given the climatic scenario assumed, we use our ecological understanding, knowledge, model, or theory – whatever expression we like to use here – to infer the ecological scenario, i.e., a projection of the ecological phenomena occurring under the assumed climatic scenario (Fig. 8.4).

In this chapter, process-based ecophysiological modelling and related experimental studies are discussed in order to assess the effects of the projected climate change on the seasonality of boreal and temperate trees. The main emphasis is on the effects on overwintering discussed in the following Sect. 8.3.

8.3 The Frost Damage Hypothesis

8.3.1 Deducing the Frost Damage Hypothesis

8.3.1.1 A Conceptual Frost Damage Hypothesis

One would assume that as climatic warming reduces the freezing stress, it should also reduce the incidence of frost damage in boreal and temperate trees. However, as air temperature is a major environmental factor regulating the annual cycle of these trees (Chaps. 3, 4, 5, and 6), a more complicated picture emerges. Cannell (1985) presented the paradoxical hypothesis that climatic warming may cause an increase, not a decrease, in the incidence of frost damage in boreal and temperate trees. The reason is that climatic warming causes warmer, longer and more frequent mild periods to occur in future winters than occur at present. In those intermittent mild periods, trees may deharden and in extreme cases even start to grow and then get damaged in subsequent frost periods. This frost damage hypothesis has been addressed in several later studies, both with computer simulations and experimentally. It was first addressed in the temperate and maritime conditions of Scotland (Cannell and Smith 1986; Murray et al. 1989). However, since the patterns found in those conditions were more complicated than those found later in boreal conditions, the studies will not be discussed chronologically in what follows; rather, those addressing boreal conditions will be discussed first.

8.3.1.2 The Frost Damage Hypothesis for Boreal Conditions

Motivated by Cannell's (1985) and Cannell and Smith's (1986) studies, Hänninen (1987, 1990a) carried out preliminary scenario simulations addressing the frost damage hypothesis in the boreal conditions of central Finland. These preliminary studies provided support for the hypothesis, so that Hänninen (1991) then carried out a simulation study addressing the hypothesis more thoroughly. He simulated the timing of the growth onset of a generalised boreal tree species across 73 years, in both the current and a scenario climate. Daily mean, T_{mean}, and daily minimum air temperatures, T_{min}, were used as input for the simulations. For the current climate, data were available from Jyväskylä, central Finland. The data for the scenario climate were constructed by raising the daily values of both T_{mean} and T_{min} according to a non-uniform warming scenario corresponding to a doubled concentration of atmospheric carbon dioxide (Table 8.1).

A modified version of Sarvas's (1972, 1974) model, i.e., a sequential model with the air temperature responses formulated by Hänninen (1990b), was used in the simulations.[1] The onset of rest, where the simulations were initiated for each annual cycle, was deliberately set at 1 September each year. For each annual cycle starting

Table 8.1 A non-uniform warming scenario for central Finland, corresponding to a doubled level of concentration of atmospheric carbon dioxide in comparison with the preindustrial era (Bach 1987; Kettunen et al. 1987)

Month	ΔT (°C)
January	6.2
February	5.7
March	5.1
April	4.4
May	3.3
June	2.1
July	1.6
August	2.1
September	3.2
October	4.3
November	5.2
December	5.9

ΔT = increase in monthly mean temperature. In the simulations reviewed in the present volume, both the daily mean temperature, T_{mean}, and the daily minimum air temperature, T_{min}, were raised by the amount indicated by ΔT (Redrawn from Hänninen 1991; with the permission of John Wiley & Sons, Inc.)

[1] The three sub-models forming the overall model are described in detail in Sects. 3.2.1.4, 3.3.2.2, and 3.4.1.1.

on that day, the date of growth onset was calculated. In order to address the frost damage connected with premature growth onset, the minimum post-dormant air temperature, T_{PD}, was determined as the lowest of the daily minimum air temperatures occurring between the simulated date of growth onset and 31 July. The value of T_{PD} indicates the heaviest freezing stress the tree will be exposed to in the active growth phase in the particular year examined. Furthermore, the year-to-year minimum of T_{PD}, $T_{PD\text{-}min}$, was determined as the lowest of the 73 annual values of T_{PD} obtained in the simulations. The value of $T_{PD\text{-}min}$ indicates the heaviest freezing stress the tree will be exposed to in the active growth phase during the entire simulation period of 73 years.

In the modelling approach taken, the theoretical tree genotype is identified with the combination of its chilling requirement of rest completion, C_{crit}, and the high temperature requirement of growth onset, H_{crit}.[2] The simulations were initiated with a genotype $C_{crit} = 30$ CU and $H_{crit} = 5.3$ HU, since these are typical values for central Finnish trees according to empirical findings (Hänninen 1990b). Over the 73 years in the current climate, the simulated growth onset of this theoretical tree genotype took place on 17 May on the average, the range being 37 days, i.e., from 27 April to 2 June (Fig. 8.5a). These results accord well with the timing of the bud burst of *Betula* observed over 36 years in Jyväskylä (Fig. 3.4 in Sect. 3.2.1.1). The simulated values of minimum post-dormant air temperature, T_{PD}, ranged from −4.1 to +4.5 °C, indicating that no severe frost damage was predicted for the current climate (Fig. 8.6a).

In the simulations with the scenario climate, however, the results changed drastically. Not only did the average date of growth onset accelerate by 41 days to 6 April, but more importantly, the range widened to 109 days, extending now from 24 January to 12 May (Fig. 8.5b). The distribution of the dates of growth onset was strongly skewed to the left, so that in some rare years the projected growth onset took place much earlier than in most years (Fig. 8.5b). The distribution of T_{PD} was correspondingly skewed to the left in the scenario climate (Fig. 8.6b). In most years, T_{PD} was in the range of −5 to +5 °C, indicating that no severe frost damage was projected for those years. However, in years with exceptionally early growth onset, the value of T_{PD} was below −10 °C, and in the most extreme case it was −26.6 °C (Fig. 8.6b). The actively growing tissues of trees do not tolerate temperatures as low as that (Sakai and Larcher 1987; Repo 1992). Altogether, then, the simulations projected that in the long run the theoretical tree genotype with $C_{crit} = 30$ CU and $H_{crit} = 5.3$ HU would be killed in the scenario climate due to frost damage caused by premature growth onset even if the genotype would acclimate to the scenario climate assumed for most years.

In order to examine the background of the results obtained in the simulations over the 73 years with the tree genotype with $C_{crit} = 30$ CU and $H_{crit} = 5.3$ HU (Figs. 8.5 and 8.6), the simulations carried out for one annual cycle are presented in Fig. 8.7. The winter of 1972–1973 was exceptionally mild in Jyväskylä. The daily

[2] The approach is similar to that presented in Sect. 6.3.4, where the adaptation of different tree genotypes to the current climate was examined.

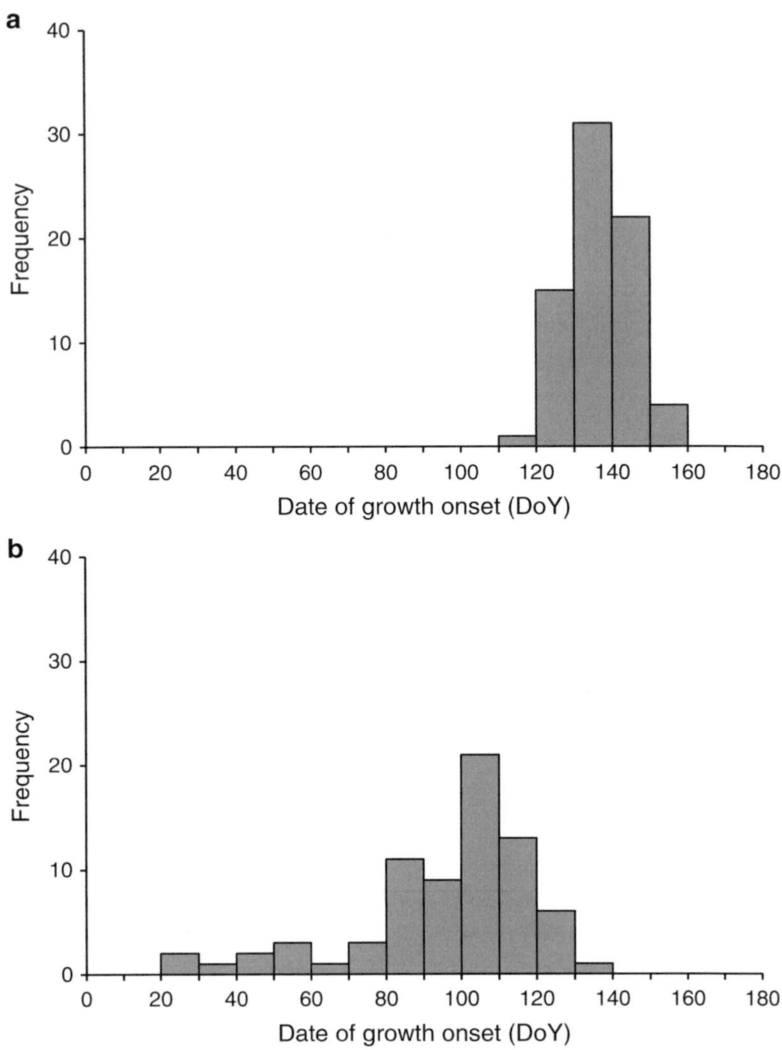

Fig. 8.5 Frequency distributions of the simulated dates of growth onset across 73 years for an average genotype of a generalised boreal tree species in Jyväskylä, central Finland. (**a**) Model prediction for the current climate. (**b**) Model projection for a scenario climate where the observed daily mean and minimum air temperatures were elevated according to the scenario presented in Table 8.1. *DoY* Day of Year (Modified from Hänninen 1991; published with the permission of John Wiley & Sons, Inc.)

mean temperatures were frequently above zero after the simulated rest completion on 7 November. These air temperatures were sufficiently high to cause ontogenetic development in the simulation, so that considerable ontogenetic development towards growth onset was predicted to occur in the winter, as indicated by the increasing value of the state of ontogenetic development, S_o (Fig. 8.7a).

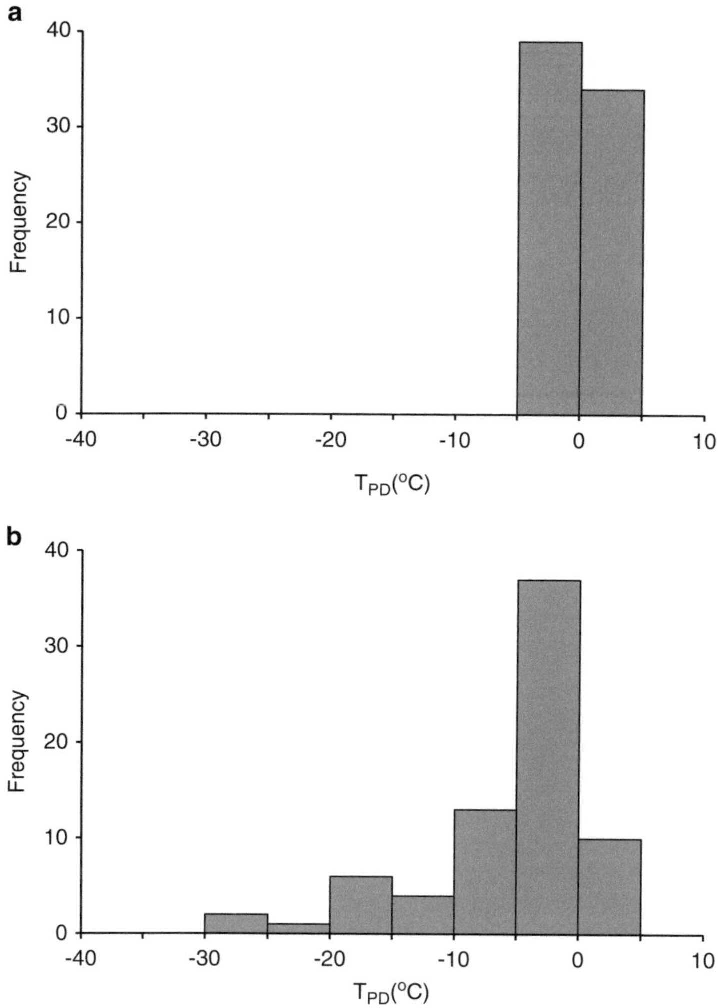

Fig. 8.6 Frequency distributions of the simulated minimum post-dormant air temperature, T_{PD}, across 73 years for an average genotype of a generalised boreal tree species in Jyväskylä, central Finland. T_{PD} indicates the heaviest freezing stress the tree will be exposed to while it is in the susceptible growth phase in the particular year examined. (**a**) Model prediction for the current climate. (**b**) Model projection for a scenario climate where the observed daily mean and minimum air temperatures were elevated according to the scenario presented in Table 8.1 (Modified from Hänninen 1991; published with the permission of John Wiley & Sons, Inc.)

Accordingly, the simulated growth onset took place on 9 May (Fig. 8.7a), i.e., 8 days earlier than average (Fig. 8.5a). The value of T_{PD} was +0.8 °C, so that no frost damage was predicted.

The results changed considerably when the simulations were carried out in the scenario climate for the annual cycle corresponding to the 1972–1973 one, i.e., with

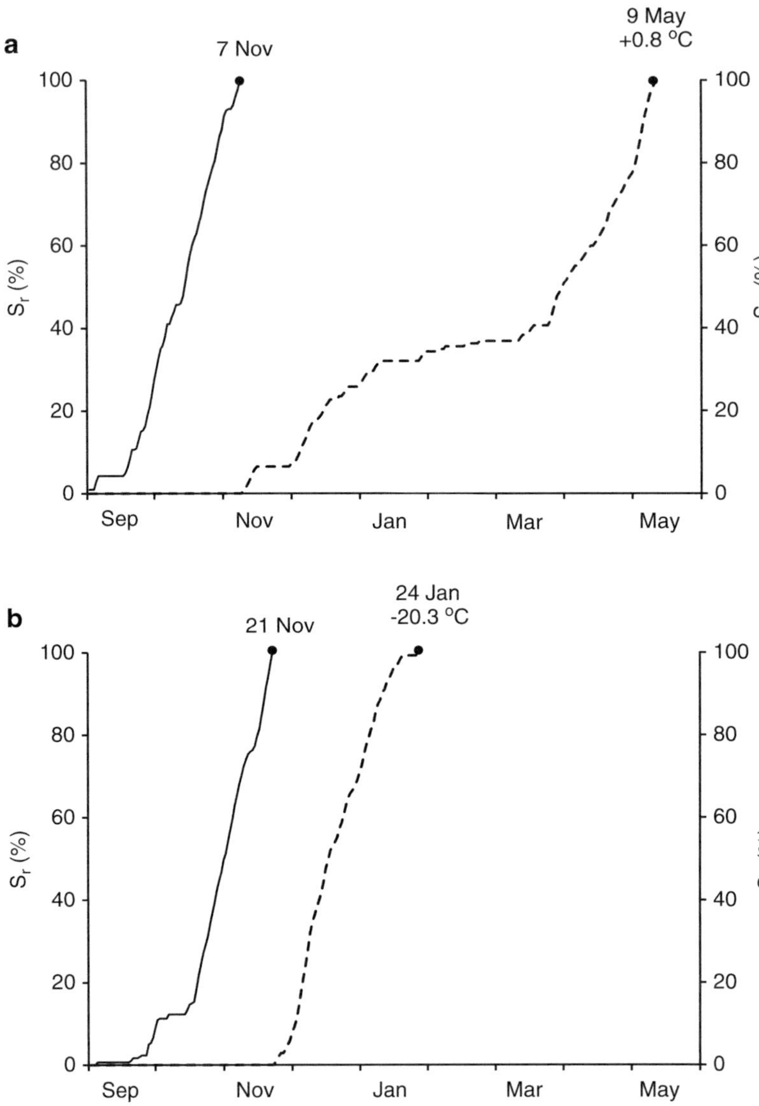

Fig. 8.7 Simulated rest break and simulated ontogenetic development towards growth onset in the buds of an average genotype of a generalised boreal tree species in Jyväskylä, central Finland. The continuous line indicates the state of rest break (S_r, *left-hand vertical axis*), with the date of rest completion indicated at the end of the line. The dashed line indicates the state of ontogenetic development (S_o, *right-hand vertical axis*), with the date of growth onset and the minimum post-dormant air temperature, T_{PD}, indicated at the end of the line. (**a**) Simulation for the current climate in 1972–1973. (**b**) Simulation for a scenario climate where the daily mean and minimum air temperatures of 1972–1973 were elevated according to the scenario presented in Table 8.1 (Modified from Hänninen 1991; published with the permission of John Wiley& Sons, Inc.)

input data where both the T_{mean} and T_{min} of 1972–1973 were raised according to the warming scenario assumed (Table 8.1). Rest completion was projected to occur on 21 November (Fig. 8.7b), so that because of the higher air temperatures in the scenario climate, the chilling requirement was met 14 days later than it was in the corresponding simulation with the current climate (Fig. 8.7a). However, this lag was subsequently more than compensated for, when the daily mean air temperatures, now considerably above zero on most days, promoted rapid ontogenetic development towards growth onset, as indicated by the rapidly increasing value of the state of ontogenetic development, S_o, in December and January (Fig. 8.7b). The projected growth onset occurred on 24 January, after which the daily minimum air temperatures dropped as low as the value of $T_{PD} = -20.3$ °C, so that the genotype $C_{crit} = 30$ CU and $H_{crit} = 5.3$ HU was projected to suffer lethal frost damage under the assumed scenario climate.

In the scenario climate, premature growth onset with subsequent lethal frost damage occurred only when the winter was exceptionally warm, even for the scenario climate. This explains why premature growth onset (Fig. 8.5b) with subsequent lethal frost damage (Fig. 8.6b) was projected to occur in the scenario climate in relatively few years only.

Simulations were then carried out for all the 73 years similarly with 40 theoretical tree genotypes represented by various combinations of the chilling requirement of rest completion, C_{crit}, and the high temperature requirement of growth onset, H_{crit}. C_{crit} was varied in the range of 5–50 CU and H_{crit} in the range of 1.8–7.0 HU (Fig. 8.8). The ranges of both traits were set so as to exceed the variation found in real central Finnish trees at both ends of the scale. At the low end, this also showed up in the simulation results, for with the low values of the two genotype-specific parameters, the projected year-to-year minimum, $T_{PD\text{-}min}$, of the annual values of the minimum post-dormant air temperature, T_{PD}, implied lethal frost damage even in the current climate (Fig. 8.8), indicating that such genotypes could not grow in the Jyväskylä region under the current climate. For instance, with the lowest value of $H_{crit} = 1.8$ HU, the $T_{PD\text{-}min}$ was below -20 °C in the current climate regardless of the C_{crit} value assumed in the simulation (Fig. 8.8a); quite clearly, then, genotypes with $H_{crit} = 1.8$ HU could not survive in the Jyväskylä region under the current climatic conditions.

The genotypes surviving in the Jyväskylä region under the current climate are delineated in Fig. 8.8, where it is assumed for simplicity's sake that the threshold for the survival of the trees is $T_{PD\text{-}min} = -10$ °C. Due to the joint effect of C_{crit} and H_{crit} in determining the timing of growth onset, the value of C_{crit} required for survival in the current climate decreases with increasing values of H_{crit} (Fig. 8.8b–d).

The generality of the conclusion about lethal frost damage in the assumed scenario climate obtained with the genotype $C_{crit} = 30$ CU and $H_{crit} = 5.3$ HU (Figs. 8.5, 8.6, and 8.7) was tested by examining the value of $T_{PD\text{-}min}$ found in the scenario climate for all of the genotypes surviving in the Jyväskylä region under the current climate (Fig. 8.8b–d). With nearly all of these genotypes, the value of $T_{PD\text{-}min}$ in the scenario climate was below -10 °C, indicating lethal frost damage in the scenario climate for genotypes surviving in the current climate. However, the

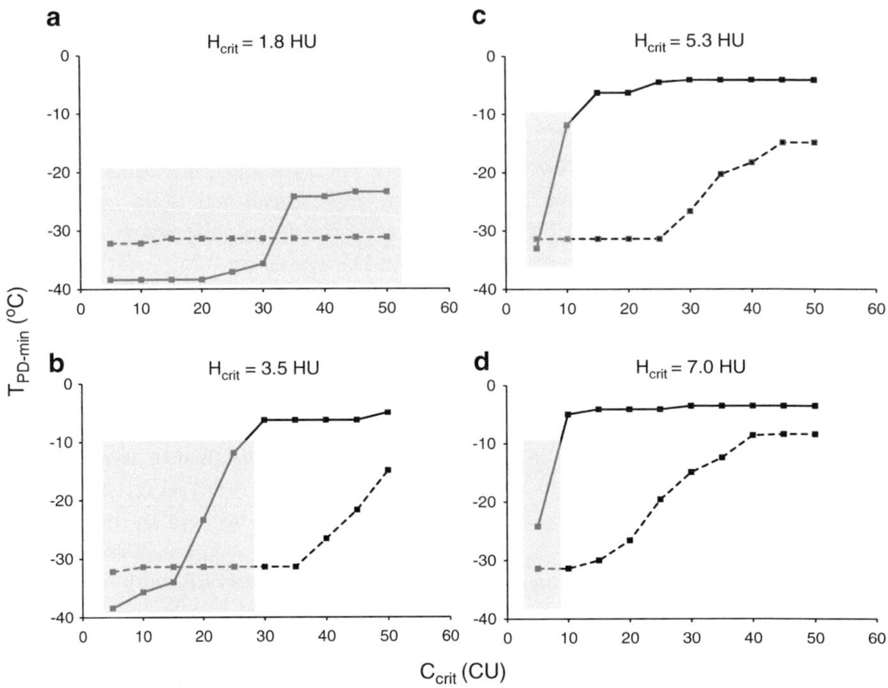

Fig. 8.8 Simulated post-dormant freezing stress in 40 genotypes of a generalised boreal tree species in Jyväskylä, central Finland. The freezing stress is indicated by the year-to year minimum, $T_{PD\text{-}min}$, of the annual values of the minimum post-dormant air temperature, T_{PD}. The $T_{PD\text{-}min}$, represented on the *vertical axes*, indicates the heaviest freezing stress the tree will be exposed to while it is in the susceptible growth phase during the entire simulation period. Each dot corresponds to one genotype defined by the combination of its chilling requirement of rest completion, C_{crit} (on the *horizontal axes*, CU = arbitrary chilling unit), and by its high temperature requirement of growth onset, H_{crit} (panels **a**–**d**, HU = arbitrary high temperature unit). The values of H_{crit} are presented in decimal fractions because they have been transformed from the original values expressed in the forcing unit scale. See the discussion in Sect. 3.2.1.4. The simulations were carried out across 73 years and both for the current climate (*continuous curves*) and for a scenario climate (*dashed curves*). The scenario climate was constructed by elevating the observed daily mean and minimum air temperatures according to the scenario presented in Table 8.1. In each panel, the grey area indicates genotypes that do not survive in the current Jyväskylä climate and are thus irrelevant for the examination of the effects of climatic warming (Modified from Hänninen 1991; published with the permission of John Wiley & Sons, Inc.)

genotypes with the highest values of both requirements, i.e., $H_{crit} = 7.0$ HU and $C_{crit} \geq 40$ CU, were an exception. For them, the value of $T_{PD\text{-}min}$ was slightly above -10 °C in the scenario climate (Fig. 8.8d). The simulation suggests, then, that they might survive in the scenario climate but would get severely damaged in the most critical years. However, such tree genotypes evidently do not exist or are rare in the tree populations of central Finland (Hänninen 1990b).

In a further study of the frost damage hypothesis in the conditions of central Finland, Hänninen et al. (1996) used the Kellomäki et al. (1992, 1995) integrated frost hardiness model discussed in Sect. 5.3.2.2. For input in the simulations, they used the same air temperature data from the Jyväskylä region as had been used in the Hänninen (1991) study discussed in the previous paragraphs. According to the prediction of the model, the annual cycle of frost hardiness was well synchronised with the annual cycle of daily minimum air temperature in the current climate (Fig. 8.9a), so that practically no frost damage was predicted for the current climate. This accords with the general finding that native tree species are well adapted to the climatic conditions prevailing at their growing site.

In the scenario climate, constructed by elevating the daily T_{mean} and T_{min} values according to the scenario also used by Hänninen (1991) (Table 8.1; Figs. 8.5, 8.6, 8.7, and 8.8), the synchronisation was lost (Fig. 8.9b). Consequently, in exceptionally mild winters there was premature dehardening in the simulations, leading to heavy frost damage during subsequent periods of frost when the daily minimum temperature dropped again (Fig. 8.9b).[3] This was no surprise as such because the Kellomäki et al. (1992, 1995) integrated frost hardiness model used in the simulations includes as a submodel the model that had projected the premature growth onset (Figs. 8.5, 8.6, 8.7, and 8.8) in the simulations carried out earlier by Hänninen (1991). This simulation study (Hänninen et al. 1996), however, provided support for the frost hardiness hypothesis, especially in the case of *Pinus sylvestris*, for which species the frost hardiness model had passed an independent test with needle frost hardiness data (Hänninen et al. 1996; Fig. 5.14a in Sect. 5.3.2.2).

In conclusion, Hänninen's (1991) and Hänninen et al.'s (1996) scenario simulations support Cannell's (1985) frost damage hypothesis as regards boreal trees growing in thermal conditions such as those prevailing in central Finland. In regard to the universal methodological principle presented in Fig. 8.4, the results of these simulations show that if the models applied in the simulations (Sarvas 1972, 1974; Koski and Sievänen 1985; Hänninen 1990b, 1991; Kellomäki et al. 1992, 1995) provide a sufficiently realistic description of the regulation of the annual cycle in boreal trees, then climatic warming such as presented in Table 8.1 will lead to catastrophic frost damage to boreal trees growing in thermal conditions similar to those prevailing in central Finland, which will probably lead to the removal of many native tree species from the flora.

The adverse effects of the climate change on ecosystems and human beings are quite often related to the increased incidence of extreme climatic events (IPCC 2013, 2014a, b). Here it is instructive to notice that if the concept of extreme climatic event is defined as increased variation in meteorological quantities such as

[3] In these diagnostic simulations, carried out at the whole-tree level only, the effects of the projected frost damage on the subsequent development of the trees were not addressed. Instead, a new annual cycle of frost hardiness was simulated even in cases where lethal frost damage had been projected. This diagnostic approach is mainly followed throughout this chapter. It goes without saying that when frost hardiness models are used as sub-models of large stand and ecosystem models, mortality and other, less drastic effects of frost damage on the subsequent development of the trees need to be addressed explicitly (Kellomäki et al. 1995; Kramer and Hänninen 2009; Chap. 7, Sect. 8.5.2).

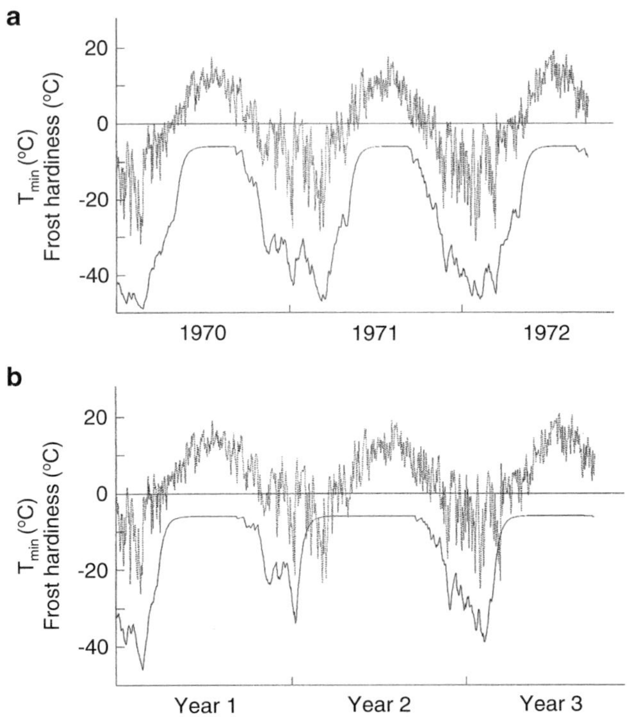

Fig. 8.9 Daily minimum air temperatures (T_{min}, uppermost curves in both panels) and the simulated frost hardiness (smooth curves) of trees in Jyväskylä, central Finland. The frost hardiness was simulated with Kellomäki et al.'s (1992, 1995) model, originally developed for a generalised boreal tree species but later tested with data on needle frost hardiness in *Pinus sylvestris* (Fig. 5.14a in Sect. 5.3.2.2). (**a**) Simulation for the current climate in 1970–1972. (**b**) Simulation for a scenario climate where the daily mean and minimum air temperatures of 1970–1972 were elevated according to the scenario presented in Table 8.1 (Hänninen et al. 1996; redrawn with the permission of the Finnish Forest Research Institute, currently the Natural Resources Institute Finland, and the Finnish Society of Forest Science)

air temperature, then the projected increased incidence of frost damage to boreal trees in the scenario climate (Hänninen 1991; Hänninen et al. 1996) is not caused by the increased incidence of extreme climatic events. This is because the scenario climate used in the simulations was constructed by raising the observed values of the daily mean and the daily minimum temperature for each day by an increment that was constant in any given month (Table 8.1). Thus, in any given month the amount of day-to-day variation in air temperature was exactly the same in the current and the scenario climate.

Therefore the increased risk of frost damage was not caused by increased day-to-day variation in air temperature but by the fact that in the warmest winters of the scenario climate, the air temperatures rose to a level causing simulated dehardening (Fig. 8.9b). This reasoning reveals the essential non-linearity of the frost hardiness hypothesis. According to the hypothesis, fluctuations of air temperature in winter are not risky to the trees as long as the air temperature does not, even in the warmest spells, rise to a level causing dehardening in the trees.

8.3.1.3 The Frost Damage Hypothesis for Temperate Conditions

In the temperate, maritime conditions of Scotland, Cannell and Smith (1986) and Murray et al. (1989) found a complicated pattern of the effects of climatic warming on the timing of the bud burst of trees. They found the pattern by using Cannell and Smith's (1983) alternating model (Fig. 8.10).[4] Warming increases the accumulation

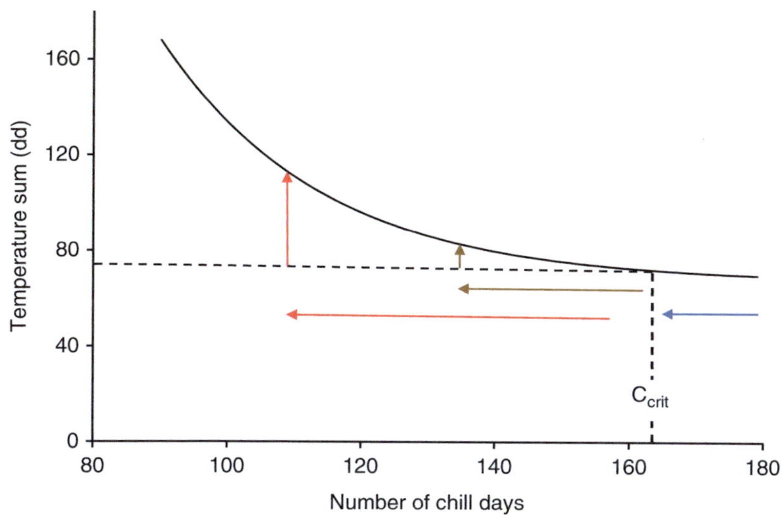

Fig. 8.10 The principle of using Cannell and Smith's (1983) alternating model for projecting the effects of climatic warming on the timing of bud burst in temperate trees (Cannell and Smith 1986; Murray et al. 1989). The model predicts that bud burst occurs when the combination of the accumulated temperature sum and the number of chill days meets the exponentially declining curve. When the timing of bud burst under climatic warming is projected, the increased accumulation of day degrees (i.e., the increased potential rate of ontogenetic development; see Sect. 3.4.1.1) caused by climatic warming is compared with the possible increase in the day degree requirement of bud burst (i.e., the possible decrease in ontogenetic competence, C_o; see Sect. 3.4.1.1) caused by the reduced accumulation of chilling. The vertical dashed line represents the chilling requirement of rest completion, C_{crit}, as indicated by the levelling off of the exponential curve. The *horizontal dashed line* represents the minimum day degree requirement corresponding to the meeting of the chilling requirement. The coloured *horizontal arrows* indicate three hypothetical cases of reduction of chilling caused by climatic warming, and the *vertical arrows* of the same colour indicate the corresponding increase in the day degree requirement caused by insufficient chilling under climatic warming (See text for details)

[4] Cannell and Smith's (1983) model is discussed in Sect. 3.4.6.5.

of day degree units, but because of the decrease in chilling it may also increase the number of day degree units required for bud burst (Morin et al. 2009; Fu et al. 2015a). Thus the direction of change, if any, in the timing of bud burst depends on which of these two changes is greater. This again depends on which part of the exponential curve of the alternating model the decrease of chilling occurs (Cannell and Smith 1986; Murray et al. 1989; Fig. 8.10).

First, if the reduced amount of chilling still meets the chilling requirement C_{crit} indicated by the levelling-off of the exponential curve, then the number of day degrees required for bud burst will not increase and an acceleration of bud burst is projected (Fig. 8.10, blue arrow). This is typically the case with most species in upland sites with relatively cool climates.

Second, if the reduced amount of chilling is slightly smaller than C_{crit}, then the number of day degrees required for bud burst increases slightly. This may compensate for the higher accumulation of day degrees, so that no change in the timing of bud burst is projected (Fig. 8.10, brown arrows). This can be the case in lowland sites with relatively warm climates and species with a relatively low chilling requirement.

Finally, if the reduced amount of chilling drops far below C_{crit}, then the number of day degrees required will increase drastically, and a delay in bud burst is projected (Fig. 8.10, red arrows). This can be the case in lowland sites and species with a relatively high C_{crit}.

As discussed in Sect. 3.4.6.5, the alternating model and the parallel model are different presentations of the same ecophysiological phenomenon of the gradual effect of accumulated chilling on the rest break of buds. In the parallel model the gradual effect is described in terms of an increase in ontogenetic competence, C_o, with increased accumulation of chilling (Fig. 3.16b in Sect. 3.4.1.1). Since the parallel model is in line with the dynamic modelling approach adopted in the present volume, the findings of Murray et al. (1989) are discussed below in terms of the concept of ontogenetic competence, C_o, in order to facilitate comparisons with other sections and chapters. Accordingly, the increased day degree requirement, if any, caused by climatic warming (Murray et al. 1989; Fig. 8.10) is referred to as reduced ontogenetic competence; and the increased accumulation of day degree units caused by climatic warming (Murray et al. 1989) is referred to as increased potential rate of ontogenetic development (Sect. 3.4.1.1).

By using the regrowth test described in Sect. 3.3.2, Murray et al. (1989) determined the parameters of the alternating model for fifteen woody species experimentally. On the basis of similarities and differences among the species in the form of the exponentially declining curve of the alternating model, they were able to divide the 15 species into 5 groups, so that the species in each group had approximately the same form of the curve (Table 8.2). Using the five curves thus obtained, they examined the timing of bud burst by means of computer simulations both in the current climate and in three scenario climates assuming a uniform warming of 1–3 °C. The simulations were carried out for two sites, i.e., a lowland site in Edinburgh and an upland site in Braemar.

Group 1 consisted of *Fagus sylvatica* alone. It had both a great chilling requirement of rest completion, C_{crit}, and a great high temperature requirement of growth onset, H_{crit}, so that it burst bud in the current climate later than any other species

Table 8.2 Comparison of the simulated timing of bud burst in 15 woody plant species in the temperate and maritime conditions of Scotland at a warm lowland site (Edinburgh) and a cool upland site (Braemar), both in the present climate and in a scenario climate assuming a 3 °C uniform warming (Murray et al. 1989). The 15 species were divided into five groups on the basis of their temperature responses of bud burst timing (see text for details)

Group	Species	Edinburgh	Braemar
1	*Fagus sylvatica*	No change	No change C, S
2	*Picea sitchensis* *Robina pseudoacacia* *Tsuga heterophylla*	Acceleration C, S	Acceleration
3	*Betula pendula* *Corylus avellana* *Rubus idaeus* *Sorbus aucuparia*	Acceleration C	Acceleration (S)
4	*Larix decidua* *Prunus avium* *Rosa rugosa* *Salix viminalis* *Sambucus nigra*	Acceleration C	Acceleration S
5	*Crataegus monogyna* *Populus trichocarpa*	Delay C	Acceleration S

The verbal descriptions "Acceleration", "No change", and "Delay" indicate, for each group and location, the direction of change in the bud burst timing caused by warming. The location with the earlier bud burst is indicated with the letters C and S for the current and the scenario climate, respectively, and the letter in parentheses indicates a minimal difference. Table compiled on the basis of the quantitative results presented in Murray et al. (1989)

considered.[5] At the cool site in Braemar it was predicted to burst bud in the current climate slightly earlier than at the warm site in Edinburgh (Table 8.2), which is an

[5] Following the convention introduced in Sect. 3.2.1, the concept of "high temperature requirement of growth onset" is used here for the sake of uniformity even though Murray et al. (1989) specified the visible activation of the apical meristems as bud burst.

illustrative indication of the importance of the high chilling requirement for the spring phenology of trees in the climatic conditions of Scotland (Murray et al. 1989). Due to the high chilling requirement of *Fagus sylvatica*, the warming scenario reduced its ontogenetic competence, and this compensated for the increase in the potential rate of ontogenetic development, so that the projected timing of bud burst remained almost unaltered at both sites in the different warming scenarios (Table 8.2).[6]

With the exception of Group 5 at the warm site in Edinburgh, warming caused a projected acceleration of bud burst in Groups 2–5 at both sites (Table 8.2). This indicated that the increase in the potential rate of ontogenetic development caused by climatic warming overrode the decrease in ontogenetic competence also potentially caused by the warming. In Group 2 the projected acceleration was almost identical at the two sites, whereas in Groups 3 and 4 it was clearly more pronounced at the cooler site in Braemar. For this reason, climatic warming eliminated the difference in the timing of bud burst between the two locations in Group 3, and even reversed the order in Group 4, so that bud burst in the current climate was earlier in Edinburgh but in the scenario climate with 3 °C warming, it was earlier in Braemar (Table 8.2).

Group 5, i.e., *Populus trichocarpa* and *Crataegus monogyna*, displayed a deviating pattern with opposite changes at the two sites. These species have a low C_{crit}, and it is met in the current climate. Therefore these species burst bud relatively early in comparison with most other groups at both sites in the current climate. Furthermore, as the chilling requirement was met at both sites, growth onset in the current climate was predicted to occur about 1 month earlier at the warm site in Edinburgh than at the cool site in Braemar (Murray et al. 1989; results not shown).

However, the chilling requirement of Group 5 in the current climate was exceeded by far at Braemar, whereas it was barely met at Edinburgh (Murray et al. 1989). For this reason, Group 5 reacted to warming in drastically different ways at the two sites (Table 8.2). In the scenario of 3 °C warming, the bud burst at Braemar was projected to be accelerated by about 1.5 months, from the end of April to mid-March, whereas at Edinburgh it was projected to be delayed from late March to early or mid-April.

In the current climate, bud burst was predicted to be earlier in Braemar than in Edinburgh only for *Fagus sylvatica*, but after the warming this was projected to be the case for 12 species, with only the 3 species belonging to Group 2 still being earlier in Edinburgh (Table 8.2). This was despite the fact that an acceleration of bud burst as a result of warming was projected for most species-location combinations. Once again, this is an illustrative indication of the importance of chilling requirement in the maritime conditions of Scotland (Murray et al. 1989).

The insightful results of Murray et al. (1989) demonstrate the efficiency of ecophysiological dynamic modelling in assessing the effects of the projected climate change. The model applied is relatively simple, addressing the variation

[6] Murray et al. (1989) did not consider the night-length requirement of rest completion evidently existing in *Fagus sylvatica*, so that the high chilling requirement they found for this species may partially reflect the delaying effects of long nights on rest break. See Sect. 3.3.3.

among species in terms of three ecophysiological traits, i.e., the chilling requirement of rest completion, the high temperature requirement of growth onset, and the effects of the state of rest on ontogenetic competence. Despite this simplicity, the model projects drastically different implications of the climate change for the timing of bud burst in different species growing at different sites in Scotland (Table 8.2). Recently, Roberts et al. (2015) made a similar projection that depending on the tree species, climatic warming will either advance or delay the spring phenology in Central England.

Murray et al. (1989) addressed the risk of frost damage simply by considering the mean daily air temperature, T_{mean}, prevailing on the day of bud burst. Though this method does not address the risk of frost damage explicitly, it is possible to examine the risk with this indirect method, too, because of the temporal autocorrelation of air temperature. Murray et al. (1989) projected that with each species growing at both sites, the risk of frost damage decreases, i.e., the T_{mean} of the bud burst day rises because of climatic warming (results not shown). This was self-evident in cases where climatic warming either did not change the timing of bud burst (Group 1) or delayed it (Group 5 at Edinburgh). However, Murray et al. (1989) found that the risk of frost damage also decreased in those species whose bud burst was accelerated as a result of climatic warming. This indicates that in the temperate and maritime conditions of Scotland, the overall decrease caused by climatic warming in the freezing stress overrides the potential increase of frost damage risk generally associated with an earlier bud burst.

Kramer (1994) adopted the phenological approach to study the effects of climatic warming on the timing of leaf unfolding and the related risk of frost damage to trees in Central European conditions. He used long-term phenological and air temperature data to parameterize both the sequential and the alternating model for several tree species in Dutch and north-western German conditions.[7] The models were then applied to scenario studies addressing both uniform and non-uniform warming. For the uniform warming studies, a total of 11 scenarios were used, with the assumed change in air temperature ranging from a cooling by two degrees (−2 °C) to a warming by eight degrees (+8 °C) in one-degree steps. To take the sequential model and *Fagus sylvatica* as examples, climatic warming was projected to accelerate leaf unfolding slightly (Fig. 8.11a). Despite this acceleration, however, the risk of frost damage, defined in Kramer's (1994) study as the probability of occurrence of sub-zero temperatures during a period of 11 days around the simulated time of leaf unfolding, was projected to decrease on account of climatic warming (Fig. 8.11b).

Kramer (1994) found that in the air temperature range of two-degree cooling to four-degree warming, the projected risk of frost damage depended approximately linearly on the change in temperature assumed in the simulations (Fig. 8.11b). Thus the rate of change of the frost damage risk per one degree of warming was determined as the slope of a straight line fitted to the data (Fig. 8.11b). The slope

[7] These models are discussed in Sects. 3.4.1 and 3.4.6.5 and the phenological approach in Sect. 3.4.6.1.

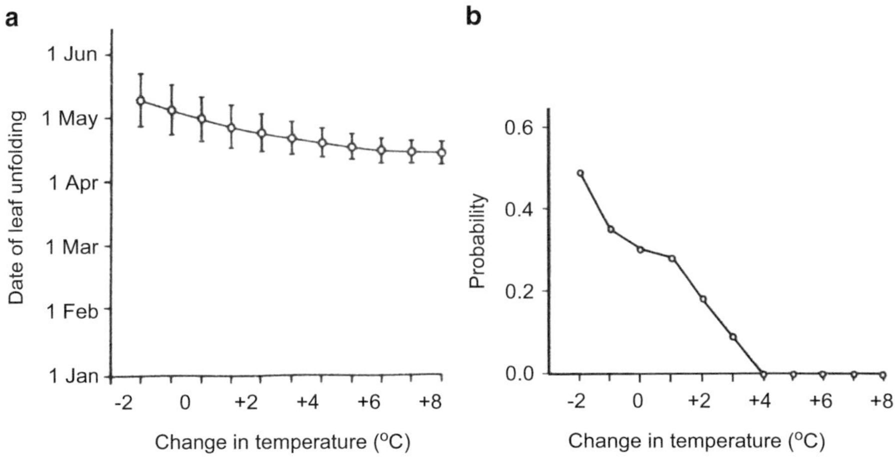

Fig. 8.11 Simulated effects of the climate change on the springtime development of *Fagus sylvatica* trees over 57 years in the Netherlands. As indicated on the horizontal axes, 11 climatic scenarios, ranging from cooling by two degrees Celsius (−2 °C) to warming by eight degrees Celsius (+8 °C), were used in the simulations. The effects of the climate change (**a**) on the date of leaf unfolding and (**b**) on the associated risk of spring frost damage as defined by the probability of sub-zero temperatures occurring during an 11-day time interval around the simulated leaf unfolding (Modified from Kramer 1994; published with the permission of John Wiley & Sons, Inc.)

was negative with all the species examined, indicating a decrease in the risk of frost damage with warming. The rate of the decrease per one degree of warming varied from 0.04 to 0.10 among the species studied (Table 8.3).

Kramer (1994) also carried out hypothetical provenance transfers by applying to Dutch conditions both the sequential model used by Hänninen (1991) for a generalised boreal tree species (Fig. 8.7) and the alternating model parameterised by Murray et al. (1989) separately for the five groups of species growing in the UK (Fig. 8.10; Table 8.2). In all simulations with the trees transferred hypothetically to the South, a decrease in the risk of frost damage was found. The rates of decrease in the frost damage risk per one degree of warming in the hypothetical transfers were similar to the ones obtained in the scenario simulations addressing the effects of climatic warming on species growing in the Netherlands and north-western Germany (Table 8.3).

The findings with the hypothetical transfers provide further support for the conclusion that in the temperate conditions of north-western Germany and the Netherlands, climatic warming causes a decrease in the incidence of spring frost damage to trees (Kramer 1994). Unfortunately, however, these considerations do not solve the methodological shortcoming inherent in the phenological approach adopted (discussed in Sect. 3.4.6.1). Nevertheless, the reliability of Kramer's (1994) conclusion is improved by the fact that it agrees with the conclusion reached by Murray et al. (1989) by using models parameterized with experimental data. In both of these studies, the overall decrease in freezing stress caused by the warming

Table 8.3 Simulated effects of the climate change on the risk of spring frost damage in tree species growing in north-western Germany (D) or in the Netherlands (NL)

Species	ΔP (°C^{-1})
Larix decidua (D)	−0.10
Betula pubescens (NL)	−0.08
Betula pubescens (D)	−0.09
Tilia platyphylla (D)	−0.06
Fagus sylvatica (NL)	−0.08
Fagus sylvatica (D)	−0.08
Tilia cordata (D)	−0.05
Quercus rubra (NL)	−0.09
Quercus robur (D)	−0.06
Quercus robur (NL)	−0.06
Fraxinus excelsior (D)	−0.05
Quercus petrea (NL)	−0.05
Picea abies (D)	−0.04
Pinus sylvestris (D)	−0.04

The frost damage risk was defined as the probability of occurrence of sub-zero air temperatures during an 11-day time interval around the simulated leaf unfolding. Several climatic scenarios were addressed in the simulations, and the negative values of ΔP indicate the rate of decrease of the frost damage risk per one degree of warming (Modified from Kramer 1994; published with the with the permission of John Wiley and Sons)

overrode the potential increase in frost damage generally associated with an earlier bud burst.

Murray et al. (1994) studied the effects of the climate change on the timing of bud burst and the related risk of spring frost damage in *Picea sitchensis* by means of two versions of the alternating model. First, the version previously parameterized by Murray et al. (1989) on the basis of their experimental data was used to address climatic warming. Second, Murray et al. (1994) re-parameterized the alternating model for the same species in order to address their experimental findings of the delaying effects of elevated concentrations of atmospheric CO_2 on the timing of bud burst. This new version of the model was used in simulations for assessing the effects of elevated concentrations of atmospheric CO_2, either with or without climatic warming, on the timing of bud burst in *Picea sitchensis*. In addition to the two sites in Scotland addressed in Murray et al.'s (1989) earlier study (Table 8.2), the simulations were also carried out for Masset, a coastal site in Queen Charlotte Islands belonging to the natural geographical range of *Picea sitchensis* in western North America.

In Murray et al.'s (1994) scenario simulations, climatic warming accelerated the timing of bud burst whereas elevated concentrations of atmospheric CO_2 delayed it (Fig. 8.12). There were, however, interesting differences among the three sites in the joint effect of climatic warming and elevated concentrations of atmospheric CO_2 in comparison with the timing of bud burst predicted for the current conditions (Murray et al. 1994). At the Scottish lowland site in Edinburgh, a two-degree

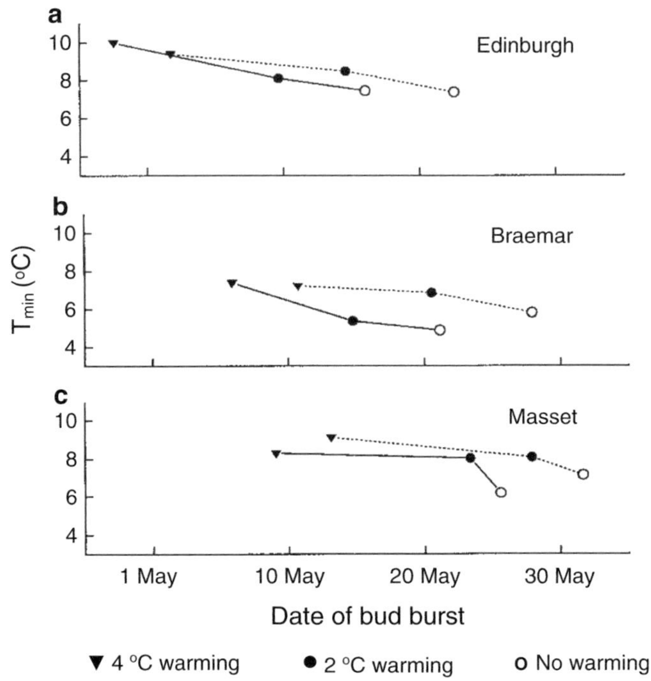

Fig. 8.12 Simulated effects of climatic warming and elevated concentrations of atmospheric carbon dioxide on the date of bud burst (*horizontal axis*) and the associated risk of spring frost damage to *Picea sitchensis*. The risk is indicated by the minimum air temperature on the date of bud burst (T_{min}, *vertical axis*). For both the date of bud burst and the T_{min}, the mean value obtained in the simulations carried out for the years 1897–1978 is indicated. The simulations were carried out for current (*continuous curves*) and elevated (*dotted curves*) concentrations of atmospheric carbon dioxide and for both the current climate and two warming scenarios, as indicated in the figure. (**a**) Simulations for Edinburgh, a lowland site in Scotland. (**b**) Simulations for Braemar, an upland site in Scotland. (**c**) Simulations for Masset, a coastal native site on the Queen Charlotte Islands in western North America (Modified from Murray et al. 1994; published with the permission of Oxford University Press)

warming overcompensated for the delaying effect of elevated CO_2, so that a slight acceleration of bud burst was projected for these conditions (Fig. 8.12a). At the Scottish upland site in Braemar, the overcompensation was so slight that the timing of bud burst was projected to remain unaltered in the case of two-degree warming (Fig. 8.12b). Finally, at the native site in Masset, the two-degree warming undercompensated for the delaying effect of elevated CO_2, so that bud burst was delayed despite the two-degree warming (Fig. 8.12c). However, similarly to the other two sites (Figs. 8.12a, b), a four-degree warming together with elevated concentrations of atmospheric CO_2 accelerated bud burst considerably at Masset, too (Fig. 8.12c). In all cases, the minimum air temperature on the day of bud burst rose as a result of warming or elevated CO_2, or both, indicating a decreased risk of spring frost damage (Murray et al. 1994, Fig. 8.12).

The results of Murray et al. (1994) show that projections made on the basis of climatic warming only may change drastically when the other major environmental change, i.e., the elevation of atmospheric CO_2 concentration, is also addressed. However, the evidence for the effects of elevated CO_2 on growth onset in boreal and temperate trees is contradictory (Sect. 3.4.3), so that further experimental studies addressing these potential effects are called for.

In conclusion, the scenario simulations of Murray et al. (1989, 1994) and Kramer (1994) suggest that under temperate conditions, climatic warming would decrease the incidence of tree frost damage in most cases. This is because climatic warming did not accelerate growth onset in all cases, and even when it did, the overall decrease in freezing stress, also caused by climatic warming, overrode the increase in potential frost risk generally related to an earlier growth onset. These projections for temperate conditions deviate drastically from the projections of an increased risk of frost damage found by Hänninen (1991) and Hänninen et al. (1996) for the more continental boreal conditions of Finland (Figs. 8.5, 8.6, 8.7, 8.8, and 8.9).

However, Cannell's (1985) frost damage hypothesis cannot be ruled out for all tree species in the temperate conditions, either. The hypothesis received some support already in the scenario simulations carried out by Cannell and Smith (1986) for the British conditions. In addition to the premature dehardening and growth onset addressed originally in the frost damage hypothesis, the incidence of frost damage may be increased due to climate change also during other parts of the annual cycle. This more general frost damage hypothesis has found support in simulation studies with *Eucalyptus pauciflora* growing in sub-alpine temperate regions in Australia (Woldendorp et al. 2008).

8.3.2 Tests of the Frost Damage Hypothesis

8.3.2.1 Provenance Transfer Trials

Existing long-term provenance trials provide an inexpensive and efficient way to test the frost damage hypothesis (Cannell 1985; Hänninen 1991) without any further experimentation. Beuker (1994) found that northern Finnish *Picea abies* trees not only survived long transfers to central and southern Finland but showed even higher productivity at the new sites than at their native northern sites. The transfer to central Finland roughly corresponds to the warming projected for the northern site as a result of a doubled concentration of atmospheric CO_2, and the warming associated with the transfer to southern Finland clearly exceeds the projected warming. Apparently, then, the vigorous growth of the *Picea abies* trees over several decades at the central and especially at the southern growing site falsifies the frost damage hypothesis of Cannell (1985) and Hänninen (1991).

Taking this reasoning as his starting point, Hänninen (1996) examined the evidence provided by the provenance trial experiment more specifically with reference to the frost damage hypothesis. Using calculations basically similar to those used in deducing the hypothesis for boreal conditions (Hänninen 1991; Figs. 8.5, 8.6, 8.7, and 8.8), he compared the implications of the provenance transfers with those of both the uniform and non-uniform warming projected for the original northern site. Accordingly, he simulated the timing of growth onset by means of a modified version of Sarvas's (1972, 1974) model, i.e., a sequential model with the air temperature responses formulated by Hänninen (1990b).[8] The simulations were carried out for 80 theoretical tree genotypes, i.e., different combinations of the chilling requirement of rest completion, C_{crit}, and the high temperature requirement of growth onset, H_{crit}.

The simulations were carried out with air temperature data representing five climates as input. First, the present climate at the three locations was considered: the northern Finnish native site (mean annual temperature −1.6 °C), the central Finnish site (+2.7 °C), and the southern Finnish site (+5.5 °C). Furthermore, for the northern Finnish native site, both a uniform and a non-uniform warming scenario of 4.7 °C was considered, the mean annual air temperatures being +3.1 °C in both. As regards the mean annual temperature, the transfer to central Finland roughly corresponded to the scenario climates at the northern native site; whereas the warming included in the transfer to southern Finland clearly exceeded the warming projected for the native northern site (Fig. 8.13).

For the survival of a given tree genotype in the given climate, it was required that the minimum post-dormant air temperature, T_{PD}, i.e., the minimum air temperature occurring between the growth onset and 1 August, did not drop below −8 °C during any of the simulated years.[9] This facilitated the drawing of a distinction between surviving and non-surviving genotypes for all the five climates considered and thus also the testing of the frost damage hypothesis.

According to the frost damage hypothesis, the mortality rate should be high in the case of non-uniform warming, which was assumed in the deduction of the hypothesis (Sect. 8.3.1.2). As expected, this model prediction was realised in the calculations, for the entire current *Picea abies* population was killed in the simulation corresponding to the non-uniform warming (Fig. 8.13).

The vigorous growth of the transferred trees that had been observed in the actual trial required that little if any mortality be predicted for the transferred trees. In broad terms, this requirement was met in the calculations for the transfer to central Finland, for only two of the 15 genotypes were predicted to die in the transfer (Fig. 8.13). Furthermore, the values of both C_{crit} and H_{crit} for the killed genotypes were close to their assumed minima in the population (genotype-specific results not shown here). Therefore, assuming that the values of both these traits were about normally distributed in the population, the percentage of killed individuals was

[8] The three sub-models forming the overall model are described in detail in Sects. 3.2.1.4, 3.3.2.2, and 3.4.1.1.

[9] In order to check the consistency of the results, a sensitivity analysis was carried out by using, in addition to the −8 °C, five alternative temperatures in the range of −5 to −10 °C as the threshold for killing frost. Use of the alternative threshold temperatures did not change the conclusions of the study.

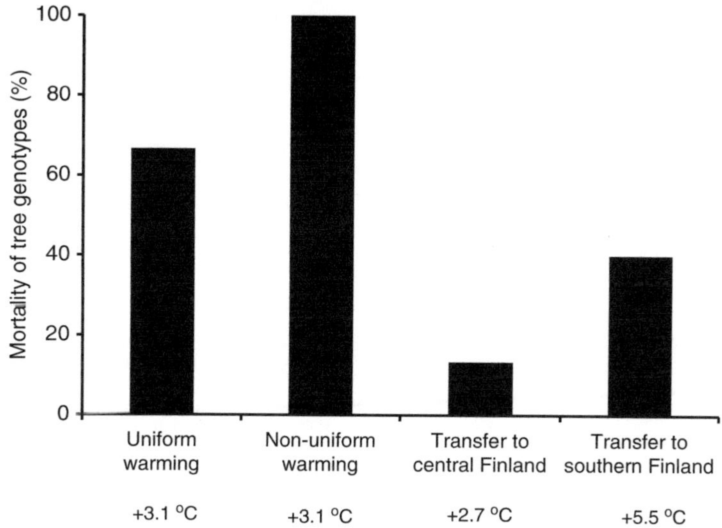

Fig. 8.13 Simulated mortality of northern Finnish *Picea abies* genotypes for two warming scenarios at the native northern site and for their transfers to central and southern Finland (Hänninen 1996). The numbers at the *bottom* indicate the mean annual temperature for the four climates considered. The mean annual temperature at the native northern site is −1.6 °C. The mortality percentages were calculated on the basis of the genotype-specific results presented in Hänninen (1996)

much less than the 13.3 % calculated for the mortality rate of the different genotypes (Figs. 8.13). In all, then, the results for the transfer to central Finland did not falsify the frost damage hypothesis.

For the transfer to southern Finland, however, much higher mortality was predicted. Now 40 % of the genotypes were predicted to die (Fig. 8.13), and even though the percentage of killed individuals was probably smaller because of the assumed normal distributions of C_{crit} and H_{crit}, the projected percentage of killed individuals was so great in any case that in order to support the frost damage hypothesis, quite high mortality should have been seen in the transferred tree population. However, no great mortality has been documented for the transfer to southern Finland, so that the results of the trial were contradictory to the hypothesis. Thus, though the results are less than conclusive because of the general methodological problems involved in using provenance transfer trials as model systems in climate change research (see Sect. 8.2.2.3), they call for further critical tests of the frost damage hypothesis (Hänninen 1996).

Methodologically, Hänninen's (1996) results show that the mean annual temperature is not a sufficient index for testing ecophysiological hypotheses such as the frost damage hypothesis (Cannell 1985; Hänninen 1991) with climatic data from provenance transfer trials. According to a comparison of the mean annual temperatures, the transfer to central Finland almost corresponded to the warming hypothesised to cause dramatic frost damage. However, practically no frost damage was implied for the transfer by the model predicting catastrophic frost damage for non-uniform warming (Fig. 8.13). Furthermore, though the same rise of mean annual air temperature was assumed in both uniform and non-uniform warming, the projected damage in the former was not as complete as it was in the latter (Fig. 8.13), where the warming was especially pronounced in winter.

These comparisons show that the rise of the mean annual temperature realised in provenance transfers is not relevant for the phenomena concerned in the frost damage hypothesis. This is because only intermittent warming during mild spells in winter is relevant for the hypothesis, whereas rising air temperatures during the growing season have no relevance for it at all (however, see the discussion on quantitative dormancy induction in Sect. 3.4.5). This reasoning shows that when provenance transfer experiments are used as model systems for the effects of the projected climate change, the suitability of the trials for this purpose should be addressed with calculations specific to the ecophysiological hypothesis in question, not with simple climatic indices such as the mean annual air temperature (Hänninen 1996).

8.3.2.2 Whole-Tree Chamber Experiments

Hänninen (1995a) tested the frost damage hypothesis in a whole-tree chamber (WTC) experiment with *Pinus sylvestris* saplings in eastern Finland over two years. A factorial 2×2 design with two levels, ambient and elevated, for both the air temperature and the concentration of atmospheric CO_2 was used. Furthermore, two additional treatments were used where the air temperature was elevated but the concentration of carbon dioxide was not.

In accordance with the hypothetico-deductive research strategy introduced in Sect. 2.1, the elevated air temperature treatments were designed on the basis of the simulations whereby the frost damage hypothesis was deduced for boreal conditions (Hänninen 1991; Sect. 8.3.1.2). In those simulations, only the warmest winters of the assumed scenario climate caused the projected premature growth onset leading to catastrophic frost damage, so that in order to test the hypothesis, the air temperature treatments had to correspond to the warmest winters of the scenario simulations. Had the standard method of WTC experiments been used here, too, with the elevated temperature treatment set as equal to the ambient air temperature plus the scenario increment, then with an exceptionally cold winter occurring at the experimental site, the air temperatures in the elevated air temperature treatment would have corresponded to an approximately average winter in the scenario climate, and no test of the frost damage hypothesis would actually have been carried out.[10]

[10] The contribution of Heikki Smolander is acknowledged here. He made this important point though he did not otherwise participate in the study.

With the above reasoning in mind, the elevated air temperatures were kept above zero most of the time in winter, too. The ambient temperature was usually well below zero, with the daily minimum air temperatures approaching −40 °C in some cases (Hänninen et al. 1993). Contrary to the air temperature pattern in the scenario climate, however, heavy frosts were generally not allowed in the elevated temperature treatments. This was because the experimental saplings were needed for other studies during the growing season; the intention was therefore to keep them alive even if they dehardened prematurely as predicted by the frost damage hypothesis. In occasional periods of exceptionally heavy frost, however, the control of the air temperature in the chambers was not sufficient to prevent air temperatures below −10 °C in the elevated temperature treatments.

The sequential model that was used to deduce the frost damage hypothesis for boreal conditions predicted that depending on the degree of temperature elevation, the *Pinus sylvestris* saplings should commence height growth sometime in January in the elevated temperature treatments. Should this happen in the experiment, then the results would support the frost damage hypothesis even if no frost damage occurred, for the saplings were generally not subjected to damaging or killing frosts in the chambers with elevated temperatures.

No clear effect of elevated CO_2 concentration on the timing of height growth onset in the *Pinus sylvestris* saplings was observed in the WTC experiment (Hänninen 1995a), whereas the elevation of the air temperature had a clear effect: height growth onset took place between 11 and 14 May in the treatments with ambient temperatures and between 25 March and 19 April in the treatments with elevated temperatures (Fig. 8.14a). However, the elevations of air temperature accelerated the height growth onset much less than predicted by the sequential model, the difference between the observation and the model prediction being about 70 days on average (Fig. 8.14a). Working at the same site with new types of WTCs, Kilpeläinen et al. (2006) later confirmed these findings for *Pinus sylvestris*.

The findings of Hänninen (1995a) and Kilpeläinen et al. (2006) show that despite its relatively strong empirical support (Sects. 3.2.1.3 and 3.3.2.1), the model applied to the deducing of the frost damage hypothesis (Hänninen 1991) does not address all essential aspects of the environmental regulation of height growth onset in *Pinus sylvestris* saplings. These results of the WTC experiment also suggest that climatic warming causes much less, if any, frost damage than was predicted in Hänninen's (1991) simulation study for boreal conditions (Figs. 8.6 and 8.8).

By combining the various sub-models for the three ecophysiological aspects concerned (Sect. 3.4.1.1), Hänninen (1995a) also tested 95 other models besides the model applied to the deducing of the frost damage hypothesis. Some of the models were able to predict the timing of height growth onset in the elevated temperature treatments, too (Fig. 8.14b). However, the results did not enable the validation of any of the models, since models including different ecophysiological assumptions implied approximately the same degree of accuracy.

What is more, the night length model, which was found to be the most accurate in the WTC test (Fig. 8.14b), was actually falsified in a separate experiment, where

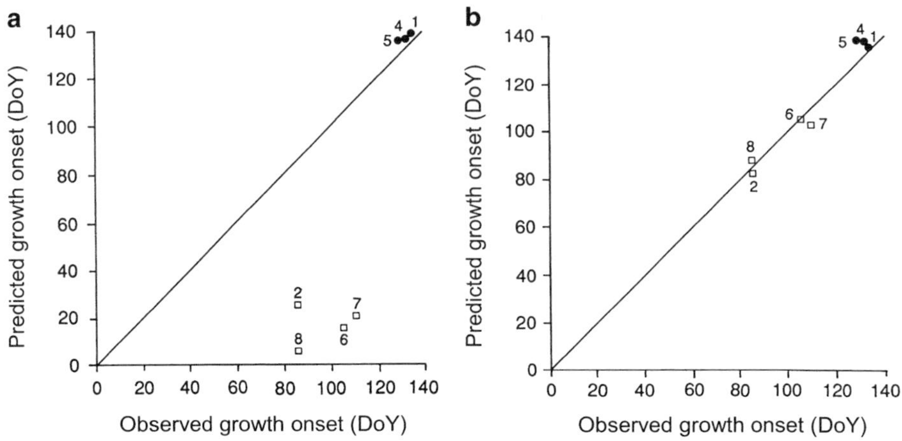

Fig. 8.14 Predicted and observed dates of height growth onset in a whole-tree chamber experiment designed for testing the frost damage hypothesis with *Pinus sylvestris* saplings. The *black circles* and the *white squares* indicate results obtained with ambient and elevated air temperatures, respectively. DoY denotes Day of Year. (**a**) The prediction of the sequential model that was used to deduce the frost damage hypothesis for boreal conditions (Sect. 8.3.1.2). (**b**) The prediction of the most accurate model among the 96 models tested. In this model, rest break was assumed to be regulated by night length (Modified from Hänninen 1995a; published with the permission of NRC Research Press)

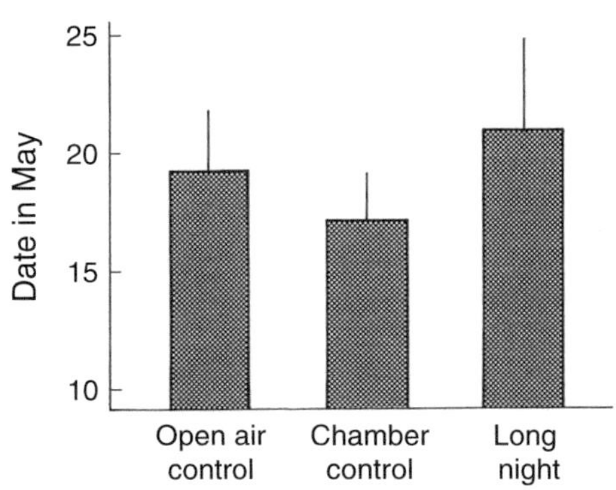

Fig. 8.15 Timing of growth onset in *Pinus sylvestris* saplings in a field experiment designed for testing the night length model of rest completion. The saplings were growing at their natural growing site, either in natural conditions (*Open-air control*) or covered with a transparent (*Chamber control*) or opaque chamber (*Long night*) for 19 h daily. The 19-h night length was selected for the experiment as it is approximately the natural night length prevailing on winter solstice at the growing site (Modified from Hänninen 1995a; published with the permission of NRC Research Press)

the night length was manipulated experimentally (Fig. 8.15). As discussed in Sect. 3.4.6.1, findings of this kind call for an ecophysiological approach, i.e., an approach where the ecophysiological aspects concerned are addressed one at a time with experiments specifically designed for testing each aspect. The WTC experiment simulating future natural conditions, though an experimental approach as such, does not fulfil the requirements of the ecophysiological approach defined in the above way.

Repo et al. (1996), using the same WTC experiment to examine the needle frost hardiness of *Pinus sylvestris* saplings, found no clear effect of elevated concentration of atmospheric CO_2 on the annual cycle of needle frost hardiness. The elevation of air temperature delayed the autumn hardening by 19 days and accelerated the spring dehardening by 29–46 days. No clear difference in hardiness could be seen between the treatments with ambient and elevated air temperatures in winter (Repo et al. 1996). This finding was contrasted by later findings with *Pseudotsuga menziesii* (Guak et al. 1998) and *Betula pubescens* ssp. *czerepanovii* (Taulavuori et al. 2004). In both of these studies the seedlings in the elevated temperature treatment were less hardy than the seedlings in the ambient temperature treatment in winter, too.

Hänninen et al. (1996) used the frost hardiness data collected by Repo et al. (1996) to test Kellomäki et al.'s (1992, 1995) frost hardiness model, which projects dramatic frost damage as a result of climatic warming (Fig. 8.9b). According to the model's prediction, elevated autumn, winter and spring temperatures would cause a complete failure in the annual cycle of frost hardiness of saplings. In autumn 1993, for instance, the saplings were predicted to harden only to −20 °C, approximately, and to subsequently deharden to the low hardiness typical for the active growth phase (Fig. 8.16). In reality, however, the experimental saplings displayed quite a normal annual cycle of needle frost hardiness

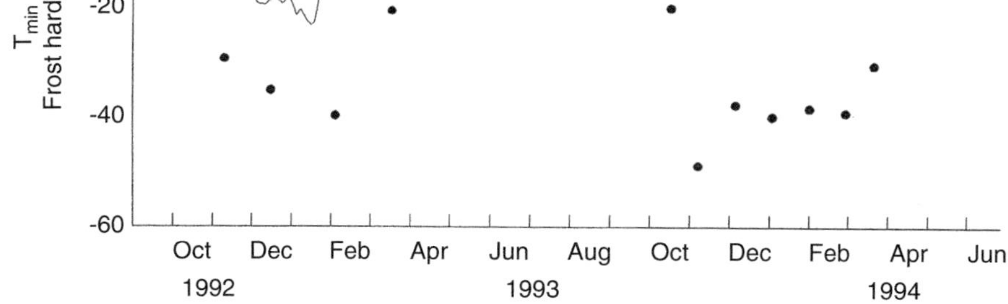

Fig. 8.16 The predicted (*smooth curve*) and measured (*black circles*) needle frost hardiness of *Pinus sylvestris* saplings under elevated air temperatures in a whole-tree chamber experiment designed for testing the frost damage hypothesis. The *uppermost curve* represents the daily minimum air temperature, T_{min} (Hänninen et al. 1996; Repo et al. 1996; redrawn with the permission of the Finnish Forest Research Institute, currently the Natural Resources Institute Finland, and the Finnish Society of Forest Science)

(Fig. 8.16), though the hardening was delayed and the dehardening accelerated (Repo et al. 1996), as discussed in the previous paragraph.

This test of the frost damage hypothesis, carried out directly with experimental frost hardiness data (Hänninen et al. 1996), supported the conclusion drawn earlier in Hänninen's (1995a) study, where the frost hardiness hypothesis was tested indirectly with experimental data on height growth onset (Fig. 8.14a). The results of both these studies suggest that climatic warming will not cause such catastrophic frost damage to boreal trees as was projected in the simulation studies discussed in Sect. 8.3.1.2.

The failure of Kellomäki et al.'s (1992, 1995) model to predict the needle frost hardiness of the *Pinus sylvestris* saplings in the elevated temperature conditions (Fig. 8.16) is in striking contrast with the accuracy the model displayed in predicting the frost hardiness of the saplings growing in natural conditions (Fig. 5.14a in Sect. 5.3.2.2). Therefore, though the test in the elevated air temperature conditions clearly falsifies the model as such, the falsification should not be considered to imply a rejection of the model from any further use in research. Rather, considering not only the predictive power of the model in natural conditions but also the large amount of empirical data supporting the various sub-models included in the overall model, discussed in Chaps. 3 and 5, one must conclude that the model accounts for several aspects of the environmental regulation of frost hardiness in boreal trees but that something essential is still either missing or misrepresented in the model. To use the terminology introduced in Sect. 2.4, Kellomäki et al.'s (1992, 1995) frost hardiness model has a degree of realism, then, but the realism is reduced by its insufficient coverage. That is why the model gives false projections for the effects of climate change on the frost hardiness of and frost damage to boreal coniferous trees.

Furthermore, it is crucial to notice that the insufficient coverage was revealed only in a test explicitly addressing the elevated temperature conditions (Fig. 8.16), not in tests carried out in current natural conditions (Fig. 5.14a in Sect. 5.3.2.2). This finding is generally important for studies addressing the ecological effects of climate change rather than being restricted to studies addressing the frost damage hypothesis of boreal and temperate trees. The models used for projecting these effects have often been tested only in current climatic conditions, so that their trustworthiness in projecting the ecological effects of the climate change remains questionable (Hänninen 1995b).

In 2002–2004, Slaney et al. (2007) examined the effects of elevated air temperatures and elevated concentrations of atmospheric carbon dioxide on the bud burst of 40-year-old *Picea abies* trees growing in WTCs (Fig. 8.3 in Sect. 8.2.2.4). A factorial 2×2 design was used in this study, too. The elevated air temperature treatments were constructed in the standard way by setting the elevated air temperature to equal the ambient air temperature plus the scenario increment. For the increment, a non-uniform scenario was used, so that it varied from 2.8 °C in July to 5.6 °C in December. Accordingly, the elevated temperature treatments were less extreme than in the earlier WTC experiment with *Pinus sylvestris* (Fig. 8.16). However, as the ambient air temperature was well above zero during one period of about 2 weeks in January 2002, the elevated air temperatures were consequently near 10 °C during that period. As a considerable quantity of high temperature units was meteorologically accumulated during that period, the testing of the ecophysiological models of bud burst was facilitated.

In Slaney et al.'s (2007) study, elevated concentrations of carbon dioxide had no effect on the timing of growth onset in spring, whereas elevated air temperatures accelerated it by 2–3 weeks. Using these experimental results, Hänninen et al. (2007) tested models of rest break and ontogenetic development towards growth onset. For rest break, three models were addressed. First, the chilling model presented by Sarvas (1974) (Fig. 3.13a in Sect. 3.3.2.1) and second, the corresponding model presented by Landsberg (1974), where frost temperatures, too, contribute to rest break similarly to the most efficient chilling temperatures (Fig. 3.13b in Sect. 3.3.2.1). In the third model, rest completion was assumed to take place on a fixed calendar day (Häkkinen et al. 1998). In Hänninen et al.'s (2007) test, the calendar model was found to be more accurate than either of the chilling models considered. Furthermore, the residual error of the model was minimised by assuming rest completion to take place relatively late in spring, i.e., approximately in mid-March (Fig. 8.17).

The findings of WTC experiments suggest that for both *Pinus sylvestris* (Figs. 8.14a and 8.16) and *Picea abies* (Fig. 8.17), the meteorological accumulation of high temperature units in long-lasting elevated temperature conditions over the winter months is not physiologically effective, i.e., does not cause dehardening and ontogenetic development towards growth onset, until relatively late in spring. This notion is in accordance with the alternative conception discussed in Sect. 3.4.4.1, stating that rest completion, and thereby the attainment of ontogenetic competence, takes place relative late in spring. If this is the case for real trees also, then climatic warming will cause less, if any, frost damage to boreal trees than was projected by

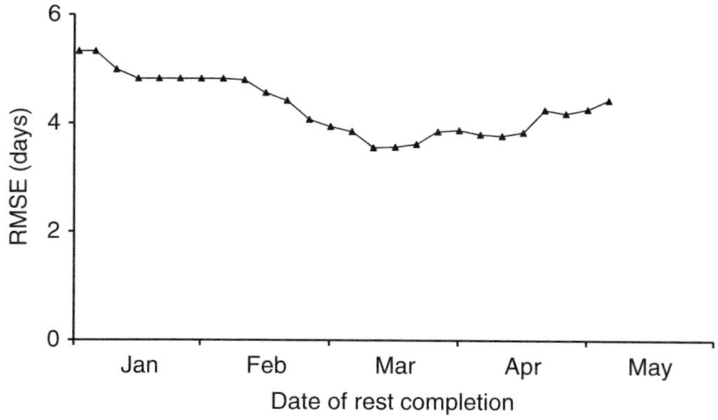

Fig. 8.17 The root mean squared error (RMSE) of the calendar model fitted to bud burst data on 40-year-old *Picea abies* trees growing in whole-tree chambers under both ambient and elevated air temperatures (for details of the data, see Slaney et al. 2007). In the model, rest completion was assumed to take place on a fixed calendar day, indicated on the *horizontal axis*, each year (Modified from Hänninen et al. 2007; published with the permission of Oxford University Press)

the models based on the traditional conception of rest completion taking place relatively early in autumn (Hänninen 1991; Kellomäki et al. 1992, 1995).

Similarly to the results obtained with boreal conifers discussed above (Hänninen 1995a; Kilpeläinen et al. 2006; Hänninen et al. 2007), Fu et al. (2012, 2013) more recently found a moderate acceleration of bud burst in warming experiments carried out with three deciduous tree species in temperate conditions. Together with the simulation studies discussed in Sect. 8.3.1.3, Fu et al.'s (2012, 2013) findings suggest a low risk of frost damage in temperate conditions as a result of climatic warming.

8.3.2.3 Findings from Observational Studies

Results from the whole-tree chamber experiments discussed in Sect. 8.3.2.2 suggest that climate warming will probably not cause such catastrophic frost damage as was projected in the early simulation studies discussed in Sect. 8.3.1.2. Even so, the hypothesis of the increased incidence of frost damage caused by the climate change cannot be ruled out for all boreal and temperate tree species. Available experimental evidence remains limited (see also Wolkovich et al. 2012), and further uncertainties are revealed by the simulation studies discussed in Sect. 8.3.3 below.

The most concrete evidence for the frost damage hypothesis is provided by observational field studies. It is now well documented that the springtime phenological development of boreal and temperate trees has advanced during the last decades (Myneni et al. 1997; Menzel and Fabian 1999; Menzel et al. 2006; Linkosalo et al. 2009; see, however, Fu et al. 2015b). In several cases, tree frost damage has been documented in natural conditions after unseasonally warm periods in spring. Such a phenomenon was documented as early as 1989 by van der Kamp and Worrall (1990) for several coniferous species growing in interior British Columbia. More recently, large-scale frost damage in several tree species growing in eastern United States was documented after the 2007 spring freeze by Gu et al. (2008). Exceptional frost damage occurring after unseasonally warm periods has been documented not only in trees but also in the dwarf shrub *Empetrum hermaphroditum* growing in sub-arctic conditions in northern Scandinavia (Bokhorst et al. 2009).

Recently, Augspurger (2013) examined spring frost damage in several tree species growing in Illinois. By combining observations of frost damage from 1993 to 2012 with meteorological observations from the same period, she identified the meteorological conditions that coincided with the damage. Using this information in an examination of a long-term (1889–1992) temperature record, she found that the frequency of the temperature patterns that coincided with the frost damage had increased during 1889–1992. In all, then, she concluded that the risk of spring frost damage in the trees is increasing.

The studies reviewed above show that the phenomenon addressed in the frost damage hypothesis may in fact already happen in the present climatic conditions. For this reason the possibility of large-scale frost damage to trees in the future cannot be ruled out when the trees are projected to be exposed to more frequent unseasonally warm periods alternating with freezing temperatures.

In addition to boreal and temperate trees, addressed in the present volume, other plant life forms at high latitudes may also run an increased risk of frost damage as a result of climatic warming. In addition to frost stress, climatic warming will also increase the incidence of other wintertime stresses to the plants, such as anoxia, ice encasement, and infections of various pathogens (Cooper 2014; Rapacz et al. 2014). In all, then, warming winters cause a major uncertainty to the functioning of high latitude ecosystems in the future (Kreyling 2010).

8.3.3 Ecophysiological Traits Critical for the Frost Damage Hypothesis

8.3.3.1 Model Comparisons to Identify the Critical Traits

In this section, the two integrated models of frost hardiness introduced in Sect. 5.3.2 are used in scenario simulations in order to identify the ecophysiological traits that are critical for the hypothesis that climatic warming causes increased incidence of frost damage in boreal and temperate trees. These two models will be referred to as $Model_1$ (Kellomäki et al. 1992, 1995) and $Model_2$ (Leinonen 1996). The section includes both reviews of published studies and reports on the results of new simulations. In one case, new simulations were carried out in order to elucidate one critical factor identified earlier by modelling the timing of bud burst (Linkosalo et al. 2000).

Both the original versions and modified versions of both integrated models are used. In each case, the principle of identification is based on the hypothetico-deductive research strategy outlined in Sect. 2.1, so that the projections obtained with the original model are compared with those obtained with the modified model. If one model projects and the other does not project increased incidence of frost damage as a result of climatic warming, then the ecophysiological trait in which the two models differ is a critical one (Hänninen 2006). The designs of the comparisons are presented in Table 8.4.

All of the new simulations were carried out with the same long-term air temperature data of 92 years from Jyväskylä, central Finland, as has been used in several of the earlier chapters. In each case, the daily mean and minimum air temperatures were raised according to the scenario (Table 8.1) used to deduce the frost damage hypothesis for boreal conditions (Hänninen 1991; Hänninen et al. 1996; Figs. 8.5, 8.6, 8.7, 8.8, and 8.9 in Sect. 8.3.1.2). The reviewed studies of frost hardiness had also addressed the frost damage hypothesis deduced for Jyväskylä and had used the same air temperature scenario (Table 8.1) in their simulations.

In each simulation, the incidence of frost damage was quantified by two indices (Hänninen 2006). First, the number of critical annual cycles, i.e., those cycles during which the daily minimum air temperature dropped at least once below the frost

Table 8.4 The designs of the model comparisons carried out to identify ecophysiological traits critical for the hypothesis that climatic warming causes increased incidence of frost damage in boreal and temperate trees. Frost hardiness and frost damage were projected in scenario simulations carried out over 92 years in Jyväskylä, central Finland. Each time, the projections of one of the original models, $Model_1$ (Kellomäki et al. 1992, 1995) or $Model_2$ (Leinonen 1996), were compared with those of a corresponding modified model, where one ecophysiological aspect of the annual cycle was altered

Comparison	Original model	Modification	Results	Reference
I	$Model_1$	$R_o(T)$ from $Model_2$	Fig. 8.19a, b Fig. 8.20a Fig. 8.21a, b	Hänninen (2006)
II	$Model_2$	$R_o(T)$ from $Model_1$	Fig. 8.19c, d Fig. 8.20b Fig. 8.21a, b	Hänninen (2006)
III	$Model_1$	Sequential sub-model replaced with calendar sub-model	Fig. 8.22a Table 8.6	
IV	$Model_1$	Sub-model for post-rest introduced	Fig. 8.22b Table 8.6	

$R_o(T)$ = air temperature response of the rate of ontogenetic development towards growth onset

hardiness projected by the model, was counted. Second, the mean annual needle loss caused by frost damage was assessed. Regardless of the model used for simulating frost hardiness, the needle loss was calculated in each case by comparing the projected daily frost hardiness with the daily minimum air temperature according to Leinonen's (1996) method. The effects of the projected frost damage on the subsequent development of the trees were not addressed in these diagnostic simulations.

8.3.3.2 The Air Temperature Response of the Rate of Ontogenetic Development

Leinonen (1996) used his integrated frost hardiness model to assess the effects of climatic warming on the frost hardiness of and frost damage to the needles of *Pinus sylvestris* trees in central Finland. Unlike the older integrated model (Kellomäki et al. 1992, 1995), Leinonen's (1996) model did not project increased incidence of frost damage as a result of climatic warming.[11] Thus the models of Kellomäki

[11] Unfortunately, Leinonen's (1996) model has not been subjected to any such test in elevated temperature conditions as falsified Kellomäki et al.'s (1992, 1995) model (Fig. 8.16). However, a preliminary version of Leinonen's (1996) model was found to predict the needle frost hardiness quite accurately in elevated temperature conditions, too (Leinonen et al. 1996). This provides some support for the projection of Leinonen's (1996) model, i.e., for the projection that climatic warming will not increase the incidence of frost damage in boreal coniferous trees. It goes without saying, though, that ultimately the projections of Leinonen's (1996) model, too, should be examined by testing the model in elevated temperature conditions.

et al. (1992, 1995) and Leinonen (1996) give drastically different projections for the effects of climatic warming though they have numerous similarities, above all the shared integrative principle (Sect. 5.3.2).

This led Hänninen (2006) to examine which of the ecophysiological differences between $Model_1$ (Kellomäki et al. 1992, 1995) and $Model_2$ (Leinonen 1996) caused the difference in their projections. Since the two models share the same modular structure, the examination was technically carried out by simulating the frost hardiness and frost damage in the scenario climate not only with the original two integrated models but also with novel models, where one sub-model at a time was reciprocally exchanged between the two integrated models. Accordingly, if the trait described by a given sub-model was the critical one responsible for the different projections, then the reciprocal exchange of the sub-model would remove the dramatic frost damage from the projection of $Model_1$ and introduce it into the projection of $Model_2$.

Whenever frost damage was projected for climatic warming in Hänninen's (2006) simulations, it was always associated with premature dehardening and growth onset. This finding suggests that neither the hardening in late summer and early autumn nor the fluctuations of hardiness in winter are critical for the risk of frost damage under climate warming.

Of the simulations carried out by Hänninen (2006), only those identifying the critical trait are discussed here in detail (Table 8.4, Comparisons I and II). Hänninen (2006) found that the critical ecophysiological trait causing the different projections of $Model_1$ and $Model_2$ was the air temperature response of the rate of ontogenetic development towards growth onset (Table 8.5). The former model uses the relatively steep PU curve of Sarvas (1972) as a sub-model, whereas the latter makes use of the more gently sloping curve determined by Kramer (1994) (Fig. 8.18). For air temperatures between zero and 10 °C, commonly occurring during intermittent mild periods in winter under the scenario climate, more rapid ontogenetic development is predicted with the steep response than with the gently sloping response. Thus the steep response, whether used as a sub-model in $Model_1$ or in $Model_2$, projects (Fig. 8.19a, d) and the gently sloping response does not project (Fig. 8.19b, c) a premature growth onset in the scenario climate for several years (Hänninen 2006).

The two air temperature responses of the rate of ontogenetic development towards growth onset (Fig. 8.18) also have a major effect on the annual cycle of frost hardiness (Fig. 8.20) and the incidence of frost damage (Fig. 8.21) projected for climatic warming (Hänninen 2006). Whether used as a sub-model in $Model_1$ or in $Model_2$, the steep air temperature response projects premature dehardening, associated with premature growth onset, (Figs. 8.19a, d and 8.20), under the scenario climate for many years. Thus, when the air temperature drops again after a mild spell that has caused dehardening and premature growth onset, heavy frost damage is projected to occur. In the illustrative simulation, this was projected to happen in February (Figs. 8.20a, b, red curves). Accordingly, regardless of the integrated model in which used as a sub-model, the steep air temperature response of the rate of ontogenetic development projects catastrophic incidence of frost

Table 8.5 The ecophysiological traits and phenomena critical for the hypothesis that climatic warming will cause increased incidence of frost damage in boreal and temperate trees

Trait or phenomenon	Incidence of frost damage	Evidence
$R_o(T)$		
Steep	High	Computer simulations
Gently sloping	Low	
Rest completion		
In autumn	High	Computer simulations
In spring	Low	
Phase of post-rest		
Existing	Low	Computer simulations
Non-existing	High	
Quantitative dormancy induction		
Existing	Low	Conceptual inference
Non-existing	High	
Effects of the maternal environment		
Existing	Low	Conceptual inference
Non-existing	High	

$R_o(T)$ = air temperature response of the rate of ontogenetic development towards growth onset. See text for details and references

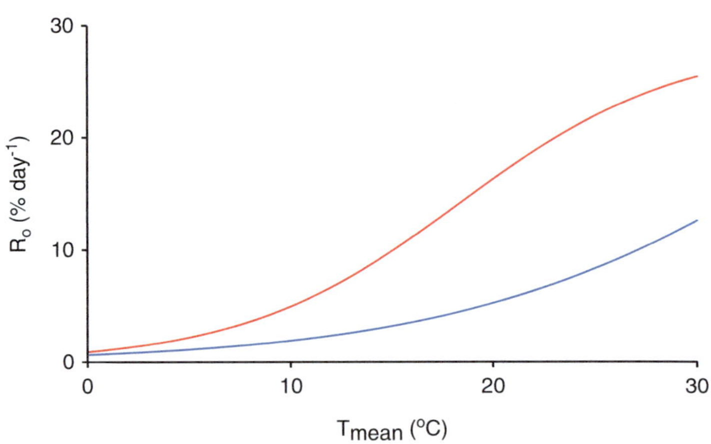

Fig. 8.18 Two air temperature responses of the rate of ontogenetic development towards growth onset, R_o, in boreal and temperate trees. T_{mean} = daily mean air temperature. *Red curve*: the steep response presented by Sarvas (1972) for several boreal tree species, as formulated for an average genotype of a generalised boreal tree species growing in thermal conditions similar to those of central Finland (Hänninen 1990b, 2006). *Blue curve*: the gently sloping response presented by Kramer (1994) for central European *Pinus sylvestris*. Both *curves* are presented using the ecophysiological percentage scale adopted in the present volume (Sect. 3.2.1.4; Hänninen and Kramer 2007). The steep response is originally included as a sub-model in Kellomäki et al.'s (1992, 1995) integrated frost hardiness model ($Model_1$, Table 8.4), and the gently sloping response is correspondingly included in Leinonen's (1996) integrated frost hardiness model ($Model_2$, Table 8.4) (Modified from Hänninen 2006; published with the permission of Oxford University Press)

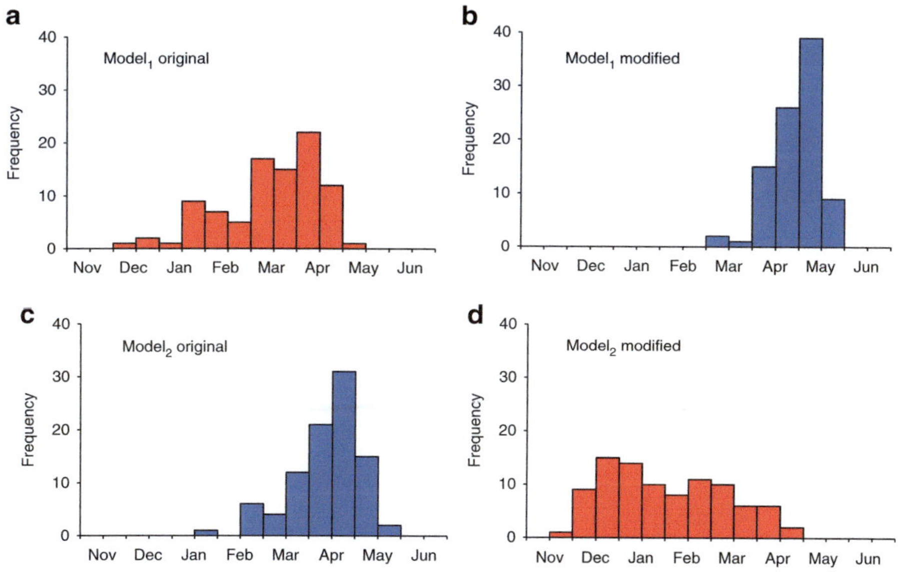

Fig. 8.19 Frequency distributions of the dates of growth onset in trees subjected to climatic warming in central Finland (Table 8.1), as projected for 92 years by ecophysiological models of the annual cycle. Simulations with (**a**, **b**) Kellomäki et al.'s (1992, 1995) integrated model ($Model_1$, Comparison I in Table 8.4); and (**c**, **d**) Leinonen's (1996) integrated model ($Model_2$, Comparison II in Table 8.4). Original = simulations with the original integrated model, modified = simulations with a modified integrated model, where the sub-model for the air temperature response of the rate of ontogenetic development has been replaced with the corresponding sub-model of the other integrated model (Table 8.4). The diagrams with red and blue bars indicate simulations with the two corresponding air temperature responses marked with the corresponding colour in Fig. 8.18 (Modified from Hänninen 2006; published with the permission of Oxford University Press)

damage during the 92-year simulation period of climatic warming (Fig. 8.21, red histograms).

With the gently sloping air temperature response (Fig. 8.18), in contrast, growth onset is postponed to later spring (Figs. 8.19b, c). Thus, regardless of the integrated model in which it is used as a sub-model, the gently sloping air temperature response projects considerable rehardening to take place in February during cold spells occurring after mild spells, and therefore no frost damage is projected (Figs. 8.20a, b, blue curves). Accordingly, again regardless of the integrated model in which it is used as a sub-model, the gently sloping air temperature response of the rate of ontogenetic development projects low incidence of frost damage for the entire 92-year simulation period of climatic warming (Fig. 8.21, blue histograms).

The steep air temperature response is based on Sarvas's (1972) experimental results concerning the ontogenetic development taking place in the flower buds of several boreal tree species (Fig. 3.7 in Sect. 3.2.1.3). Furthermore, as discussed in Sect. 3.2.1.3, though the shape of the curve remained unchanged in Sarvas's (1972)

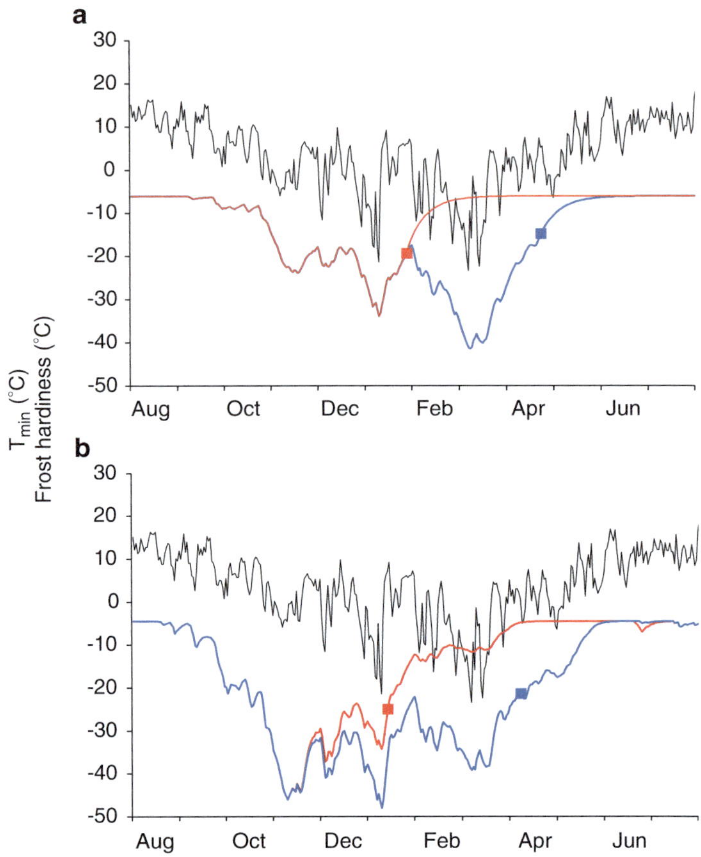

Fig. 8.20 The annual cycle of frost hardiness (*coloured curves*) and the timing of growth onset (*coloured squares*) in trees subjected to climatic warming in central Finland, as projected by models of the annual cycle. In the simulations, the observed daily mean and minimum air temperatures of 1970–1971 were elevated according to the non-uniform scenario presented in Table 8.1. The uppermost *black curves* in both panels indicate the daily minimum air temperatures in the scenario climate. Simulations with (**a**) Model_1 (Kellomäki et al. 1992, 1995; Comparison I in Table 8.4) and (**b**) Model_2 (Leinonen 1996; Comparison II in Table 8.4). The *red* and the *blue symbols* indicate simulations with the two air temperature responses marked with the corresponding colour in Fig. 8.18. At the beginning and the end of the simulation, the projections of the two models overlap; therefore, only the curve representing the original model (*red* for Model_1 and *Blue* for Model_2) is indicated (Modified from Hänninen 2006; published with the permission of Oxford University Press)

experiments, its level varied according to the high temperature requirement of the species and the developmental phenomenon in question. In Hänninen's (2006) study (Fig. 8.18), the high temperature requirement was selected to correspond to that of an average boreal tree species growing in central Finland. The gently sloping curve (Fig. 8.18) was formulated by Kramer (1994) by fitting a sigmoidal air temperature response into long-term historical data on growth onset and air

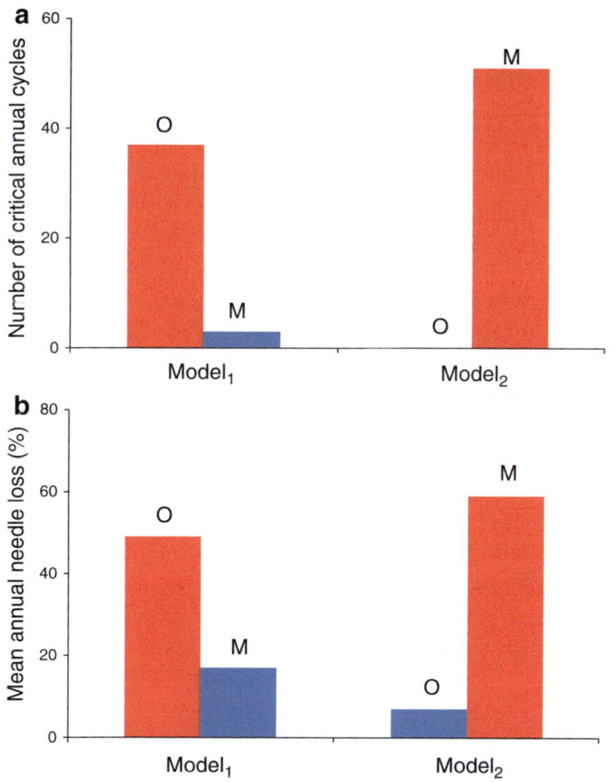

Fig. 8.21 Incidence of frost damage in trees subjected to climatic warming in central Finland (Table 8.1), as projected for 92 years by models of the annual cycle. (**a**) The number of critical annual cycles, i.e., those annual cycles where the daily minimum air temperature dropped below the simulated frost hardiness at least once. (**b**) Mean annual needle loss caused by the frost damage. Simulations with Kellomäki et al.'s (1992, 1995) integrated model (Model$_1$, Comparison I in Table 8.4); and Leinonen's (1996) integrated model (Model$_2$, Comparison II in Table 8.4). Original = simulations with the original integrated model, Modified = simulations with a modified integrated model, where the sub-model for the air temperature response of the rate of ontogenetic development has been replaced with the corresponding sub-model of the other integrated model. The diagrams with red and blue bars indicate simulations with the two corresponding air temperature responses marked with the corresponding colour in Fig. 8.18 (Modified from Hänninen 2006; published with the permission of Oxford University Press)

temperature. Though Kramer's (1994) study was carried out in temperate conditions, Leinonen's (1996) integrated model, which has the gently sloping curve as a sub-model, predicted the needle frost hardiness of *Pinus sylvestris* growing under natural conditions in central Finland accurately (Leinonen 1996; Fig. 5.14b in Sect. 5.3.2.3).

In conclusion, Hänninen's (2006) findings demonstrate the importance of the air temperature response of the rate of ontogenetic development towards growth onset for the risk of tree frost damage under climate warming (Figs. 8.18, 8.19, 8.20, and 8.21). This conclusion calls for a thorough review of all available experimental data published about the response in different boreal and temperate tree species.

As discussed in Sect. 3.2.1.3, the response can be readily determined with a simple experimental design. Still, relatively few experimental studies are available for this crucial tree trait (Campbell and Sugano 1975, 1979; Campbell 1978; Caffarra 2007). Furthermore, the use of various arbitrary developmental units (Hänninen 1990b; Myking 1997) and scaling factors (Caffarra and Donnelly 2011; Caffarra et al. 2011) in the published results often hamper a comparison of the responses. This is because the various developmental units and results scaled with different scaling factors are not comparable with each other. As discussed at length in Sect. 3.2.1, the only comparable scale is the one obtained as the reciprocal of the quantity originally measured in the experiment, i.e., as the reciprocal of the developmental time, Δt (Fig. 8.18; Eq. 3.2 in Sect. 3.2.1.3). With the exception of Sarvas (1972) (Table 3.2 in Sect. 3.2.1.3), results for Δt are often not reported in the studies applying the arbitrary developmental units and scaling factors.

In order to get access to the data originally measured for Δt, cooperation among the authors of the published studies is needed for a comprehensive review comparing the experimentally measured air temperature responses of the rate of ontogenetic development towards growth onset. For this reason, no attempt for such a review was made in the present volume.

8.3.3.3 Rest Completion

Linkosalo et al. (2000) used long-term historical phenological and air temperature field data from central Finland to parameterize two models predicting the timing of bud burst in *Betula* species. First, they used a sequential model where rest completion was assumed to be caused by accumulated chilling. Second, they used a calendar model where rest completion was assumed to take place on a given calendar day. They adopted the intermediate approach discussed in Sect. 3.4.6.1, i.e., fixed forms of the air temperature responses were used, but the three critical model parameters were estimated by fitting the models into the historical data. These parameters included the chilling requirement of rest completion, C_{crit}, in the sequential model, the critical calendar day, t_{crit}, causing rest completion in the calendar model, and the high temperature requirement of growth onset, H_{crit}, used in both models. Similarly to several other studies using long-term historical field data (Sect. 3.4.4.1), the calendar model was found to fit to the data more accurately than the sequential model (Linkosalo et al. 2000).

Linkosalo et al. (2000) used both of the parameterized models in scenario simulations to assess the effects of climatic warming on the timing of bud burst and the associated risk of spring frost damage in *Betula* spp. trees growing in central Finland. In all warming scenarios, the calendar model projected much less frost damage, if any, than the sequential model. This was because in most cases the simulated rest completion took place in December according to the sequential model, whereas according to the calendar model it was postponed till March or early April (Linkosalo et al. 2000). Thus, as in Hänninen's (1991) simulations for the generalised boreal tree species (Fig. 8.7b in Sect. 8.3.1.2), the sequential model, now fitted to empirical field data on *Betula* spp., projected rapid ontogenetic development in the warmest winters of the scenario climate, which led to premature bud burst and frost damage. According to the fitted calendar model, in contrast, the high air temperatures did not cause ontogenetic development until March or early April, so that the model did not generally project premature bud burst with concomitant frost damage (Linkosalo et al. 2000).

The difference between the projections of the sequential model and the calendar model found by Linkosalo et al. (2000) in simulations addressing the timing of bud burst is further elaborated here by means of scenario simulations explicitly addressing the frost hardiness of and frost damage to the trees. To this end, Kellomäki et al.'s (1992, 1995) frost hardiness model ($Model_1$) was used in scenario simulations. Besides the original $Model_1$, frost hardiness was also simulated with a modified model, where the sequential sub-model included in the original model was replaced with a calendar model that assumed rest completion to take place at spring equinox on 20 March each year (Comparison III, Table 8.4).

Like the original $Model_1$, the modified model that included a calendar model as the sub-model for rest completion also projected considerable dehardening to take place during intermittent mild spells under the scenario conditions in winter (Fig. 8.22a). However, due to the postponing of rest completion until 20 March, the modified model projected much later growth onset than the original $Model_1$ did (Fig. 8.22a). For this reason, the modified model projected considerable rehardening for February and early March, when the air temperature drops, so that unlike the original model, the modified model did not project heavy frost damage (Fig. 8.22a, violet dashed curve). Thus, contrary to the projection of the original model, the incidence of frost damage projected by the modified model remained low through the entire simulation period of 92 years (Table 8.6).

The fact that the calendar model provided a better fit to the historical long-term air temperature and bud burst data than the sequential model did (Linkosalo et al. 2000) supports the projection of the calendar model, suggesting that the incidence of frost damage to *Betula* spp. in central Finnish conditions will not be increased considerably by climatic warming (Linkosalo et al. 2000; Fig. 8.22a; Table 8.6). This suggestion also accords with the results of the whole-tree chamber experiments with *Pinus sylvestris* and *Picea abies* discussed in Sect. 8.3.2.2 (Hänninen 1995a; Hänninen et al. 1996, 2007).

However, as emphasised in the same section, models used for assessing the ecological effects of the climate change should ultimately be also tested in scenario conditions, not only in current environmental conditions. Therefore, as the calendar model applied by Linkosalo et al. (2000) to the case of *Betula* spp. lacks such tests in scenario conditions, its projections remain inconclusive. Most importantly, Linkosalo et al.'s (2000) finding, further elucidated in Fig. 8.22a, reveals the crucial

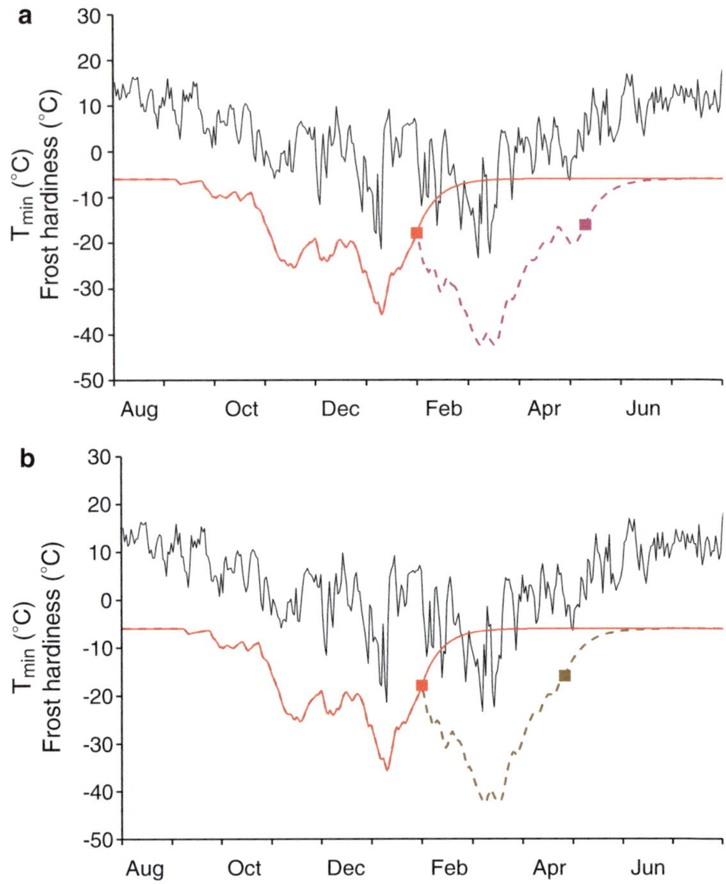

Fig. 8.22 As Fig. 8.20, but in both panels, the continuous *red line* and the *red quadrat* indicate simulations with Kellomäki et al.'s (1992, 1995) original integrated model and the *violet* or *brown symbols* and the *dashed curve*, simulations with a modified version of the model. (**a**) In the modified model (*violet symbols*, *dashed curve*), the original sequential sub-model was replaced with a calendar sub-model that assumed rest completion to take place on 20 March each year (Comparison III, Table 8.4). (**b**) The post-rest model introduced in Sect. 3.3.1.2 on the basis of Vegis's (1964) theory was introduced into the modified model (*brown symbols*, *dashed curve*) by assuming the air temperature response of the rate of ontogenetic development to change as indicated in Fig. 8.23 (Comparison IV, Table 8.4)

importance of rest completion for the frost damage hypothesis (Table 8.5). A late rest completion related to the calendar model efficiently decreases the risk of increased incidence of frost damage in a warming climate (Table 8.6). This reasoning calls for the solving of the paradox of rest completion discussed in Sect. 3.4.4.

Table 8.6 Incidence of frost damage projected in simulations with integrated models of frost hardiness over 92 years for a scenario climate (Table 8.1) in Jyväskylä, central Finland. The simulations were carried out with Kellomäki et al.'s (1992, 1995) original frost hardiness model ($Model_1$) and with two modifications of it

Model	Number of critical annual cycles[a]	Mean annual needle loss (%)
$Model_1$ as such	33[b]	42.6[b]
$Model_1$ with the sequential sub-model replaced with a calendar sub-model	1	12.7
$Model_1$ with a sub-model for post-rest introduced	3	14.3

[a]A critical annual cycle = a cycle during which the daily minimum air temperature dropped below the projected frost hardiness at least once
[b]These values, calculated for the present volume, are slightly different from the corresponding values published earlier by Hänninen (2006) (Fig. 8.21). This difference is caused by a difference in the input air temperature data used. See footnote 5 in Sect. 6.2.3

8.3.3.4 The Phase of Post-Rest

Vegis's (1964) theory of gradual changes in the dormancy status of trees, discussed in Sects. 3.3.1.2 and 3.4.4.2, is also potentially critical for the frost damage hypothesis. If the air temperature response of the rate of ontogenetic development towards growth onset changes as predicted by the post-rest model based on Vegis's (1964) theory (Fig. 8.23a), then air temperatures a few degrees above zero, projected to be common in future winters due to climatic warming, would not cause ontogenetic development towards growth onset during that time of the year. Thus, contrary to the assumption made when the frost damage hypothesis for boreal conditions was deduced (Fig. 8.7b in Sect. 8.3.1.2), air temperatures slightly above zero would not cause ontogenetic development towards growth onset until relatively late in spring, and boreal and temperate trees would therefore run a lower risk, if any, of increased incidence of frost damage in the scenario climate.

This notion, based on conceptual inference, is examined here quantitatively in the light of computer simulations with a procedure similar to the one used above for the other two ecophysiological traits (Hänninen et al. 2013). The scenario simulations made use of both Kellomäki et al.'s (1992, 1995) original $Model_1$ and a modified version of it that included the post-rest model (Table 8.4, comparison IV). In the modified model it was assumed that at rest completion, the air temperature response of ontogenetic development is located at relatively high temperatures, as shown in Fig. 8.23a. It was further assumed that during post-rest the curve shifts towards lower air temperatures as a result of chilling, attaining the location based on Sarvas's (1972) empirical results at the end of post-rest (=onset of quiescence) (Fig. 8.23a).

The introduction of the post-rest model into the modified integrated model delayed growth onset, allowed rehardening when the air temperature dropped after a mild spell in winter, and in this way prevented the considerable frost damage projected by the original model (Fig. 8.22b; Table 8.6). Therefore the phase of

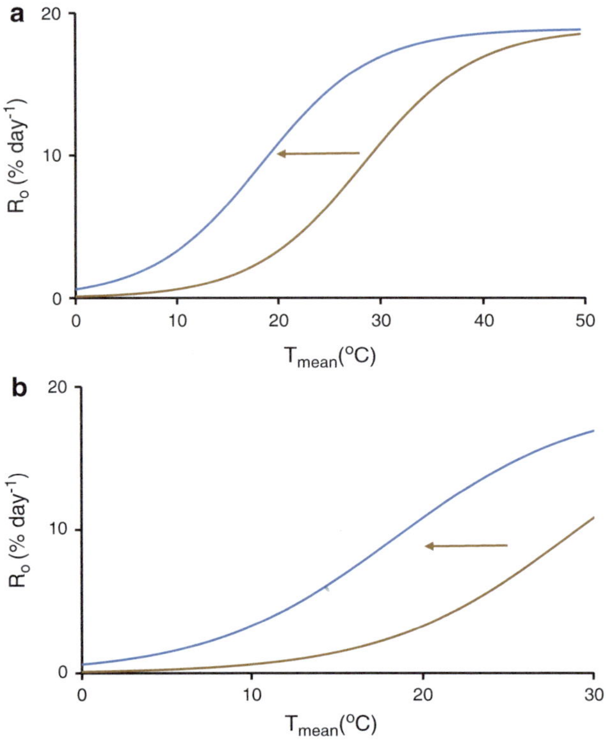

Fig. 8.23 The formulation of the post-rest model based on Vegis's (1964) theory (see Sect. 3.3.1.2), as introduced into Kellomäki et al.'s (1992, 1995) integrated frost hardiness model in the present volume. Post-rest is assumed to occur between rest completion and the onset of quiescence, and chilling is assumed to be its driving force just like for rest. The chilling requirement for progressing from the onset of post-rest to the onset of quiescence is assumed to be twice that of the chilling requirement of rest completion, C_{crit}. T_{mean} = daily mean air temperature, and R_o = rate of ontogenetic development. The *brown* and the *blue curve* represent the air temperature response of the rate of ontogenetic development towards growth onset at the beginning of post-rest (=at rest completion) and at the end of post-rest (= at the onset of quiescence), respectively. The *arrow* indicates the change in the response taking place during the post-rest phase. (**a**) *Curves* drawn on a broad air temperature scale, showing that by the present definition of post-rest, the form of the *curve* remains unaltered while its location moves to lower air temperatures during post-rest. (**b**) The previous *curves*, drawn in an ecologically meaningful air temperature scale

post-rest is added to the list of ecophysiological traits critical for the frost damage hypothesis (Table 8.5).

8.3.3.5 A Comparison of the Three Critical Traits

It is remarkable that in comparison with the projection of the catastrophic frost damage originally provided by Kellomäki et al.'s (1992, 1995) integrated frost hardiness model, the three ecophysiological aspects found in computer simulations to be critical (Table 8.5) all imply mutually strikingly similar changes in the projected frost hardiness pattern (Figs. 8.20 and 8.22). Though addressing different ecophysiological traits, all these critical aspects affect the rate of ontogenetic development and, through its changes, the incidence of frost damage in the scenario climate. In the calendar model the rate of ontogenetic development is simply zero at any temperature until the assumed date of rest completion, whereas the other two concern quantitative differences in the air temperature response (Figs. 8.18 and 8.23).

When mapped onto an ecologically meaningful air temperature scale, the change in the air temperature response implied by the post-rest model (Fig. 8.23b) actually resembles the difference found by Hänninen (2006) to explain the different projections of the two original integrated models of frost hardiness (Fig. 8.18). Despite this graphical resemblance, these two results address completely different phenomena. Hänninen's (2006) study addresses two hypothetical genotypes, each of them having a constant genotype-specific air temperature response (Fig. 8.18), whereas the post-rest model addresses a hypothetical change in the air temperature response of a given genotype (Fig. 8.23; see also the comparison of the parallel and the post-rest model in Sect. 3.4.1.3).

8.3.3.6 Long-Term Phenomena Calling for New Modelling Principles

There is increasing empirical evidence to support the theory of quantitative dormancy induction (Sect. 3.4.5). High air temperatures during dormancy induction have been found to cause deepening of the dormancy by increasing either the chilling requirement of rest completion or the high temperature requirement of growth onset, or both. In any case, growth onset is delayed, so that the risk of premature growth onset with the concomitant risk of increased incidence of frost damage as a result of climatic warming is decreased (Heide 2003; Table 8.5). It is therefore crucial to examine how common and how strong the effect is among various boreal and temperate tree species.

There is also substantial empirical evidence to support the theory of the effects of the maternal environment on trees. The environment in which the tree seeds develop has been found to affect the properties of the annual cycle of the offspring considerably (Sect. 6.5). For *Picea abies* the empirical evidence is conclusive, but for other tree species it remains fragmentary and inconclusive. Due to the effects of the maternal environment, climatic warming will tune the ecophysiological traits of trees towards those typical for more southern provenances. Besides several other effects of the maternal environment, higher air temperatures during flowering and seed development will delay the growth onset of the offspring (Sect. 6.5). This reduces the risk of premature growth onset occurring with concomitant increased incidence of frost damage as a result of climatic warming. It is therefore important in future empirical work to examine how common this phenomenon is among boreal and temperate tree species (Table 8.5).

In the ecophysiological modelling approach adopted in the present volume, the ecophysiological traits of the trees are described in a simplified but quantitative way with the equations and parameter values included in the models. Thus, unlike the dynamic ecophysiological output variables predicted by the models, such as needle frost hardiness, the values of parameters such as the chilling requirement of rest completion or the high temperature requirement of growth onset are classically assumed to be genotype-specific constants (Hänninen and Kramer 2007). The findings supporting the theory of quantitative dormancy induction suggest, however, that rather than being genotype-specific constants, the ecophysiological traits represented by the values of these parameters also acclimate to the environment in which the tree is growing. This line of reasoning calls for a hierarchical model structure with a specific sub-model developed for each parameter affected by such acclimation. The output of these models, i.e., the values of the parameters, is then used in the models of the annual cycle similarly to the way they are used currently.

The findings supporting the theory of the effects of the maternal environment call for a somewhat analogous revision of the current modelling approach. However, rather than acclimating to the environmental conditions prevailing during each annual cycle, as is the case in quantitative dormancy induction, the ecophysiological traits evidently stay affected by the maternal environment through the entire life cycle of the offspring.

For instance, the high temperature requirement of a given tree individual can be regarded as constant, but rather than being strictly specific to the genotype, it also depends on the environmental conditions prevailing during the flowering and the seed development. Thus the theory of the effects of the maternal environment, too, calls for a new hierarchical model structure, but in this case the new model needed to determine the value of the parameter is to be used only before the birth of the new individual tree.

The experimental evidence for both quantitative dormancy induction and the effects of the maternal environment may be still too meagre for novel sub-models addressing these aspects to be introduced into the ecophysiological models of rest completion and growth onset. Even so, the implications of these phenomena for the timing of growth onset and the associated risk of frost damage can be hypothetically assessed by means of computer simulations. This can be carried out even in the absence of any experimental data, as shown by the simulations addressing the hypothetical post-rest model (Figs. 8.22b and 8.23). By using the hypothetico-deductive research strategy adopted in the present volume, the results of such theoretical studies can subsequently be used for guiding the planning of efficient experimental designs for empirical studies (Hänninen and Lundell 2007; Sect. 2.1).

8.3.3.7 Experimental Designs for the Critical Traits

Each of the five critical ecophysiological traits or phenomena (Table 8.5) can be addressed with experimental designs specific to each trait or phenomenon, some more readily than others. The air temperature response of the rate of ontogenetic development towards growth onset (Fig. 8.18) can be readily determined with the simple experimental design described in Sect. 3.2.1.3. The same design can be used to test the post-rest model, but to do that, the experimental procedure needs to be repeated for the different phases of dormancy so as to check whether the air temperature response changes during dormancy as predicted by the theory (Fig. 8.23).

According to most studies, the chilling requirement of rest completion is met during autumn. Accordingly, it has been suggested that the potential postponing of rest completion from autumn to spring (Table 8.5) is caused by additional environmental factors other than the accumulation of chilling (Sect. 3.4.4). The possible additional factors required for rest completion can be readily addressed with designs in which such additional factors are regulated. Night length and the quality of light, for instance, can be readily manipulated experimentally to test if these factors affect rest break and rest completion (Linkosalo and Lechowicz 2006; Basler and Körner 2012). Similarly, quantitative dormancy induction (Table 8.5) can be readily examined with experimental designs in which the air temperature during dormancy induction is manipulated (Sect. 3.4.5).

Finally, the effects of the maternal environment (Table 8.5) can also be readily studied experimentally by first producing seeds in field trials under varying climatic conditions and then examining the traits of the annual cycle of the offspring with standard ecophysiological experimental designs used for assessing, e.g., the chilling requirement of rest completion and the high temperature requirement of growth onset. It goes without saying, however, that such studies require rather long-lasting projects.

It is to be noted that all the critical ecophysiological traits (Table 8.5) are related to the annual phenological cycle of boreal and temperate trees. This is because the alternation of the main phases of growth and dormancy has a major effect on the annual cycle of frost hardiness, as discussed in Chap. 5. This was one of the main reasons why the annual phenological cycle, and especially the timing of growth onset, was discussed more thoroughly in Chap. 3 than the other aspects of the annual cycle in other chapters. As discussed in Chap. 3, the modelling of the annual phenological cycle would be greatly facilitated if observations on developmental phenomena in the buds invisible to the naked eye were available. Such observations would also be useful for examining the traits that are critical for the risk of frost damage in a changing climate (Table 8.5). This reasoning calls for further cooperation between researchers taking the whole-tree modelling approach discussed in the present volume and those addressing the developmental phenomena at the anatomical, cellular, and molecular level (Cooke et al. 2012; Viherä-Aarnio et al. 2014).

Microscopic observations on anatomical development inside the buds are obtained at the organisational level next to the whole-tree level, so that they offer especially promising prospects for improving the whole-tree models (Sutinen et al. 2009, 2012; Viherä-Aarnio et al. 2014). As discussed in detail in Sect. 3.2.1.3, such microscopic observations were already used by Sarvas (1972, 1974) to determine the air temperature response of the rate of ontogenetic development

(Fig. 8.18), which later on has been found critical for assessing the ecological effects of the climate change (Table 8.5).

Though the experimental designs needed for addressing the critical ecophysiological traits are readily available in principle, the realisation of the experiments in practice is hampered by the need to also address mature trees, not only small seedlings. The regulation of the annual cycle changes in several respects as the trees get older (Ununger et al. 1988; Vitasse 2013). This is a problem of economy, however, rather than principle. It is technically possible to develop efficient experimental designs for the study of the annual cycle of adult trees, too. This can be done by using large whole-tree chambers or even by using big machinery to transfer large adult trees, with their root systems and all, into large greenhouses. It goes without saying, however, that such approaches are extremely expensive, so that the researchers of the annual cycle need to settle for more modest experimentation in most cases, if not always. This is an unfortunate shortcoming because, as documented in this volume, informative experimental designs are absolutely necessary for creating trustworthy models for the assessment of the ecological effects of the climate change.

8.4 Effects on the Seasonality of Growth

When the effects of the climate change on boreal and temperate trees are considered in the present volume, the main emphasis is on the hypothesis of tree overwintering and frost damage discussed in detail in Sect. 8.3. It goes without saying that climate change affects these trees in many ways during the growing season, too. These effects are discussed in detail elsewhere (Ceulemans and Mousseau 1994; Saxe et al. 2001; Ainsworth and Long 2005; Morison and Morecroft 2006; Zaehle et al. 2014), so that only a brief summary of the potential effects on the seasonality of growth is provided here. This is done by recalling the discussions presented earlier in this chapter in connection with the frost damage hypothesis (Sect. 8.3) and the discussions presented in Chap. 3 on the environmental regulation of the annual phenological cycle.

The rising of air temperature in spring is crucial for growth onset in all boreal and temperate plants, including trees. Contrary to several other aspects of the environmental regulation of the annual cycle, this high temperature requirement of growth onset is a universal phenomenon among all boreal and temperate trees (Sect. 3.2.2). As a result of climate warming, the high temperature requirement is going to be fulfilled earlier than it is at present, so that growth onset is generally projected to be accelerated because of climate warming (Kilpeläinen et al. 2006; Slaney et al. 2007).

There are, however, two major phenomena related to the dormancy of trees that may counteract the projected accelerating effect of climate warming on growth onset in spring (Hänninen and Tanino 2011). First, climate warming may delay rest completion by delaying and reducing the chilling accumulation (Sect. 8.3.1.3).

Second, higher air temperatures in late summer may cause deeper dormancy manifested in a higher chilling requirement of rest completion and/or a higher high temperature requirement of growth onset (Sect. 3.4.5). These phenomena may reduce, and in some cases even reverse, the accelerating effect of springtime warming on growth onset.

The effects of night length on rest break in boreal and temperate trees present further complications to the projections of growth onset under climate warming. It has been known for a long time that in many tree species short nights compensate for lack of chilling in causing rest break (Sect. 3.3.3). This effect may reduce or even remove the delaying effects on growth onset caused by reduced chilling. In such cases, growth onset may be accelerated as projected on the basis of earlier fulfilment of the high temperature requirement of growth onset. However, there is increasing evidence for the notion that in some tree species the effects of night length on rest break are not restricted to the compensation of lack of chilling (Sects. 3.3.3 and 3.4.4). In such cases long nights may reduce, or even prevent, the accelerating effects of climate warming on growth onset. The overall effects of climate change on growth onset may be further complicated by potential effects of other environmental factors, such as elevated concentrations of atmospheric carbon dioxide (Sect. 3.4.3).

Increasing night length is a major environmental factor causing growth cessation during late summer in most boreal and temperate trees (Sect. 3.5.2). Climate change does not alter the annual cycle of nigh length, so that one would expect the effect of climate change on growth cessation to be smaller as a rule than its effect on growth onset. However, as discussed in Sect. 3.5, air temperature interacts with night length in many species, and growth cessation is regulated by air temperature alone in some species. These phenomena bring out a complicated picture in which climate warning obviously accelerates growth cessation in some species but delays it in some others (Hänninen and Tanino 2011).

There are considerable genetic differences in the environmental regulation of both growth onset and growth cessation, not only among boreal and temperate tree species but also among provenances in many species (Hänninen and Tanino 2011; Sect. 6.3). These genetic differences should be addressed when the effects of climate warming on the timing of growth onset and growth cessation are projected. Under climatic warming, the meteorologically defined growing season favourable for growth will be prolonged by both an earlier start in spring and a later cessation in autumn, but because of the genetic differences, the season when growth actually occurs will not be prolonged similarly in all tree species and provenances.

The ecophysiological modelling approach adopted in the present volume is well suited for assessing the effects of climate change on the seasonality of growth in boreal and temperate trees. Still, the approach should be applied with caution. One should not jump into species-specific conclusions on the basis of scenario simulations with generalised models, such as the ones used by Hänninen (1991) and Kellomäki et al. (1992, 1995), for deducing generalised hypotheses concerning the effects of climate change. Rather, for species- and provenance-specific projections, the models need to be tested with corresponding species- and provenance-specific data. This means, once again, that rather than providing ready-to-use tools for assessing the effects of climate change, the present modelling approach provides a framework for developing such tools.

8.5 Implications to the Stand and the Ecosystem Level

8.5.1 Process-Based Forest Growth Modelling in Climate Change Research

The ecological effects of the climate change on various ecosystems, including boreal and temperate forests, have frequently been assessed by means of large process-based ecosystem models such as those discussed in Chap. 7. It has generally been concluded in these large-scale modelling studies that the climate change will considerably enhance the productivity of boreal and temperate forests (Kellomäki et al. 1997; Bergh et al. 2003). Accordingly, in comparison with the catastrophic projections, e.g., desertification, presented for several more southern ecosystems (IPCC 2014a, b), the effects of the climate change on boreal and temperate forests may even be regarded as positive in several cases.[12]

However, it has also been concluded that climate change may increase the incidence of various forms of damage to boreal and temperate trees. These include biotic damage caused by new species of insect pests (Hlásny and Turčáni 2009) and pathogenic fungi (Desprez-Loustau et al. 2007), and abiotic damage caused for instance by heavy storms (Peltola et al. 1999), or by waterlogging and the subsequent anoxia developed in the soil during warm periods in winter (Wang et al. 2013). Thus, though the projection of increased productivity of boreal and temperate forests is the main line prevailing in the literature, there is still considerable uncertainty about the projection.

In recent years, the models of the annual cycle of boreal and temperate trees discussed throughout the present volume have increasingly been used as sub-models in large process-based ecosystem models when the effects of the climate change on boreal and temperate forests have been assessed. Among the three attributes of the annual cycle addressed in the present volume, the models of the annual phenological cycle (Chap. 3) and the annual cycle of photosynthesis (Chap. 4) have been used as sub-models in the stand and ecosystem models particularly frequently, whereas the models of the annual cycle of frost hardiness (Chap. 5) have only rarely been used for that purpose. The studies using ecophysiological models of various aspects of the annual cycle in large-scale models were

[12] The valuation of environmental changes is ultimately a philosophical, not a scientific issue. Many environmentalists consider any human-induced environmental change negative by definition.

reviewed in detail by Kramer and Hänninen (2009); the discussion below is restricted to two illustrative cases.

8.5.2 *Photosynthetic Production in* Pinus sylvestris *Stands*

Mäkelä et al. (2004) introduced their model of the annual cycle of photosynthetic capacity as a sub-model into their process-based main model simulating the photosynthetic production of a boreal coniferous stand.[13] The main model first simulates the attenuation of light going down the canopy and then the effects of four environmental factors on the rate of shoot photosynthesis at different heights in the canopy (Fig. 8.24). Hänninen et al. (2005) further introduced a sub-model of frost damage into the main model, thus creating an overall model consisting of three components (Fig. 8.24). The sub-model for frost damage offers three options: it is not used at all, i.e., frost hardiness and frost damage are not addressed in the simulation, or the needle loss caused by frost damage is addressed by simulating frost hardiness with either Kellomäki et al.'s (1992, 1995) integrated model or Leinonen's (1996) corresponding integrated model. Following the convention adopted in Sect. 8.3.3, these will be referred to briefly as $Model_1$ and $Model_2$, respectively.

The outputs of the two sub-models are used as input for the main model as follows (Fig. 8.24): first, the changes in photosynthetic capacity change the photosynthetic light curve as indicated in Fig. 4.2d in Sect. 4.1.2.1; thus, the rate of photosynthesis at a given light level increases with increasing photosynthetic capacity; second, the output of the optional frost damage model is used in two ways in the main model: besides decreasing the rate of shoot photosynthesis, the needle loss caused by frost damage also decreases shading, and thus also the light attenuation in the canopy (Fig. 8.24).

Using the modular overall model (Fig. 8.24) in a simulation study, Hänninen et al. (2005) examined how the dynamics of photosynthetic capacity and the needle loss potentially caused by frost damage affect the photosynthetic production of *Pinus sylvestris* under climatic warming. For comparison, simulations for the current climate were carried out first. As the main model had been parameterized for a *Pinus sylvestris* stand located at the SMEAR II station in Hyytiälä, southern Finland (Hari et al. 2008); the simulations for the current climate were carried out by using half-hourly meteorological data measured at SMEAR II in 1998–2003. In the meteorological data for the scenario climate, the half-hourly air temperature records were elevated according to the non-uniform scenario (Table 8.1) that was also used in the present volume when the frost damage hypothesis was discussed in Sect. 8.3.

[13] The model of the annual cycle of photosynthetic capacity is discussed in Sect. 4.2.1.2.

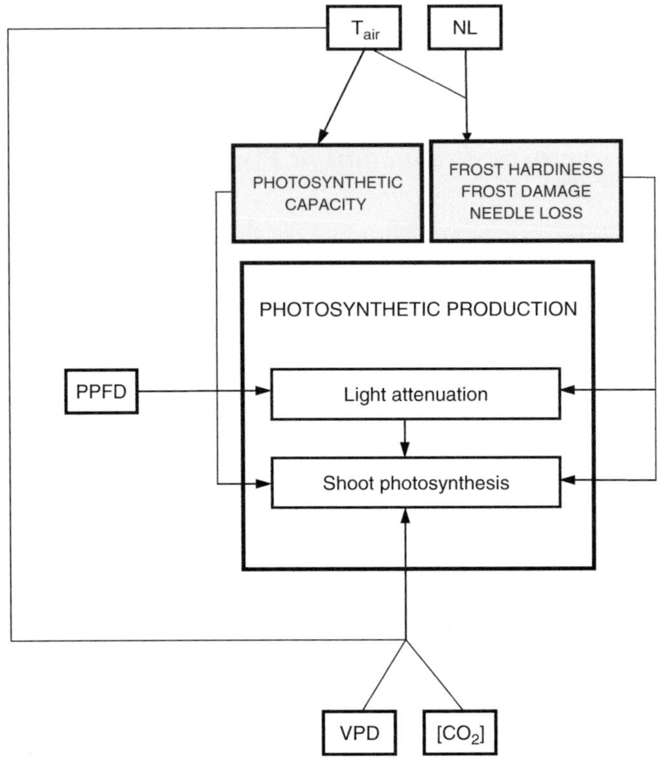

Fig. 8.24 A schematic presentation of a model of photosynthetic production in boreal coniferous trees, including two models of the annual cycle as sub-models (the *shaded boxes*). In the main model (the unshaded *large box*), light attenuation down the canopy is modelled first, and then the effects of the environmental factors on shoot photosynthesis (net CO_2 exchange) are calculated. PPFD = the photosynthetic photon flux density, VPD = the water vapour deficit of air, $[CO_2]$ = the concentration of atmospheric carbon dioxide, T_{air} = the air temperature (Mäkelä et al. 2004). The sub-model of photosynthetic capacity is driven by long-term changes in air temperature, which thus affect the simulation of photosynthetic production in the main model (Mäkelä et al. 2004). The sub-model of frost hardiness and damage (Hänninen et al. 2005) is driven by long-term changes in air temperature and night length; when the modelled frost hardiness is insufficient to tolerate the prevailing air temperature, needle loss caused by frost damage is predicted. In the main model, the needle loss decreases the photosynthetic production of the shoots and also the light attenuation down the canopy (Modified from Hänninen et al. 2005; published with the permission of NRC Research Press)

The effect of climatic warming on the annual cycle of photosynthesis mediated by photosynthetic capacity is seen in the projections where frost damage was not addressed (Fig. 8.25a). Photosynthetic production started about 1 month earlier in the scenario climate than in the current one. Subsequently, when photosynthetic production also started in the current climate, it was lower than in the scenario climate for several weeks. In the summertime there was no difference in photosynthetic production between the two climatic conditions (Figs. 8.25a). In autumn there

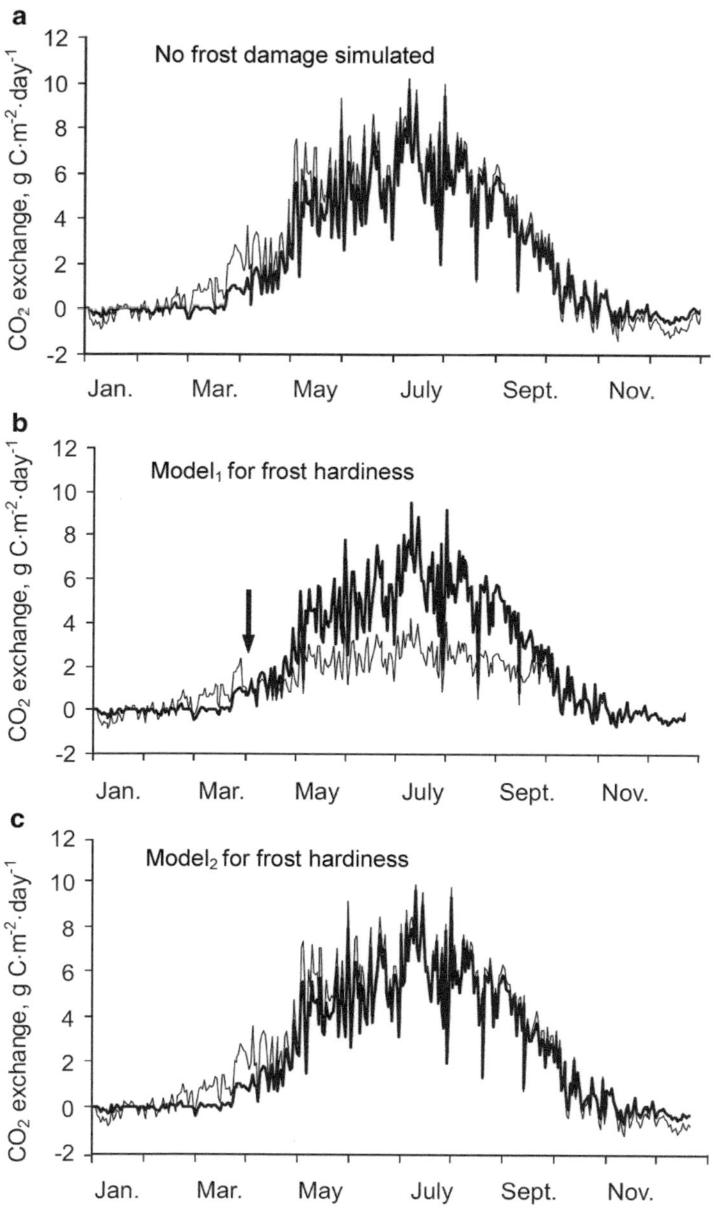

Fig. 8.25 Simulated daily net carbon dioxide exchange of shoots in a southern Finnish *Pinus sylvestris* stand, expressed in units of mass of carbon per square metre in the stand. The *thick* and the *thin curves* indicate simulations for the current (year 2000) and the scenario climate, respectively. In the latter, the half-hourly temperature records were elevated according to the non-uniform scenario presented in Table 8.1. (**a**) Simulations not addressing frost damage. Simulations addressing the needle loss due to frost damage by simulating needle frost hardiness (**b**) with $Model_1$ (Kellomäki et al. 1992; 1995) or (C) with $Model_2$ (Leinonen 1996). The *arrow* in (**b**) indicates the time of simulated severe needle loss caused by frost damage (Hänninen et al. 2005; redrawn with the permission of NRC Research Press)

was again a difference in the projected photosynthetic capacity (results not shown), but because of the limited amount of incoming solar radiation, there was only a minor difference in photosynthetic production. In all, then, in the simulations where frost damage was not addressed, the earlier recovery of photosynthesis in the scenario climate (Fig. 8.25a) implied higher photosynthetic production accumulated over the entire growing season than was projected for the current climate (Fig. 8.26). This conclusion accords with the main line prevailing in the literature, suggesting increased productivity of boreal forests as a result of the climate change (Kellomäki et al. 1997; Bergh et al. 2003).

When frost hardiness was simulated with $Model_2$, very little frost damage and needle loss were projected for the scenario climate (results not shown). The projection obtained for the annual cycle of photosynthesis (Fig. 8.25c) was similar to that obtained when frost damage was not addressed (Fig. 8.25a), and the projections for the photosynthetic production accumulated during the entire growing season did not differ markedly between these two simulations (Fig. 8.26).

However, a different projection was obtained when frost hardiness was simulated according to $Model_1$ (Fig. 8.25b). Here again, photosynthetic production was projected to start earlier in the scenario climate than in the current one. However, unlike $Model_2$, $Model_1$ projected considerable frost damage in spring for the scenario climate (results not shown). Accordingly, after the occurrence of the

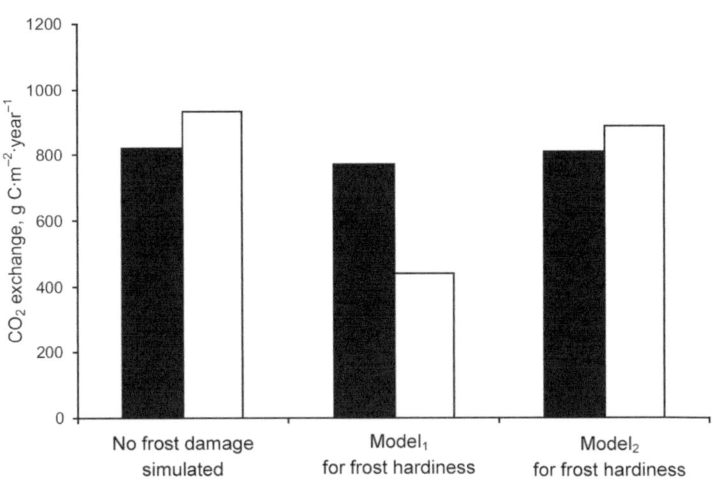

Fig. 8.26 Simulated average annual net carbon dioxide exchange of shoots in a southern Finnish *Pinus sylvestris* stand, expressed in units of mass of carbon per square metre in the stand. The *black* and the *white bars* indicate simulations for the current (years 1998–2003) and the scenario climate, respectively. In the latter, the half-hourly temperature records were elevated according to the non-uniform scenario presented in Table 8.1. The simulations were carried out by not addressing frost damage (the *bars* on the left-hand side) and by calculating the needle loss caused by frost damage according to the frost hardiness projected by Kellomäki et al.'s (1992, 1995) integrated model ($Model_1$) and by Leinonen's (1996) integrated model ($Model_2$) (Hänninen et al. 2005; redrawn with the permission of NRC Research Press)

needle loss caused by frost damage in spring under the scenario climate, photosynthetic production remained at a level considerably lower in the scenario climate than in the current one for the rest of the growing season (Fig. 8.25b). Consequently, the photosynthetic production accumulated over the entire growing season was lower for the scenario climate than for the current one in the simulation with $Model_1$ (Fig. 8.26). This was the case even though the tree mortality caused by frost damage (Kellomäki et al. 1995) was not addressed in the simulations.

In qualitative terms, the difference found by Hänninen et al. (2005) between the implications of the two frost hardiness models for the projected photosynthetic production was actually obvious a priori. This was because it was known a priori that Kellomäki et al.'s (1992, 1995) model projected and Leinonen's (1996) model did not project heavy frost damage for climatic warming (Fig. 8.21 in Sect. 8.3.3.2). However, the illustration of the potential effects of frost damage on photosynthetic production was necessary, for the potentially occurring frost damage is not addressed in most process-based forest ecosystem models. Furthermore, as concluded in Sect. 8.3, the risk of increased incidence of frost damage to boreal and temperate trees cannot be ruled out even though limited data from empirical tests with *Picea abies* and *Pinus sylvestris* suggest a falsification of the particular models, including Kellomäki et al.'s (1992, 1995) integrated model of frost hardiness, that project catastrophic frost damage as a result of climatic warming. In all, then, the results shown in Fig. 8.26 demonstrate that in addition to the usually acknowledged uncertainties discussed in Sect. 8.5.1, the risk of frost damage caused by premature dehardening and growth onset should also be addressed when the effects of the climate change on the productivity of boreal and temperate forest stands are projected (Hänninen et al. 2005).

8.5.3 Productivity in Picea abies *Stands*

Rammig et al. (2010) studied the effects of climate change on the productivity of *Picea abies* stands in Swedish conditions. They took a large-scale modelling approach, combining a regional climate model (RCM) with the generalised ecosystem model LPJ-GUESS. As a novelty, they introduced into LPJ-GUESS the effects of frost damage to the trees. To this end, they used Jönsson et al.'s (2004) frost hardiness model and Kellomäki et al.'s (1995) frost damage model as sub-models in LPJ-GUESS. Using the ecosystem model thus refined in scenario simulations, they projected interesting regional differences in the productivity of *Picea abies* stands in future conditions. As a result of climatic warming, the effect of frost damage on productivity was mainly projected to decrease, especially in northern Sweden, but in some cases the incidence of frost damage was projected to increase, thus decreasing the productivity of the *Picea abies* stands by up to 25 %.

Unfortunately, the main issue of this chapter, i.e., the critical testing of the models used in projecting the ecological effects of climate change, needs to be brought forward in the case of Rammig et al.'s (2010) study, too. No tests were presented for the frost hardiness model, and the frost damage model was tested by comparing its output with observed needle damage documented in 1999–2005. In the light of the reasoning presented in Sect. 8.3.2.2, it is highly uncertain how much the model's ability to predict such observations in the current climate tells us about its predictive power for the changed climatic conditions projected to prevail in the future.

Still, this criticism should not obscure the obvious merits of Rammig et al.'s (2010) pioneering study. They introduced, evidently for the first time, tree frost damage into a regional ecosystem model, in this case LPJ-GUESS, and with the refined ecosystem model they demonstrated some potential effects that tree frost damage may have on tree productivity in the changed climate. In a sense, then, the obvious shortcomings in Rammig et al.'s (2010) study are actually useful because they highlight the urgent need to develop realistic and accurate models of tree frost hardiness and damage for use in large-scale ecosystem modelling.

References

Ainsworth, E. A., & Long, S. P. (2005). What have we learned from 15 years of free-air CO_2 enrichment (FACE)? A meta-analytic review of the responses of photosynthesis, canopy properties and plant production to rising CO_2. *New Phytologist, 165*, 351–372.

Augspurger, C. K. (2013). Reconstructing patterns of temperature, phenology, and frost damage over 124 years: Spring damage risk is increasing. *Ecology, 94*, 41–50.

Bach, W. (1987). Development of climatic scenarios: A. From general circulation models. In M. L. Parry, T. R. Carter, & N. T. Konijn (Eds.), *The impact of climatic variations on agriculture* (Assessment in cool temperate and cold regions, Vol. 1, pp. 125–157). Dordrecht: Kluwer Academic Publishers.

Bakkenes, M., Alkemade, J. R. M., Ihle, F., Leemans, R., & Latour, J. B. (2002). Assessing effects of forecasted climate change on the diversity and distribution of European higher plants for 2050. *Global Change Biology, 8*, 390–407.

Bartelt, P., & Lehning, M. (2002). A physical SNOWPACK model for the Swiss avalanche warning. Part I: Numerical model. *Cold Regions Science and Technology, 35*, 123–145.

Barton, C. V. M., Ellsworth, D. S., Medlyn, B. E., Duursma, R. A., Tissue, D. T., Adams, M. A., Eamus, D., Conroy, J. P., McMurtrie, R. E., Parsby, J., & Linder, S. (2010). Whole-tree chambers for elevated atmospheric CO_2 experimentation and tree scale flux measurements in south-eastern Australia: The Hawkesbury Forest Experiment. *Agricultural and Forest Meteorology, 150*, 941–951.

Basler, D., & Körner, C. (2012). Photoperiod sensitivity of bud burst in 14 temperate forest tree species. *Agricultural and Forest Meteorology, 165*, 73–81.

Bergh, J., Freeman, M., Sigurdsson, B., Kellomäki, S., Laitinen, K., Niinistö, S., Peltola, H., & Linder, S. (2003). Modelling the short-term effects of climate change on the productivity of selected tree species in Nordic countries. *Forest Ecology and Management, 183*, 327–340.

Berglund, B. E. (Ed.). (1986). *Handbook of Holocene palaeoecology and palaeohydrology*. Chichester: Wiley. 869 p.

Beuker, E. (1994). Long-term effects of temperature on the wood production of *Pinus sylvestris* L. and *Picea abies* (L.) Karst. in old provenance experiments. *Scandinavian Journal of Forest Research, 9*, 35–45.

Bokhorst, S. F., Bjerke, J. W., Tømmervik, H., Callaghan, T. V., & Phoenix, G. K. (2009). Winter warming events damage sub-arctic vegetation: Consistent evidence from an experimental manipulation and a natural event. *Journal of Ecology, 97*, 1408–1415.

Bradley, B. A. (2009). Regional analysis of the impacts of climate change on cheatgrass invasion shows potential risk and opportunity. *Global Change Biology, 15*, 196–208.

Caffarra, A. (2007). *Quantifying the environmental drivers of tree phenology*. Thesis submitted for the degree of Doctor of Philosophy. School of Natural Sciences, Trinity College, University of Dublin.

Caffarra, A., & Donnelly, A. (2011). The ecological significance of phenology in four different tree species: Effects of light and temperature on bud burst. *International Journal of Biometeorology, 55*, 711–721.

Caffarra, A., Donnelly, A., Chuine, I., & Jones, M. B. (2011). Modelling the timing of *Betula pubescens* budburst. I. Temperature and photoperiod: A conceptual model. *Climate Research, 46*, 147–157.

Campbell, R. K. (1978). Regulation of bud-burst timing by temperature and photoregime during dormancy. In C. A. Hollis & A. E. Squillace (Eds.), *Proceedings: Fifth north American forest biology workshop* (pp. 19–33). Gainesville: Forestry Department, University of Florida.

Campbell, R. K., & Sugano, A. I. (1975). Phenology of bud burst in Douglas-fir related to provenance, photoperiod, chilling, and flushing temperature. *Botanical Gazette, 136*, 290–298.

Campbell, R. K., & Sugano, A. I. (1979). Genecology of bud-burst phenology in Douglas-fir: Response to flushing temperature and chilling. *Botanical Gazette, 140*, 223–231.

Cannell, M. G. R. (1985). Analysis of risks of frost damage to forest trees in Britain. In P. M. A. Tigerstedt, P. Puttonen, & V. Koski (Eds.), *Crop physiology of forest trees* (pp. 153–166). Helsinki: Helsinki University Press.

Cannell, M. G. R., & Smith, R. I. (1983). Thermal time, chill days and prediction of budburst in *Picea sitchensis*. *Journal of Applied Ecology, 20*, 951–963.

Cannell, M. G. R., & Smith, R. I. (1986). Climatic warming, spring budburst and frost damage on trees. *Journal of Applied Ecology, 23*, 177–191.

Ceulemans, R., & Mousseau, M. (1994). Effects of elevated atmospheric CO_2 on woody plants. *New Phytologist, 127*, 425–446.

Cooke, J. E. K., Eriksson, M. E., & Junttila, O. (2012). The dynamic nature of bud dormancy in trees: Environmental control and molecular mechanisms. *Plant, Cell and Environment, 35*, 1707–1728.

Cooper, E. J. (2014). Warmer shorter winters disrupt arctic terrestrial ecosystems. *Annual Review of Ecology, Evolution, and Systematics, 45*, 271–295.

Desprez-Loustau, M.-L., Robin, C., Raynaud, G., Déqué, M., Badeau, V., Piou, D., Husson, C., & Marçais, B. (2007). Simulating the effects of a climate-change scenario on the geographical range and activity of forest-pathogenic fungi. *Canadian Journal of Plant Pathology, 29*, 101–120.

Fu, Y. H., Campioli, M., Deckmyn, G., & Janssens, I. A. (2012). The impact of winter and spring temperatures on temperate tree budburst dates: Results from an experimental climate manipulation. *PLoS One, 7*(10), e47324. doi:10.1371/journal.pone.0047324.

Fu, Y. H., Campioli, M., Deckmyn, G., & Janssens, I. A. (2013). Sensitivity of leaf unfolding to experimental warming in three temperate tree species. *Agricultural and Forest Meteorology, 181*, 125–132.

Fu, Y. H., Piao, S., Vitasse, Y., Zhao, H., De Boeck, H. J., Liu, Q., Yang, H., Weber, U., Hänninen, H., & Janssens, I. A. (2015a). Increased heat requirement for leaf flushing in temperate woody species over 1980–2012: Effects of chilling, precipitation and insolation. *Global Change Biology, 21*, 2687–2697.

Fu, Y. H., Zhao, H., Piao, S., Peaucelle, M., Peng, S., Zhou, G., Ciais, P., Huang, M., Menzel, A., Peñuelas, J., Song, Y., Vitasse, Y., Zeng, Z., & Janssens, I. A. (2015b). Declining global warming effects on the phenology of spring leaf unfolding. *Nature, 526*, 104–107.

Gu, L., Hanson, P. J., Post, W. M., Kaiser, D. P., Yang, B., Nemani, R., Pallardy, S. G., & Meyers, T. (2008). The 2007 eastern US spring freeze: Increased cold damage in a warming world? *BioScience, 58*, 253–262.

Guak, S., Olsyzk, D. M., Fuchigami, L. H., & Tingey, D. T. (1998). Effects of elevated CO_2 and temperature on cold hardiness and spring bud burst and growth in Douglas-fir (*Pseudotsuga menziesii*). *Tree Physiology, 18*, 671–679.

Häkkinen, R., Linkosalo, T., & Hari, P. (1995). Methods for combining phenological time series: Application to bud burst in birch (*Betula pendula*) in Central Finland for the period 1896–1955. *Tree Physiology, 15*, 721–726.

Häkkinen, R., Linkosalo, T., & Hari, P. (1998). Effects of dormancy and environmental factors on timing of bud burst in *Betula pendula*. *Tree Physiology, 18*, 707–712.

Hänninen, H. (1987). Metsäpuiden vuosirytmistä. *Metsäntutkimuslaitoksen Tiedonantoja, 249*, 12–21.

Hänninen, H. (1990a). Modeling dormancy release in trees from cool and temperate regions. In R. K. Dixon, R. S. Meldahl, G. A. Ruark, & W. G. Warren (Eds.), *Process modeling of forest growth responses to environmental stress* (pp. 159–165). Portland: Timber Press.

Hänninen, H. (1990b). Modelling bud dormancy release in trees from cool and temperate regions. *Acta Forestalia Fennica, 213*, 1–47.

Hänninen, H. (1991). Does climatic warming increase the risk of frost damage in northern trees? *Plant, Cell and Environment, 14*, 449–454.

Hänninen, H. (1995a). Effects of climatic change on trees from cool and temperate regions: An ecophysiological approach to modelling of bud burst phenology. *Canadian Journal of Botany, 73*, 183–199.

Hänninen, H. (1995b). Assessing ecological implications of climatic change: Can we rely on our simulation models? *Climatic Change, 31*, 1–4.

Hänninen, H. (1996). Effects of climatic warming on northern trees: Testing the frost damage hypothesis with meteorological data from provenance transfer experiments. *Scandinavian Journal of Forest Research, 11*, 17–25.

Hänninen, H. (2006). Climate warming and the risk of frost damage to boreal forest trees: Identification of critical ecophysiological traits. *Tree Physiology, 26*, 889–898.

Hänninen, H., & Kramer, K. (2007). A framework for modelling the annual cycle of trees in boreal and temperate regions. *Silva Fennica, 41*, 167–205.

Hänninen, H., & Lundell, R. (2007). Dynamic models in plant ecophysiology. In E. Taulavuori & K. Taulavuori (Eds.), *Physiology of northern plants under changing environment* (pp. 157–175). Trivandrum, Kerala: Research Signpost.

Hänninen, H., & Tanino, K. (2011). Tree seasonality in a warming climate. *Trends in Plant Science, 16*, 412–416.

Hänninen, H., Kellomäki, S., Laitinen, K., Pajari, B., & Repo, T. (1993). Effect of increased winter temperature on the onset of height growth of Scots pine: A field test of a phenological model. *Silva Fennica, 27*, 251–257.

Hänninen, H., Leinonen, I., Repo, T., & Kellomäki, S. (1996). Overwintering and productivity of Scots pine in a changing climate. *Silva Fennica, 30*, 229–237.

Hänninen, H., Kolari, P., & Hari, P. (2005). Seasonal development of Scots pine under climatic warming: Effects on photosynthetic production. *Canadian Journal of Forest Research, 35*, 2092–2099.

Hänninen, H., Slaney, M., & Linder, S. (2007). Dormancy release of Norway spruce under climatic warming: Testing ecophysiological models of bud burst with a whole-tree chamber experiment. *Tree Physiology, 27*, 291–300.

Hänninen, H., Lundell, R., & Junttila, O. (2013). A fifty-year-old conceptual plant dormancy model provides new insights into dynamic phenology modelling. In R. Sievänen, E. Nikinmaa, C. Godin, A. Lintunen, & P. Nygren (Eds.), *Proceedings of the 7th international conference on functional-structural plant models,* Saariselkä, 9–14 June 2013. http://www.metal.fi/fspm2013/proceedings. ISBN 978-951-651-408-9. p. 176.

Hari, P., Nikinmaa, E., Vesala, T., Pohja, T., Siivola, E., Lahti, T., Aalto, P., Hiltunen, V., Ilvesniemi, H., Keronen, P., Kolari, P., Grönholm, T., Palva, L., Pumpanen, J., Petäjä, T.,

Rannik, Ü., & Kulmala, M. (2008). SMEAR network. In P. Hari & L. Kulmala (Eds.), *Boreal forest and climate change* (pp. 46–56). Dordrecht: Springer Science+Business Media B.V.

Heide, O. M. (2003). High autumn temperature delays spring bud burst in boreal trees, counterbalancing the effect of climatic warning. *Tree Physiology, 23*, 931–936.

Henry, G. H. R., & Molau, U. (1997). Tundra plants and climate change: The International Tundra Experiment (ITEX). *Global Change Biology, 3*(Suppl. 1), 1–9.

Hijmans, R. J., & Graham, C. H. (2006). The ability of climate envelope models to predict the effect of climate change on species distributions. *Global Change Biology, 12*, 2272–2281.

Hlásny, T., & Turčáni, M. (2009). Insect pests as climate change driven disturbances in forest ecosystems. In K. Střelcová, C. Mátyás, A. Kleidon, M. Lapin, F. Matejka, M. Blaženec, J. Škvarenina, & J. Holécy (Eds.), *Bioclimatology and natural hazards* (pp. 165–177). Dordrecht: Springer Science+Business Media B.V.

IPCC. (2013). T. F. Stocker, D. Qin, G.-K. Plattner, M. Tignor, S. K. Allen, J. Boschung, A. Nauels, Y. Xia, V. Bex, & P. M. Midgley (Eds.), *Climate change 2013: The physical science basis. Contribution of working group I to the fifth assessment report of the intergovernmental panel on climate change*. Cambridge/New York: Cambridge University Press. 1535 p.

IPCC. (2014a). C. B. Field, V. R. Barros, D. J. Dokken, K. J. Mach, M. D. Mastrandrea, T. E. Bilir, M. Chatterjee, K. L. Ebi, Y. O. Estrada, R. C. Genova, B. Girma, E. S. Kissel, A. N. Levy, S. MacCracken, P. R. Mastrandrea, & L. L. White (Eds.), *Climate change 2014: Impacts, adaptation, and vulnerability. Part a: Global and sectoral aspects. Contribution of working group II to the fifth assessment report of the intergovernmental panel on climate change*. Cambridge/New York: Cambridge University Press. 1132 p.

IPCC. (2014b). V. R. Barros, C. B. Field, D. J. Dokken, M. D. Mastrandrea, K. J. Mach, T. E. Bilir, M. Chatterjee, K. L. Ebi, Y. O. Estrada, R. C. Genova, B. Girma, E. S. Kissel, A. N. Levy, S. MacCracken, P. R. Mastrandrea, & L. L. White (Eds.), *Climate change 2014: Impacts, adaptation, and vulnerability. Part B: Regional aspects. Contribution of working group II to the fifth assessment report of the intergovernmental panel on climate change*. Cambridge/New York: Cambridge University Press. 688 p.

Jönsson, A. M., Linderson, M.-L., Stjernquist, I., Schlyter, P., & Bärring, L. (2004). Climate change and the effect of temperature backlashes causing frost damage in *Picea abies*. *Global and Planetary Change, 44*, 195–207.

Kalela, A. (1938). Zur Synthese der experimentellen Untersuchungen über Klimarassen der Holzarten. *Communicationes Instituti Forestalis Fenniae, 26*(1), 1–445.

Kellomäki, S., & Väisänen, H. (1997). Modelling the dynamics of the forest ecosystem for climate change studies in the boreal conditions. *Ecological Modelling, 97*, 121–140.

Kellomäki, S., Väisänen, H., Hänninen, H., Kolström, T., Lauhanen, R., Mattila, U., & Pajari, B. (1992). A simulation model for the succession of the boreal forest ecosystem. *Silva Fennica, 26*, 1–18.

Kellomäki, S., Hänninen, H., & Kolström, M. (1995). Computations on frost damage to Scots pine under climatic warming in boreal conditions. *Ecological Applications, 5*, 42–52.

Kellomäki, S., Karjalainen, T., & Väisänen, H. (1997). More timber from boreal forests under changing climate? *Forest Ecology and Management, 94*, 195–208.

Kellomäki, S., Wang, K.-Y., & Lemettinen, M. (2000). Controlled environment chambers for investigating tree response to elevated CO_2 and temperature under boreal conditions. *Photosynthetica, 38*, 69–81.

Kettunen, L., Mukula, J., Pohjonen, V., Rantanen, O., & Varjo, U. (1987). The effects of climatic variations on agriculture in Finland. In M. L. Parry, T. R. Carter, & N. T. Konijn (Eds.), *The impact of climatic variations on agriculture* (Assessment in cool temperate and cold regions, Vol. 1, pp. 513–614). Dordrecht: Kluwer Academic Publishers.

Kilpeläinen, A., Peltola, H., Rouvinen, I., & Kellomäki, S. (2006). Dynamics of daily height growth in Scots pine trees at elevated temperature and CO_2. *Trees, 20*, 16–27.

Kolari, P., & Hari, P. (2008). Photosynthetic production of a tree. In P. Hari & L. Kulmala (Eds.), *Boreal forest and climate change* (pp. 313–325). Dordrecht: Springer Science + Business Media B.V.

Koski, V., & Selkäinaho, J. (1982). Experiments on the joint effect of heat sum and photoperiod on seedlings of Betula pendula. *Communicationes Instituti Forestalis Fenniae, 105*, 1–34.
Koski, V., & Sievänen, R. (1985). Timing of growth cessation in relation to the variations in the growing season. In P. M. A. Tigerstedt, P. Puttonen, & V. Koski (Eds.), *Crop physiology of forest trees* (pp. 167–193). Helsinki: Helsinki University Press.
Kramer, K. (1994). A modelling analysis of the effects of climatic warming on the probability of spring frost damage to tree species in The Netherlands and Germany. *Plant, Cell and Environment, 17*, 367–377.
Kramer, K. (1996). *Phenology and growth of European trees in relation to climate change*. Thesis, Landbouw Universiteit Wageningen, the Netherlands, 210 p.
Kramer, K., & Hänninen, H. (2009). The annual cycle of development of trees and process-based modelling of growth to scale up from the tree to the stand. In A. Noormets (Ed.), *Phenology of ecosystem processes* (pp. 201–227). Dordrecht: Springer Science+Business Media LLC.
Kramer, K., Leinonen, I., Bartelink, H. H., Berbigier, P., Borghetti, M., Bernhofer, C., Cienciala, E., Dolman, A. J., Froer, O., Gracia, C. A., Granier, A., Grünwald, T., Hari, P., Jans, W., Kellomäki, S., Loustau, D., Magnani, F., Markkanen, T., Matteucci, G., Mohren, G. M. J., Moors, E., Nissinen, A., Peltola, H., Sabaté, S., Sanchez, A., Sontag, M., Valentini, R., & Vesala, T. (2002). Evaluation of six process-based forest growth models using eddy-covariance measurements of CO_2 and H_2O fluxes at six forest sites in Europe. *Global Change Biology, 8*, 213–230.
Kreyling, J. (2010). Winter climate change: A critical factor for temperate vegetation performance. *Ecology, 91*, 1939–1948.
Landsberg, J. J. (1974). Apple fruit bud development and growth; analysis and an empirical model. *Annals of Botany, 38*, 1013–1023.
Leinonen, I. (1996). A simulation model for the annual frost hardiness and freeze damage of Scots pine. *Annals of Botany, 78*, 687–693.
Leinonen, I., Repo, T., & Hänninen, H. (1996). Testing of frost hardiness models for *Pinus sylvestris* in natural conditions and in elevated temperature. *Silva Fennica, 30*, 159–168.
Linkosalo, T., & Lechowicz, M. (2006). Twilight far-red treatment advances leaf bud burst of silver birch (*Betula pendula*). *Tree Physiology, 26*, 1249–1256.
Linkosalo, T., Häkkinen, R., & Hari, P. (1996). Improving the reliability of a combined phenological time series by analyzing observation quality. *Tree Physiology, 16*, 661–664.
Linkosalo, T., Carter, T. R., Häkkinen, R., & Hari, P. (2000). Predicting spring phenology and frost damage risk of *Betula* spp. under climatic warming: A comparison of two models. *Tree Physiology, 20*, 1175–1182.
Linkosalo, T., Häkkinen, R., Terhivuo, J., Tuomenvirta, H., & Hari, P. (2009). The time series of flowering and leaf bud burst of boreal trees (1846–2005) support the direct temperature observations of climatic warming. *Agricultural and Forest Meteorology, 149*, 453–461.
Mäkelä, A., Hari, P., Berninger, F., Hänninen, H., & Nikinmaa, E. (2004). Acclimation of photosynthetic capacity in Scots pine to the annual cycle of temperature. *Tree Physiology, 24*, 369–376.
Marion, G. M., Henry, G. H. R., Freckman, D. W., Johnstone, J., Jones, G., Jones, M. H., Lévesque, E., Molau, U., Mølgaard, P., Parsons, A. N., Svoboda, J., & Virginia, R. A. (1997). Open-top designs for manipulating field temperature in high-latitude ecosystems. *Global Change Biology, 3*(Suppl. 1), 20–32.
McCarroll, D., Tuovinen, M., Campbell, R., Gagen, M., Grudd, H., Jalkanen, R., Loader, N. J., & Robertson, I. (2011). A critical evaluation of multi-proxy dendroclimatology in northern Finland. *Journal of Quaternary Science, 26*, 7–14.
Medhurst, J., Parsby, J., Linder, S., Wallin, G., Ceschia, E., & Slaney, M. (2006). A whole-tree chamber system for examining tree-level physiological responses of field-grown trees to environmental variation and climate change. *Plant, Cell and Environment, 29*, 1853–1869.
Menzel, A., & Fabian, P. (1999). Growing season extended in Europe. *Nature, 397*, 659.

Menzel, A., & Sparks, T. (2006). Temperature and plant development: Phenology and seasonality. In J. I. L. Morison & M. D. Morecroft (Eds.), *Plant growth and climate change* (pp. 70–95). Oxford: Blackwell Publishing Ltd.

Menzel, A., Sparks, T. H., Estrella, N., Koch, E., Aasa, A., Ahas, R., Alm-Kübler, K., Bissolli, P., Braslavská, O., Briede, A., Chmielewski, F. M., Crepinsek, Z., Curnel, Y., Dahl, Å., Defila, C., Donnelly, A., Filella, Y., Jatczak, K., Måke, F., Mestre, A., Nordli, Ø., Peñuelas, J., Pirinen, P., Remišová, V., Scheifinger, H., Striz, M., Susnik, A., van Lieth, A. J. H., Wielgolaski, F.-E., Zach, S., & Zust, A. (2006). European phenological response to climate change matches the warming pattern. *Global Change Biology, 12*, 1969–1976.

Mooney, H. A., & Ehleringer, J. R. (1997). Photosynthesis. In M. J. Crawley (Ed.), *Plant ecology* (2nd ed., pp. 1–27). Oxford: Blackwell Science.

Moore, J. (2011). *Wood properties and uses of Sitka spruce in Britain* (Forestry Commission Research Report). Edinburgh: Forest Commission. 48 p.

Morin, X., Lechowicz, M. J., Augspurger, C., O'Keefe, J., Viner, D., & Chuine, I. (2009). Leaf phenology in 22 North American tree species during the 21st century. *Global Change Biology, 15*, 961–975.

Morison, J. I. L., & Morecroft, M. D. (Eds.). (2006). *Plant growth and climate change*. Oxford: Blackwell Publishing Ltd. 213 p.

Murray, M. B., Cannell, M. G. R., & Smith, R. I. (1989). Date of budburst of fifteen tree species in Britain following climatic warming. *Journal of Applied Ecology, 26*, 693–700.

Murray, M. B., Smith, R. I., Leith, I. D., Fowler, D., Lee, H. S. J., Friend, A. D., & Jarvis, P. G. (1994). Effects of elevated CO_2, nutrition and climatic warming on bud phenology in Sitka spruce (*Picea sitchensis*) and their impact on the risk of frost damage. *Tree Physiology, 14*, 691–706.

Myking, T. (1997). Effects of constant and fluctuating temperature on time to budburst in *Betula pubescens* and its relation to bud respiration. *Trees, 12*, 107–112.

Myneni, R. B., Keeling, C. D., Tucker, C. J., Asrar, G., & Nemani, R. R. (1997). Increased plant growth in the northern high latitudes from 1981 to 1991. *Nature, 386*, 698–702.

Nikkanen, T., Karvinen, K., Koski, V., Rusanen, M., & Yrjänä-Ketola, L. (1999). Kuusen ja männyn siemenviljelykset ja niiden käyttöalueet. *Metsäntutkimuslaitoksen tiedonantoja* 730. 203 p.

Parmesan, C., & Yohe, G. (2003). A globally coherent fingerprint of climate change impacts across natural systems. *Nature, 421*, 37–42.

Peltola, H., Kellomäki, S., & Väisänen, H. (1999). Model computations of the impact of climatic change on the windthrow risk of trees. *Climatic Change, 41*, 17–36.

Persson, B., & Beuker, E. (1997). Distinguishing between the effects of changes in temperature and light climate using provenance trials with *Pinus sylvestris* in Sweden. *Canadian Journal of Forest Research, 27*, 572–579.

Rammig, A., Jönsson, A. M., Hickler, T., Smith, B., Bärring, L., & Sykes, M. T. (2010). Impacts of changing frost regimes on Swedish forests: Incorporating cold hardiness in a regional ecosystem model. *Ecological Modelling, 221*, 303–313.

Rapacz, M., Ergon, Å., Höglind, M., Jørgensen, M., Jurczyk, B., Østrem, L., Rognli, O. A., & Tronsmo, A. M. (2014). Overwintering of herbaceous plants in a changing climate. Still more questions than answers. *Plant Science, 225*, 34–44.

Rasmus, S., Gustafsson, D., Lundell, R., & Saarinen, T. (2016). Observations and snow model simulations of winter energy balance terms within and between different coniferous forests in Southern Boreal Finland. *Hydrology Research, 47*, 201–216.

Repo, T. (1992). Seasonal changes of frost hardiness in *Picea abies* and *Pinus sylvestris* in Finland. *Canadian Journal of Forest Research, 22*, 1949–1957.

Repo, T., Hänninen, H., & Kellomäki, S. (1996). The effects of long-term elevation of air temperature and CO_2 on the frost hardiness of Scots pine. *Plant, Cell and Environment, 19*, 209–216.

Roberts, A. M. I., Tansey, C., Smithers, R. J., & Phillimore, A. B. (2015). Predicting a change in the order of spring phenology in temperate forests. *Global Change Biology, 21*, 2603–2611.

Robson, T. M., Rasztovits, E., Aphalo, P. J., Alia, R., & Aranda, I. (2013). Flushing phenology and fitness of European beech (*Fagus sylvatica* L.) provenances from a trial in La Rioja, Spain, segregate according to their climate origin. *Agricultural and Forest Meteorology, 180*, 76–85.

Rousi, M., Possen, B. J. H. M., Hagqvist, R., & Thomas, B. R. (2012). From the Arctic Circle to the Canadian prairies – A case study of silver birch acclimation capacity. *Silva Fennica, 46*, 355–364.

Sakai, A., & Larcher, W. (1987). *Frost survival of plants. Responses and adaptation to freezing stress*. Berlin: Springer-Verlag. 321 pp.

Sarvas, R. (1972). Investigations on the annual cycle of development of forest trees. Active period. *Communicationes Instituti Forestalis Fenniae, 76*(3), 1–110.

Sarvas, R. (1974). Investigations on the annual cycle of development of forest trees. II. Autumn dormancy and winter dormancy. *Communicationes Instituti Forestalis Fenniae, 84*(1), 1–101.

Saxe, H., Cannell, M. G. R., Johnsen, Ø., Ryan, M. G., & Vourlitis, G. (2001). Tree and forest functioning in response to global warming. *New Phytologist, 149*, 369–400.

Slaney, M., Wallin, G., Medhurst, J., & Linder, S. (2007). Impact of elevated carbon dioxide concentration and temperature on bud burst and shoot growth of boreal Norway spruce. *Tree Physiology, 27*, 301–312.

Sutinen, S., Partanen, J., Viherä-Aarnio, A., & Häkkinen, R. (2009). Anatomy and morphology in developing vegetative buds on detached Norway spruce branches in controlled conditions before bud burst. *Tree Physiology, 29*, 1457–1465.

Sutinen, S., Partanen, J., Viherä-Aarnio, A., & Häkkinen, R. (2012). Development and growth of primordial shoots in Norway spruce buds before visible bud burst in relation to time and temperature in the field. *Tree Physiology, 32*, 987–997.

Taulavuori, K. M. J., Taulavuori, E. B., Skre, O., Nilsen, J., Igeland, B., & Laine, K. M. (2004). Dehardening of mountain birch (*Betula pubescens* ssp. *czerepanovii*) ecotypes at elevated winter temperatures. *New Phytologist, 162*, 427–436.

Thomas, C. D., Cameron, A., Green, R. E., Bakkenes, M., Beaumont, L. J., Collingham, Y. C., Erasmus, B. F. N., de Siqueira, M. F., Grainger, A., Hannah, L., Hughes, L., Huntley, B., van Jaarsveld, A. S., Midgley, G. F., Miles, L., Ortega-Huerta, M. A., Peterson, A. T., Phillips, O. L., & Williams, S. E. (2004). Extinction risk from climate change. *Nature, 427*, 145–148.

Thuiller, W., Lavorel, S., Araújo, M. B., Sykes, M. T., & Prentice, I. C. (2005). Climate change threats to plant diversity in Europe. *PNAS, 102*, 8245–8250.

Ununger, J., Ekberg, I., & Kang, H. (1988). Genetic control and age-related changes of juvenile growth characters in *Picea abies*. *Scandinavian Journal of Forest Research, 3*, 55–66.

van der Kamp, B. J., & Worrall, J. (1990). An unusual case of winter bud damage in British Columbia interior conifers. *Canadian Journal of Forest Research, 20*, 1640–1647.

Vegis, A. (1964). Dormancy in higher plants. *Annual Review of Plant Physiology, 15*, 185–224.

Viherä-Aarnio, A., & Velling, P. (2008). Seed transfers of silver birch (*Betula pendula*) from the Baltic to Finland – Effect on growth and stem quality. *Silva Fennica, 42*, 735–751.

Viherä-Aarnio, A., Sutinen, S., Partanen, J., & Häkkinen, R. (2014). Internal development of vegetative buds of Norway spruce trees in relation to accumulated chilling and forcing temperatures. *Tree Physiology, 34*, 547–556.

Vitasse, Y. (2013). Ontogenic changes rather than difference in temperature cause understory trees to leaf out earlier. *New Phytologist, 198*, 149–155.

Wallin, G., Hall, M., Slaney, M., Räntfors, M., Medhurst, J., & Linder, S. (2013). Spring photosynthetic recovery of boreal Norway spruce under conditions of elevated [CO_2] and air temperature. *Tree Physiology, 33*, 1177–1191.

Wang, A.-F., Roitto, M., Lehto, T., Zwiazek, J. J., Calvo-Polanco, M., & Repo, T. (2013). Waterlogging under simulated late-winter conditions had little impact on the physiology and growth of Norway spruce seedlings. *Annals of Forest Science, 70*, 781–790.

Willis, K. J., & MacDonald, G. M. (2011). Long-term ecological records and their relevance to climate change predictions for a warmer world. *Annual Review of Ecology, Evolution, and Systematics, 42*, 267–287.

Woldendorp, G., Hill, M. J., Doran, R., & Ball, M. C. (2008). Frost in a future climate: modelling interactive effects of warmer temperatures and rising atmospheric [CO_2] on the incidence and severity of frost damage in a temperate evergreen (*Eucalyptus pauciflora*). *Global Change Biology, 14*, 294–308.

Wolkovich, E. M., Cook, B. I., Allen, J. M., Crimmins, T. M., Betancourt, J. L., Travers, S. E., Pau, S., Regetz, J., Davies, T. J., Kraft, N. J. B., Ault, T. R., Bolmgren, K., Mazer, S. J., McCabe, G. J., McGill, B. J., Parmesan, C., Salamin, N., Schwartz, M. D., & Cleland, E. E. (2012). Warming experiments underpredict plant phenological responses to climate change. *Nature, 485*, 494–497.

Zaehle, S., Medlyn, B. E., De Kauwe, M. G., Walker, A. P., Dietze, M. C., Hickler, T., Luo, Y., Wang, Y.-P., El-Masri, B., Thornton, P., Jain, A., Wang, S., Warlind, D., Weng, E., Parton, W., Iversen, C. M., Gallet-Budynek, A., McCarthy, H., Finzi, A., Hanson, P. J., Prentice, I. C., Oren, R., & Norby, R. J. (2014). Evaluation of 11 terrestrial carbon–nitrogen cycle models against observations from two temperate Free-Air CO_2 enrichment studies. *New Phytologist, 202*, 803–822.

Chapter 9
Concluding Remarks

Contents

Abstract The most substantial findings and conclusions from the eight preceding chapters are briefly revisited. The need for biological realism in ecophysiological models for the annual cycle of boreal and temperate trees is emphasised, and the need to improve the transparency of the various published models is pointed out. Different models have drastically different degrees of empirical support, and the evaluation of this support is a major theme throughout this book. Even models with sound empirical support from earlier tests may be falsified when they are further tested with more thorough tests. However, models that have been corroborated earlier but falsified later should not be abandoned once and for all, for a lot can be learnt from them. Though the present volume essentially belongs to the discipline of whole-tree ecophysiology, recent studies addressing the upscaling of the models to the ecosystem level are also briefly discussed. In addition, the downscaling of the models to the anatomical, cellular, and molecular levels is also proposed. That would facilitate a synthesis of the whole-tree modelling approach discussed in the present volume and the molecular biology of plants, which is progressing very rapidly today. As to the effects of the projected climate change, the main emphasis is on the hypothesis that climatic warming will increase the incidence of frost damage in boreal and temperate trees due to premature dehardening and growth onset during mild spells in winter and early spring. For several reasons, this hypothesis cannot be ruled out even though experimental tests up to now have not provided support for the catastrophic damage projected earlier. For further evaluation of the hypothesis, more experimental work with different tree species, provenances, and cultivars is needed. The research discussed in the present volume has pointed out not only the ecophysiological traits of the trees that are critical for the hypothesis but also the experimental designs that facilitate the determining of those traits in any tree population.

H. Hänninen, *Boreal and Temperate Trees in a Changing Climate*, Biometeorology 3,
DOI 10.1007/978-94-017-7549-6_9

Keywords Annual cycle • Boreal trees • Computer simulations • Ecological scenario • Frost hardiness • Frost damage • Model realism • Premature dehardening • Premature growth onset • Scenario simulation • Seasonality • Temperate trees • Unseasonally warm spells • Winter warming

9.1 Methodological and Interdisciplinary Considerations

In the present volume, studies addressing the ecophysiological modelling of seasonal phenomena in boreal and temperate trees under changing climate conditions were reviewed. The approach adopted here devoted special emphasis to the ecophysiological basis and realism of the models used to assess the effects of the climate change. To this end, most of the volume actually addresses the ecophysiology of the phenomena included in the annual cycle and their modelling as such, without considering the effects of climate change.

Furthermore, studies concerned with the modelling of the annual cycle were reviewed historically, so that in addition to the current state of the modelling approach, the development of the models into their present form was also discussed. The idea was to make the models as transparent as possible to all those who wish to apply them in ecological scenario studies, often by upscaling the models for use as sub-models in large stand and ecosystem models (Fig. 9.1). Transparency is needed not only for recognising which phenomena are addressed and how they are addressed in the ecophysiological seasonality models but also for recognising the degree of empirical support each model has. As emphasised throughout the present volume, there are great differences in the empirical support available for the models currently used in projecting the ecological effects of climate change, and in general, further empirical studies are needed to improve the models.

Modelling was discussed in the present volume within the discipline of whole-tree ecophysiology, i.e., whole-tree attributes of the annual cycle were addressed. In each case, the physiological status of the tree is summarised into the value(s) of one or more state variables, depending on the number of whole-tree attributes addressed. These attributes, such as needle frost hardiness, are emergent manifestations of the basic physiological properties existing and the phenomena occurring at the anatomical, cellular, and molecular levels. Thus, though the approach taken evidently seems mechanistic to the ecosystem modeller, the plant physiologist will still see the models discussed in this volume as oversimplifications of the real-life phenomena of boreal and temperate trees in several respects. This is not only because the models applied mainly address the phenomena of the annual cycle at the whole-tree level without considering the underlying lower-level phenomena but also, as discussed in Chaps. 3, 4, 5, and 6, because several of the whole-tree concepts involved in the current models need to be revisited in order to improve the physiological realism of the models.

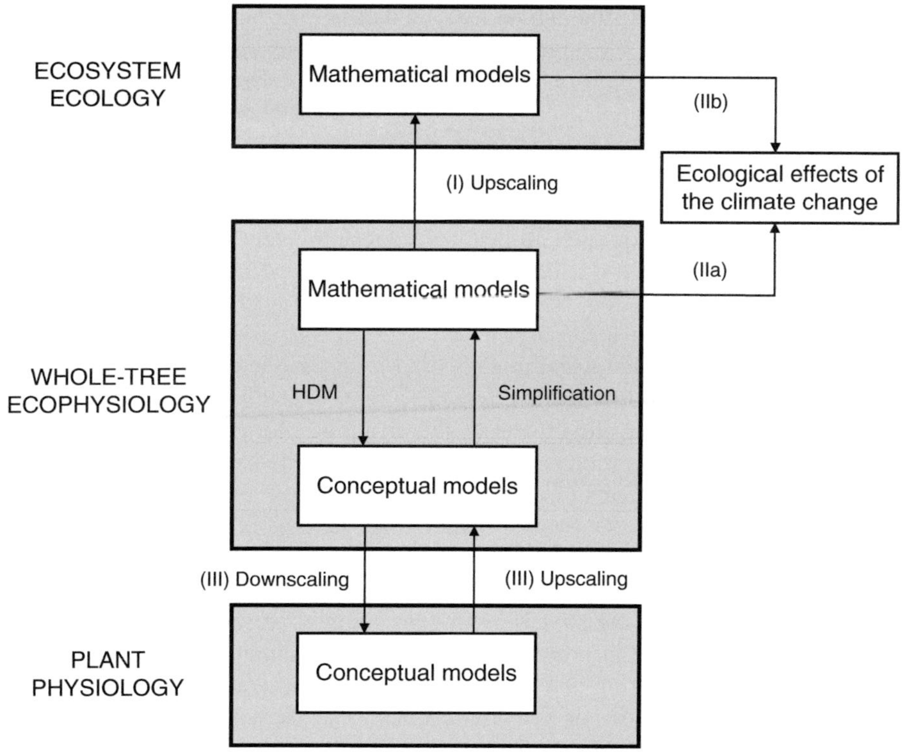

Fig. 9.1 A schematic presentation of the ecophysiological whole-tree modelling approach to the seasonality of boreal and temperate trees and its links to other related research. Within the discipline of whole-tree ecophysiology, the overall state of knowledge is represented by conceptual models. In the formulation of mathematical dynamic models on the basis of conceptual models, simplifications are needed. Dynamic models facilitate deductive studies, i.e., computer simulations used in the hypothetico-deductive (HDM) research strategy to further improve the overall state of ecophysiological knowledge represented by the conceptual models. (*I*) The knowledge represented by mathematical ecophysiological models is scaled up to the discipline of ecosystem ecology by introducing them as sub-models into large ecosystem models. (*II*) Mathematical ecophysiological models are used either as such at the whole-tree level (*IIa*) or as part of ecosystem models (*IIb*) in computer simulations to assess the ecological effects of the projected climate change. (*III*) The knowledge represented by conceptual ecophysiological models is scaled down to the discipline of plant physiology. Conceptual whole-tree models pinpoint phenomena whose anatomical, cellular, and molecular mechanisms need to be determined. The mechanistic information may then be scaled up to improve the realism of the conceptual whole-tree models. This approach also holds high potential for improving the physiological realism of the mathematical models in the discipline of whole-tree ecophysiology, though it is not yet widely used for that purpose

Improving the transparency of the seasonality models will facilitate not only the upscaling of the models to the ecosystem level but also their downscaling to the anatomical, cellular, and molecular levels (Fig. 9.1). Though cellular and molecular mechanisms were not discussed in this volume, one aim was to facilitate a synthesis of the whole-tree modelling approach taken here with approaches that address the annual cycle at the lower levels. It is evident that a more thorough consideration of these basic phenomena would improve the realism of the whole-tree models in many cases. Such a synthesis would also be beneficial to plant physiologists, as it would facilitate the use of their results in a wider context, such as studies assessing the effects of the projected climate change.

It is impossible to introduce all available physiological detail into ecophysiological models that are used for assessing the effects of the climate change at higher levels of organisation, from the individual tree to the entire globe. That is not only impossible but also purposeless. The crucial issue is the realism and coverage of the ecophysiological models used to assess the effects of the climate change. Do the models sufficiently cover the physiological whole-tree phenomena that ultimately determine the effects of the changing climatic conditions on the survival, growth, and reproduction of trees? Any effort to use ecophysiological models to project the effects of climate change on boreal and temperate trees, or on any plants or ecosystems for that matter, will stand or fall with the realism of the models applied.

The present volume is based on a hypothetico-deductive research strategy, which combines modelling closely with experimental work. Model simulations are carried out not only to project the effects of the climate change but also to facilitate the further testing and development of the models. It is crucial for the assessment of the effects of climate change that the critical testing of the models is continued after they have successfully passed the firsts tests. As shown in the studies reviewed in this volume, a seemingly realistic model that has also been proved to be accurate in tests carried out in the current natural climatic conditions may drastically lose its accuracy when tested in scenario conditions corresponding to the projected climate change: it may provide drastically erroneous projections for the ecological effects of the projected climate change. Therefore it is ultimately necessary to test any ecophysiological model used for projecting the effects of the climate change not only in the current climatic conditions but also in the scenario conditions projected for the future.

However, it was also shown in the present volume that if a model is falsified, it does not mean that the model should be abandoned once and for all. Rather, the main emphasis should be on finding out which part of the model is responsible for the falsification and improving the model accordingly. Quite often, the various parts of the model are sufficiently realistic as such, but there is something essential missing or misrepresented; to use a concept introduced in the present volume, the coverage of the model is insufficient in these cases. In this way, a lot can be learnt from models that have been falsified, strictly speaking, on account of their inability to provide predictions corresponding to observations.

9.2 Effects of Climate Change on Boreal and Temperate Trees

Overall, the effects of the climate change on a given boreal or temperate tree species may be either negative or positive. Even if the climatic scenario to be realised were known, it would be very hard to tell which one of the two potential cases would be realised with any given species in different parts of its current geographical range. This uncertainty prevails even if we just consider the seasonal phenomena of the trees addressed in the present volume: the lengthening of the growing season may either benefit the tree species or subject it to increased incidence of frost damage due to premature dehardening and growth onset.

The uncertainty is further increased by important phenomena that are outside the scope of the present volume, such as the potential effects of increased concentrations of atmospheric carbon dioxide on the photosynthetic production of the trees or the risks of biotic damage due to the immigration of new species of pest insects and pathogenic fungi. Furthermore, several phenomena related to soil conditions may also prove to be decisive for the effects of the climate change. For instance, the accelerated nutrient cycling caused by elevated temperatures may increase the fertility of the soil so as to favour some tree species at the expense of others, or the soil conditions may prove unsuitable for an alien tree species that otherwise could immigrate to the site on account of the changed climate. Furthermore, even slight differences in the ecophysiological responses among tree species may be magnified by interspecific competition in natural forests, so that a given species slightly favoured by the climate change finally prevails over a competing species whose response to the climate change is slightly negative.

Besides the methodological issues crucial for all ecological climate change research, the main emphasis in the present volume was on discussing the paradoxical hypothesis that climatic warming will increase the incidence of frost damage in boreal and temperate trees. The catastrophic frost damages projected with computer simulations about 25 years ago turned out not to be likely any longer. According to those simulations, boreal trees would deharden and even start to grow during such mild spells in winter as are commonly projected to prevail in the future climate, so that serious damage would result during subsequent periods of frost. Empirical tests designed for testing that hypothesis suggested that the trees would remain dormant and retain their hardiness during these intermittent mild periods.

Even so, the hypothesis of the increased incidence of frost damage caused by the climate change cannot be ruled out for all boreal and temperate tree species. Available experimental evidence remains limited, and theoretical work with computer simulations has shown that relatively small changes in the ecophysiological traits of trees may cause premature dehardening and growth onset during mild spells in the scenario climate. Last but not least, there have been several reports of considerable frost damage to boreal and temperate trees and other plants in natural conditions after unseasonally warm spells in winter, which show that even at present, the frost damage hypothesis is already realised to some extent.

In conclusion, to the disappointment of the end-users of ecological research results, nothing very certain can be said about the effects of the climate change on boreal and temperate trees. Though it is a cliché, it must be said that further research is required. The good news is that with a moderate amount of well-targeted research effort, the uncertainty can be essentially diminished. The research discussed in the present volume has pointed out not only the ecophysiological traits of the trees that are critical for the hypothesis of the increased incidence of frost damage to boreal and temperate trees in a changing climate but also the experimental designs that facilitate the determining of those traits in any tree population. It is to be hoped that such experiments will be carried out in the near future with a wide spectrum of boreal and temperate tree species and addressing many provenances of forest trees and many cultivars of horticultural trees. If that begins to happen, then the writing of this volume has achieved its main target.

MIX
Papier aus verantwortungsvollen Quellen
Paper from responsible sources
FSC® C105338

If you have any concerns about our products,
you can contact us on
ProductSafety@springernature.com

In case Publisher is established outside the EU,
the EU authorized representative is:
Springer Nature Customer Service Center GmbH
Europaplatz 3, 69115 Heidelberg, Germany

Printed by Libri Plureos GmbH
in Hamburg, Germany